Polymers at Cryogenic Temperatures

Preface

Polymers at cryogenic temperatures have become an important subject in polymer science and engineering, with new developments emerging in space, superconducting magnet, and electronic technologies, as well as large cryogenic engineering projects such as the International Thermonuclear Experimental Reactor (ITER). In this book, we define cryogenic temperatures as very low temperatures, such as liquid nitrogen (77 K), liquid hydrogen (20 K), and liquid helium (4 K) temperatures. To the best of our knowledge, only a few books have been published on the cryogenic properties of polymers. There has been great progress in the past four decades in this field and this book summarizes the developments so far. Twelve chapters cover various topics within this subject.

In Chap. 1, state of the art, advantages and applications for cryogenic processing are briefly surveyed. Cryogenic mechanical and physical properties of polymers and polymer nanocomposites are reviewed in Chap. 2. Chapter 3 covers the friction and wear of polymeric materials at cryogenic temperatures. This chapter gives an overview of theories and experimental studies on polymer tribology at cryogenic temperatures. Mechanical behavior of polymer composites at cryogenic temperatures is explained in Chap. 4. Chapter 5 describes the studies on the interlaminar delamination fracture and fatigue of woven glass fiber-reinforced polymer composite laminates under Mode I, Mode II, and Mode III loadings at cryogenic temperatures. Chapters 6 and 8 present the behavior of polymer-based dielectrics under cryogenic conditions. Medical applications of cryo-treated polymeric hydrogels such as poly(vinyl alcohol) cryogels are covered in Chap. 7 of this book. Chapter 9 addresses the effect of cryogenic treatment on mechanical behavior of glass fiber-reinforced plastic composite laminate. Chapter 10 covers the polyurethane and polyisocyanurate foams in external tank cryogenic insulation. Chapter 11 includes the effect of the cryogenic treatment on cutting tools and polymers. Finally, current and potential applications of cryogenic-treated polymers are covered in Chap. 12.

Since this book covers all topics related to cryogenic treatment, properties of polymers at cryogenic temperatures, and applications of cryo-treated polymer materials, we hope that it will be of interest to the academic and industrial research community.

Susheel Kalia would like to thank his students, who helped him in the editorial work. Shao-Yun Fu wishes to thank Technical Institute of Physics and Chemistry, Chinese Academy of Sciences for the opportunity to work on this subject. Financial support from the National Science Foundation of China and the Chinese Academy of Sciences is greatly appreciated. Finally, we gratefully acknowledge permissions to reproduce copyrighted materials from a number of sources.

The editors would like to express their gratitude to all contributors of this book, who have provided excellent contributions.

Solan, Himachal Pradesh, India — Susheel Kalia
Beijing, China — Shao-Yun Fu

Contents

Chapter 1
Cryogenic Processing: State of the Art, Advantages and Applications

Susheel Kalia and Shao-Yun Fu

Contents

Abstract Cryogenic processing is the one-time permanent treatment of the materials at very low temperatures to increase the physical and mechanical properties. Cryogenic processing is capable of treating a wide variety of materials such as metals, alloys, polymers, carbides, ceramics and composites. Cryogenic applications of polymers are not only limited to the fields of space, electrical and superconducting technology but also to other advanced technologies such as cryo-surgery and cryobiology in the medical field.

Keywords Cryogenic processing • Materials • Polymers • Mechanical properties

S. Kalia (✉)
Department of Chemistry, Bahra University, Waknaghat 173 234, Dist. Solan (H.P.), India
e-mail: susheel.kalia@gmail.com

S.-Y. Fu
Technical Institute of Physics and Chemistry, Chinese Academy of Sciences, Beijing 100190, China
e-mail: syfu@mail.ipc.ac.cn

S. Kalia and S.-Y. Fu (eds.), *Polymers at Cryogenic Temperatures*,
DOI 10.1007/978-3-642-35335-2_1, © Springer-Verlag Berlin Heidelberg 2013

1.1 Introduction

Cryogenic processing is the process of deep-freezing materials at cryogenic temperatures to enhance the mechanical and physical properties of materials being treated. It is a supplementary process to conventional heat treatment process. The cryogenic processing on materials increases wear resistance, hardness, and dimensional stability and reduces tool consumption and downtime for the machine tool set-up, thus leading to cost reductions [1].

The technique of cryogenic processing is an inexpensive method that improves the physical and mechanical properties of materials such as metals, plastics and composites. The word "cryogenic" comes from the Greek word "kryos", which means cold and is simply study of materials at low temperature such as 77 K. Cryo processing refines and stabilizes the crystal lattice structure and distributes carbon particles throughout the material resulting in a stronger and hence more durable material [2].

1.2 A Brief History

Scientists have been experimenting with the use of extreme cold to strengthen metals since the mid-1800s, but it was not until the advent of space travel that cryogenic processing really came into own. NASA engineers analyzed spacecraft that had returned from the cold vacuum of space and discovered that many of metal parts came back stronger than they were before spending time in space.

Cryogenics and refrigeration technologies share a common history. The most obvious difference between two is the temperature range. Cryogenics had its beginning in the mid of nineteenth century when for the first time man learned to cool objects to a temperature lower than had ever existed naturally on the surface of earth. First practical vapour compression refrigerator was invented by James Harrison in 1855. In 1872, Sir James Dewar invented the vacuum flask. The air first liquefied in 1883 by Polish scientist named Olszewski. Ten years later Olszewski and a British scientist Sir James Dewar liquefied hydrogen. In 1902, Georges Claude improved the efficiency of air liquefaction by including reciprocating expansion engine. The Dutch Physicist Kamerlingh Onnes finally liquefied helium in 1908. Thus, by the beginning of twentieth century, the door had been opened to a strange new world of experimentation [3–7].

Cold treatments were reported to have beneficial effects on tool performance as far back as 1937. Both in the United States and Europe, several reports have appeared of substantial benefits which can be realized by treating steel tools at a low temperature, around 77 K. Within the United States, claims for improvements have been expanded to include copper (Cu), carbides, nylon and some high temperature alloys [3].

The method of cryogenic processing materials at sub-zero temperatures was first acknowledged when metal parts that were transported via train had been crammed with dry ice (at −79 °C), resulting in perceptible increases in wear resistance [8]. Work is being done to confirm that cooling below the temperature of dry ice

(−79 °C), at boiling temperature of liquid nitrogen (−196 °C), would further improve the physical properties of materials [9].

1.3 Cryogenic Processing

Cryogenic processing is capable of treating a wide variety of materials such as metals, alloys, polymers, carbides, ceramics and composites. Deep cryogenics is the ultra low temperature processing of materials to enhance their desired metallurgical and structural properties. This is a temperature about 77 K. Cryogenic processing uses temperature around 77 K because this temperature is easily achieved using computer controls, a well-insulated treatment chamber and liquid nitrogen as shown in Fig. 1.1. Cryogenics is a dry process in which liquid nitrogen is converted to a gas before it enters the chamber so that it does not come into contact with the parts assuring that the dangers of cracking from too fast cooling are eliminated. The risk of thermal shock is eliminated as there is no exposure to cryogenic liquids. The whole process takes between 36 and 74 h depending upon the type and weight of material under treatment. Cryogenic processing must be done correctly in order for it to be successful. The basic steps in a cryogenic process are as follows [2, 10, 11]:

- Ramp down: cryogenics involves slow cooling of the material from room temperature to 77 K and ramp down time is in the 4–10 h range.
- Hold: the material is soaked or held at 77 K for 20–30 h which depends upon the volume of the part. This is the part of the treatment in which the micro-structural changes are realized.
- Ramp up: finally the material is brought back to room temperature. The ramp up time can be from 10 to 20 h range.

1.4 The Science of Cryogenic Processing

Cryogenics refines and stabilizes the crystal lattice structure and distributes carbon particles throughout the material resulting in a stronger and hence more durable material. All of the individual particles that make up an alloy are placed into their most stable state. These particles then are aligned optimally with surrounding particles. Also, molecular bonds are strengthened by the process [2]. Particle alignment and grain refinement combine to relieve internal stresses, which can contribute to part failure. This results in material that is optimized for durability. The extremely low temperature during cryogenic processing also slows the movement at atomic levels and increases the internal molecular bonding energy and hence promotes a pure structural balance throughout the material (Fig. 1.2) [2, 12].

As a result of cryogenic processing, material is obtained with an extremely uniform, refined and dense microstructure which ultimately leads to improvement in physical and mechanical properties.

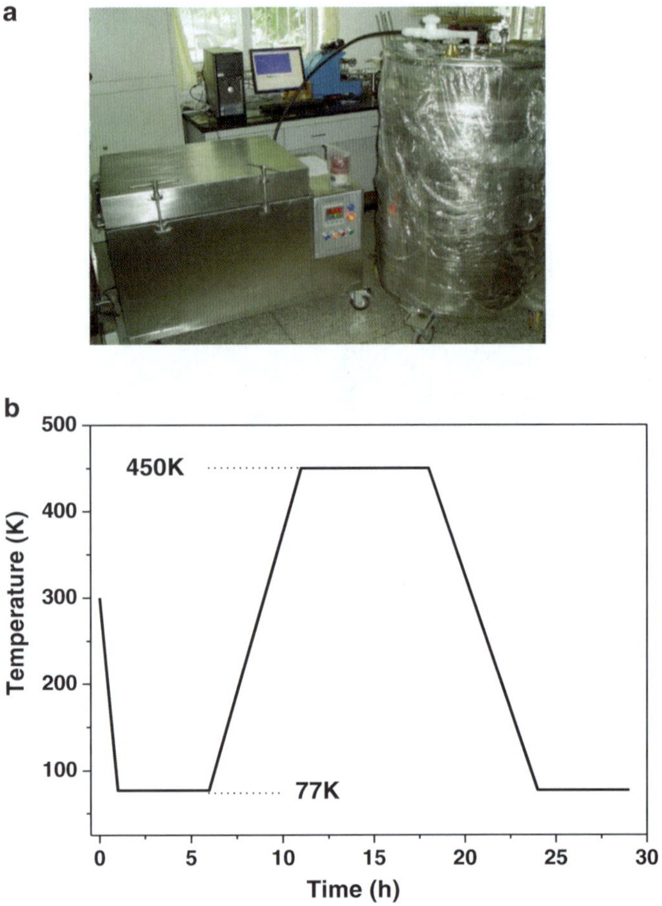

Fig. 1.1 (**a**) Set-up of cryogenic processing in the temperature of 77–450 K, which consists of a well-insulated treatment chamber (*left*), a liquid nitrogen dewar (*right*), cryogenic media transport tube, temperature controlling and data acquisition system; (**b**) temperature (77–450 K)-time curve with controllable heating and cooling speed

Fig. 1.2 Increase in bonding energy at atomic levels with decrease in temperature [2]

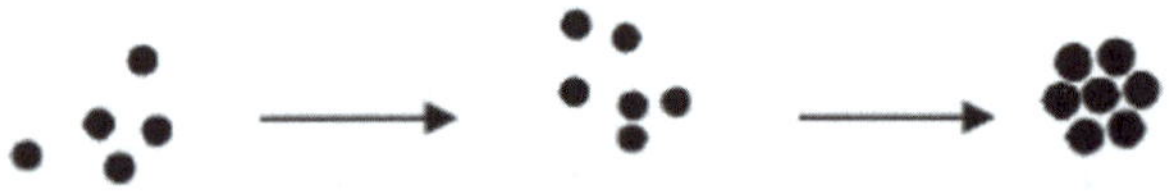

1.5 Polymers at Cryogenic Temperature

Applications of plastics in several areas are of special interest because of their excellent electrical, thermal and mechanical properties. The research work in the field of cryogenic treatment of polymers has gained significant attention in recent years. Cryogenic treatment is found to be the most effective tool to improve wear resistance capacity of polymers. In cryogenic treatment of polymers, the polymer materials are subjected to sub-zero temperature for a stipulated time using liquid nitrogen or liquid helium etc. Since during the cryogenic treatment the whole bulk is also affected, so it is not merely the surface phenomenon. During cryogenic processing, the extremely low temperature during cryogenic processing also slows the movement at atomic levels and increases the internal molecular bonding energy and hence promotes a pure structural balance throughout the material. Cryogenics technology has proven to be the best in case of metals but the need for the development of hypersonic vehicles and cryogenic storage tanks etc. has led to the evaluation of lightweight materials like polymers [2, 13].

Michael et al. [14] have carried out the tensile tests on various epoxies at room temperature and 77 K and observed that these materials are relatively brittle and show plastic behaviour at room temperature just before fracture. Whereas at 77 K no plasticity was observed, the modulus of elasticity was increased by a factor of 3. Stress–strain behaviour and fracture behaviour of polyimide were measured at 77 and 4.2 K. The modulus of elasticity and ultimate tensile strength were increased by 40 % and 60 %, respectively by cooling up to 77 K, but no further change was observed up to 4.2 K [15]. Effect of cryogenic treatment on tensile strength, elongation at break and modulus of elasticity of some polymers are listed in Table 1.1 [16].

1.6 Advantages of Cryogenic Processing

Cryogenically treated parts not only improve performance but also increase the life of materials. Main advantages include [17]:

- One of the most important advantages for cryogenic treatment of materials is an increase in wear resistance.
- The structure of materials is permanently changed, resulting in improved mechanical properties. Treated components may be ground after treatment and the benefits of treatment are retained.
- The frequency and cost of tool remanufacture are reduced. Treated worn tools require less material removal to restore a uniform cutting edge.
- Machine downtime caused by tool replacement is substantially reduced.
- Surface finishing is improved on material being manufactured with treated tooling. Treated tooling stays sharper and in tolerance longer than untreated.
- The overall durability of the treated material is increased.
- Refinishing or regrinds do not affect permanent improvements.
- Thermal shock is eliminated through a dry, computer controlled process.
- Brittleness is decreased.
- Tensile strength, toughness and stability are improved.
- Internal stresses are relaxed.
- The orderly arrangement of crystals is caused, internal bonding energy is increased, and a structural balance throughout the mass of the material is achieved.

Table 1.1 Tensile strength of various plastics at different temperatures [16]

Materials	Temperature (K)	Tensile strength (MPa)	Elongation at break (%)	Modulus of elasticity (MPa)
Plastic films & paper				
Polyimide (H film)	298	18.2	44	510
	77	32.5	16	640
	4.2	34.0	8	615
Polycarbonate (Makrolon)	298	6.4	112	
	77	13.5	4.5	
	4.2	15.6	3.0	
Polyamide (Nomex)	298	8.3	21	
	77	14.5	3.5	
	4.2	15.8	3.0	
Polyethylene terephthalate (Mylar)	298	16.5	88	450
	77	26.0	9.5	870
	4.2	26.8	5.5	880
Polyvinylchloride (rigid type)	298	2.3	156	110
	77	9.5	3.6	215
	4.2	11.6	2.9	300
Polyethylene (high density)	298	2.5	525	
	77	10.5	4.3	
	4.2	13.5	3.1	
Polytetrafluoroethylene (Teflon)	298	2.0	480	
	77	4.3	6.5	145
	4.2	5.6	3.5	217
Bulk materials				
Polyimide (Vespel)	77	18.4	3.1	
Polyethylene (high density)	77	13.8	2.3	
Polyvinylchloride (rigid)	77	11.5	11.5	

1.7 Applications of Cryogenic Processing

There are several categories of industrial applications and representative parts [12]:

- Paper and corrugated board industry: chipper knives, slitter knives, tape cutters, paper drills, trimmers, tissue perforators.
- Cutting tools: mill, cutters, ball screws, punches, drills, broaches.
- Performance vehicles: crankshafts, brake rotors, push rods, heads, pistons, blocks, cams.
- Plastics industry: trimmers, mill knives, granulating blades, dies, feed screws.
- Other applications: copper, electrode, gun barrels, sporting goods, razor blades, musical instruments.

1.8 Conclusions and Future Prospects

Cryogenic processing is an inexpensive one-time permanent treatment of the materials at very low temperature such as 77 K. This technique improves the physical and mechanical properties of the materials. Cryogenic applications of polymers are not only limited to the fields of space, electrical and superconducting technology but also to other advanced technologies such as cryosurgery and cryobiology in the medical field. More applications of cryogenic processing will take off in the near future. Applications of polymeric materials at cryogenic temperatures usually require a lot of attentions because of severe cryogenic conditions; there is limited work yet and a lot of work has to be done on the cryogenic treatment of polymers. Further newer materials have to be developed with capacity of bearing severe cryogenic conditions.

References

1. Gill SS, Singh H, Singh R, Singh J (2010) Cryoprocessing of cutting tool materials—a review. Int J Adv Manuf Tech 48:175–192
2. Kalia S (2010) Cryogenic processing: a study of materials at low temperatures. J Low Temp Phys 158:934–945
3. Foerg W (2002) History of cryogenics: the epoch of the pioneers from the beginning to the year 1911. Int J Refrig 25:283–292
4. Scurlock RG (1990) A matter of degrees: a brief history of cryogenics. Cryogenics 30:483–500
5. Steckelmacher W (1993) History and origin of cryogenics. In: Scurlock RG (ed) Monographs on cryogenics, vol 8. Oxford University Press, Oxford, p 653
6. Richardson RN, Stone HBJ (2003) The cooling potential of cryogens. Part 1 – the early development of refrigeration and cryogenic cooling technology. Ecolibrium (Australian Institute of Refrigeration Air Conditioning and Heating), pp. 10–14
7. Timmerhaus KD (1982) The cryogenic engineering conference—a record of twenty five years of low temperature progress. Advances in Cryogenic Engineering, vol 27. Plenum, New York, p 1
8. Brown J (1995) Big chill to extend gear life. Power Transm Des, pp. 59–61
9. Sweeney TP Jr (1986) Deep cryogenics: the great cold debate. Heat Treat 18:28–32
10. Singh PJ, Guha B, Achar DRG (2003) Fatigue life improvement of AISI 304 L cruciform welded joints by cryogenic treatments. Eng Fail Anal 10:1–12
11. Gulyaev A (1937) Improved heat treatment of high speed steel. Metallurgy 12:65–70
12. Singh R, Gupta S (2006) National seminar on nonferrous metals—imperative need for sustainable development. Indian Institute of Metals, Khetari Nagar Chapter, Rajasthan, India, pp 45–48
13. Pande KN, Peshwe DR, Kumar A (2012) Effect of the cryogenic treatment on polyamide and optimization of its parameters for the enhancement of wear performance. Trans Indian Inst Met 65:313–319
14. Michael PC, Aized D, Rabinowicz E, Iwasa Y (1990) Mechanical properties and static friction behaviour of epoxy mixes at room temperature and at 77 K. Cryogenics 30(9):775–786
15. Tschegg E, Humer K, Weber HW (1991) Mechanical properties and fracture behaviour of polyimide (SINTIMID) at cryogenic temperatures. Cryogenics 31:878–883
16. van de Voorde M (1976) Results of physical tests on polymers at cryogenic temperatures. Cryogenics 16:296–302
17. http://www.cryoplus.com/advantages.php

Chapter 2
Cryogenic Properties of Polymer Materials

Shao-Yun Fu

Contents

Abstract The cryogenic properties of polymer materials have received great attention with new developments in space, superconducting, electronic and defense technologies as well as large cryogenic engineering projects such as International Thermonuclear Experimental Reactor (ITER), etc. Polymer materials developed for these applications are mainly employed as electrical insulators, thermal insulators, vacuum sealants, and matrix materials for composites used in cryogenic environments. The requirements are extremely severe and complicated for polymer materials in these unique applications. The polymer materials need to possess good mechanical and physical properties at cryogenic temperatures such as liquid helium (4.2 K), liquid hydrogen (20 K), liquid nitrogen (77 K), and liquid oxygen (90 K) temperatures, etc., to meet the high requirements by the cryogenic engineering applications. Herein the cryogenic mechanical and physical properties of polymer materials will be highlighted in this chapter. Cryogenic tensile properties/behaviors are first presented in some details for various neat polymers and filled polymers. Cryogenic shear strength, impact strength, and fracture toughness are then

S.-Y. Fu (✉)
Technical Institute of Physics and Chemistry, Chinese Academy of Sciences,
Beijing 100190, China
e-mail: syfu@mail.ipc.ac.cn

S. Kalia and S.-Y. Fu (eds.), *Polymers at Cryogenic Temperatures*,
DOI 10.1007/978-3-642-35335-2_2,

discussed. Afterwards, cryogenic thermal, creep, sliding, and dielectric properties of polymers are briefly summarized. Finally, discussions about effects of water absorption and cryogenic aging on cryogenic properties of some polymers are conducted.

Keywords Polymers • Nanocomposites • Cryogenic mechanical properties • Cryogenic physical properties • Low temperatures • Cryogenic engineering

2.1 Cryogenic Tensile Properties

Polymers have wide applications in various industries and also in cryogenic engineering-related areas. Cryogenic mechanical properties of polymers are important parameters for their successful applications in extremely low temperature environments. For example, in a fiber-reinforced polymer composite, since the internal stresses will be generated as a result of the unequal coefficients of thermal expansion between fibers and matrix, the micro-cracking of the polymer matrix may take place when temperature is decreased down to cryogenic temperature [1]. Sometimes even fracture of the polymer matrix can occur if the thermal stress-induced stress intensity factor exceeds the fracture toughness of polymers [2]. The cryogenic mechanical properties including strength, modulus, ductility, fracture resistance, creep, and sliding performance, etc., of polymers play a critical role in determining their applicability in cryogenic engineering environments. Tensile behaviors at cryogenic temperatures are first discussed below for various polymers and nanocomposites.

Some polymers, that exhibit good mechanical properties at cryogenic temperatures, are studied by focusing on their cryogenic tensile behaviors. Cryogenic tensile tests are performed on aromatic polyimide films (Upilex-R and Upilex-S), which are the products of polycondensation reactions between biphenyl-tetracarboxylic dianhydride (BPDA) and various diamines, and a polypyromellitimide (PPMI)-designated Kapton film [3]. Figure 2.1 displays stress-strain curves for Upilex-R and -S films as well as Kapton film at 77 K. Although the main chains contain the BPDA moiety as a common structural unit, the tensile behaviors of the two Upilexs at 77 K are largely affected by the molecular structure of the diamine moieties. Upilex-R retains considerable ductility but exhibits much lower yield stress around 250 MPa at 77 K. In contrast, Upilex-S withstands elongations in excess of 10 % at a much higher stress level of about 500 MPa at 77 K. The stress-strain curve of Kapton film at 77 K is located between those of the two Upilexs. These observations indicate that aromatic polyimide films exhibit good mechanical properties at cryogenic temperatures.

For semicrystalline polymers, the degree of crystallinity exhibits an effect on their tensile properties at cryogenic temperatures. Figure 2.2 shows the results of tensile tests at 77 K for some biaxially stretched polyethylene-2,6-naphthalenedicarboxylate (PEN) films with different degrees of crystallinity, which is the condensation product from 2,6-naphthalenedicarboxylic acid and ethylene glycol [3].

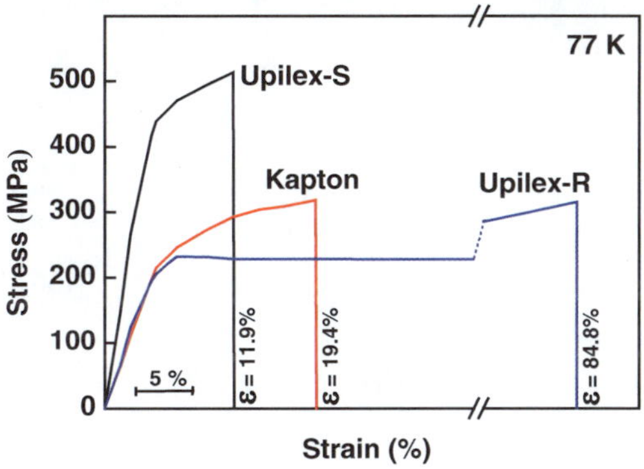

Fig. 2.1 Stress-strain curves at 77 K for Upilex-R, Upilex-S, and Kapton films [3]

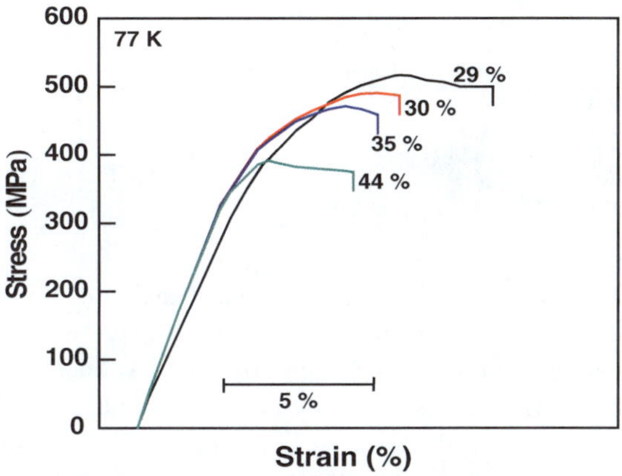

Fig. 2.2 Stress-strain curves at 77 K for biaxially stretched PEN films with various degrees of crystallinity [3]. The crystallinity degree data are shown in the figure

It appears that the strength and the elongation at break of PEN films decreases with increasing the degree of crystallinity.

Figure 2.3 exhibits the stress-strain curves of polyphenylene sulfide (PPS) film at different temperatures. The curve at RT (300 K) is typical of hard ductile polymers and shows a gradual increase in strength beyond the elastic limit. This feature remains unchanged with decreasing temperature from RT to 4.2 K. This indicates that no transition region from ductile to brittle exists between 300 and 4.2 K. Table 2.1 presents the results for the mechanical properties of PPS film at four different temperatures obtained from Fig. 2.3. Both the tensile strength and the tensile modulus gradually increase with decreasing temperature. And the elongation at break remarkably decreases with the decrease in temperature. Nonetheless, the PPS film still retains some ductility even at a very low temperature of 4.2 K.

The effect of crystallinity degree on the tensile properties is exhibited in Fig. 2.4 for the (002) plane of drawn and annealed poly(trimethylene terephthalate) (PTT) film at 18 and 300 K [4]. The results are for the drawn film (crystallinity $\chi_c = 20.2$ %) and those for the drawn (draw ratio = 4) and annealed (for 30 min at 473 K) PPT film ($\chi_c = 41$ %) at 300 K. The inclination of the line corresponds to the crystal modulus of 2.59 GPa along the chain axis. Though the drawn film and

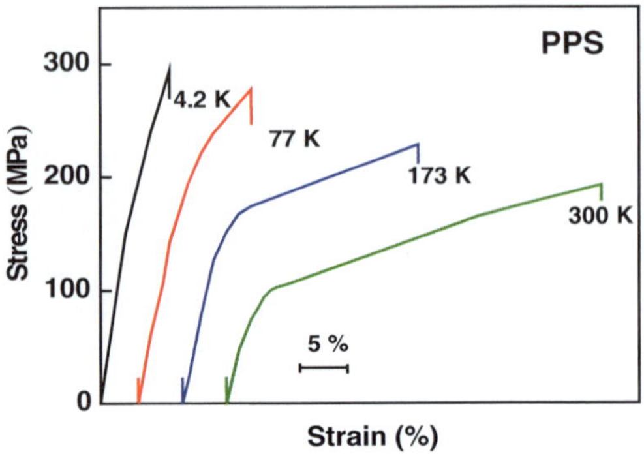

Fig. 2.3 Stress-strain curves at different temperatures for PPS film [3]

Table 2.1 Tensile properties of the polyphenylene sulfide (PPS) film at RT and cryogenic temperatures [3]

Temperature (K)	Tensile strength (MPa)	Tensile modulus (GPa)	Elongation at break (%)
4.2	295	5.2	7.9
77	260	4.8	12.2
173	215	4.2	22.8
RT (300)	188	3.8	38.2

the drawn-and-annealed film have different microstructures, their identical modulus values at 300 K are strong evidence that the stress distribution in PTT film is homogeneous. At 18 K the measured strain vs. stress gives a straight line through the origin, the slope of which corresponds to a modulus of 5.39 GPa, more than twice the modulus of PTT film at 300 K. This indicates that the modulus value of PTT film is strongly dependent on testing temperature.

Figure 2.5 exhibits the stress-strain curves of the drawn PTT film under cyclic loads at 300 and 18 K. The loading and unloading curves obtained at 300 K show hysteresis loops, but those obtained at 18 K coincide with one another up to 8 %. It is reported that the elongation at break of some polyimides is more than 10 % at 77 K [3]. Nonetheless, polyimides deform with yielding at the strain less than 5 %, after which plastic deformation is observed. Therefore, PTT film possesses a comparatively higher elasticity up to 8 % at a very low temperature of 18 K.

Addition of montmorillonite (MMT) at proper contents may enhance the cryogenic tensile strength of polyimide film as shown in Table 2.2 [5]. The tensile strength of MMT/polyimide (PI) composite films at 77 K is generally higher than that at room temperature (RT) except at 20 wt.%. This is mainly attributed to: (1) the PI strength becomes higher at 77 K, resulting in a higher composite strength than at RT; (2) the MMT/PI matrix interfacial bonding becomes stronger at 77 K than at RT because of the thermal shrinkage of PI matrix and tight clamping of the MMT by the PI matrix at cryogenic temperature, leading to a higher composite strength. Moreover, when the MMT content is 20 wt.%, the tensile strength at 77 K is dramatically decreased by the addition of MMT and is lower than that at RT. This is mainly due to the aggregations of MMT at a very high MMT content [5], which would prick up the stress concentration more severely at 77 K than at RT at the sites of MMT/PI interfaces.

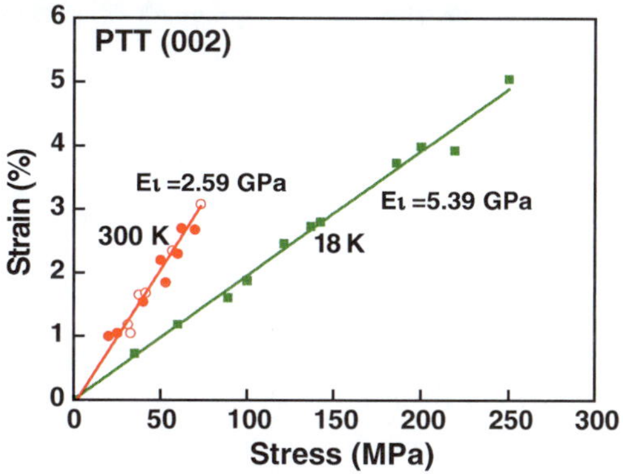

Fig. 2.4 Stress-strain curves for the (002) plane of drawn and annealed PTT films at (*filled square*) 18 K and at (*open circle*) 300 K [4]. Also shown is the plot for the drawn PTT film at 300 K (*filled circle*)

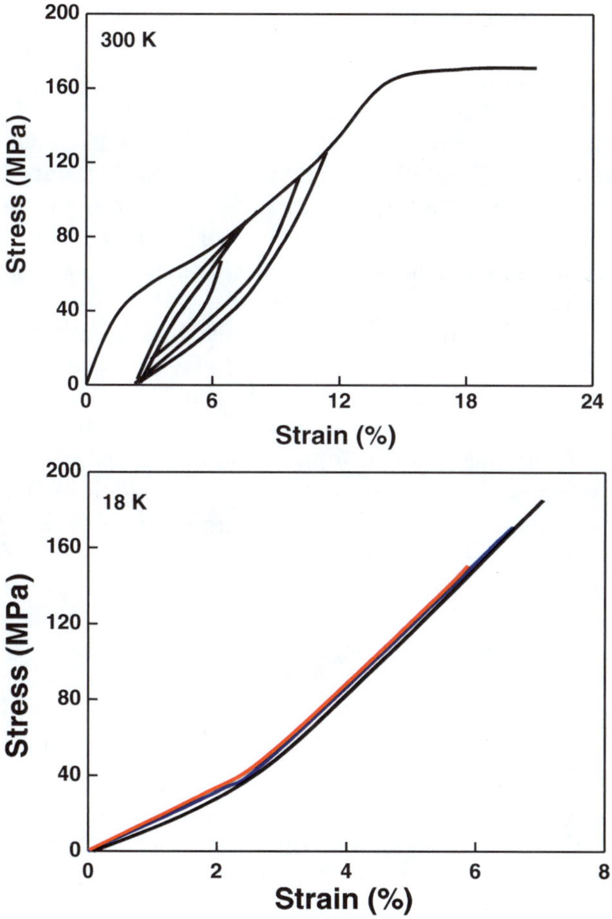

Fig. 2.5 Stress-strain curves of drawn PTT films (draw ratio = 4) under cyclic loads at 300 and 18 K [4]

Table 2.2 The tensile strength of montmorillonite (MMT)/polyimide (PI) composite films at RT and 77 K [5]

MMT content (wt.%)	0	1	3	5	10	20
RT	120	131	120	115	114	84
77 K	188	211	191	158	124	60

Except polymer films as introduced above, epoxy resins are also widely used in cryogenic engineering areas as insulators, vacuum sealants, matrix materials for composites, etc. However, epoxy resins are generally brittle at cryogenic temperatures and residual internal stresses due to thermal cycling may be high to induce fracture of epoxy resins. So, their high cryogenic mechanical properties are very important to assure that epoxy resins can be used under severe cryogenic environments. Therefore, the mechanical properties of epoxy resins at cryogenic temperatures are often enhanced using various modifiers such as carbon nanotubes, thermoplastics, MMT, and other fillers to meet the high requirements by cryogenic engineering applications.

Multi-walled carbon nanotubes (MWCNTs, Showa Denko KK) are used to reinforce a diglycidyl ether of bisphenol-F (DGEBF, DER354) epoxy resin cured with diethyl toluene diamine (DETD, ETHACURE-100) and modified by a reactive aliphatic diluent (Changshu Jiafa Chemistry Co) [6]. The software SemAfore 4.0 is used on the SEM images of carbon nanotubes to measure the diameter of carbon nanotubes as shown in Fig. 2.6 and the average diameter value is estimated to be about 87 nm for CNTs.

The tensile properties at room temperature and liquid nitrogen temperature (77 K) are studied for the epoxy matrix and epoxy nanocomposites. Table 2.3 shows the results for the tensile properties at RT and 77 K of the epoxy matrix and epoxy nanocomposites [6]. It can be seen from Table 2.3 that the tensile strength of the CNT/epoxy nanocomposites at RT is almost independent of the MWCNT addition. By contrary, a significant enhancement in the cryogenic strength is observed by the addition of MWCNTs at appropriate contents. The cryogenic tensile strength reaches the maximum at the CNT content of 0.5 wt.% and then decreases as the CNT content increases further. Moreover, it is seen from Table 2.3 that the cryogenic strength is consistently higher than the RT strength with the same composition. On the one hand, when the temperature decreases from RT to 77 K, the chemical bond and molecules will shrink and the binding forces between molecules will become strong for the epoxy matrix. Thus, a greater load will be needed to break the epoxy matrix at 77 K, leading to a higher strength of the epoxy matrix at 77 K than at RT. On the other hand, the thermal contraction of the epoxy matrix due to the temperature decrease raises the clamping force to the carbon nanotubes at 77 K, this would lead to a stronger CNT-epoxy interfacial bonding.

When there is a relatively poor bonding between the matrix and the CNTs, CNTs would be easily pulled out with smooth surfaces for the RT case as shown in Fig. 2.7a [6]. The weak bonding leads to poor stress transfer from epoxy matrix to CNTs and the CNTs would then have a low reinforcing efficiency, resulting in the insensitivity of the composite strength to the CNT content. When there is a strong

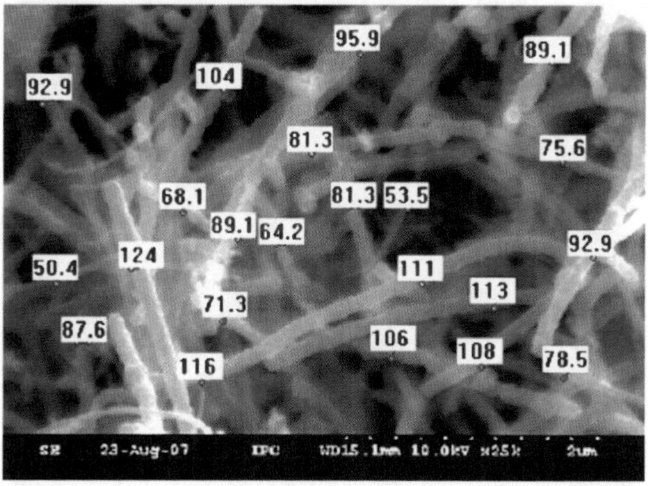

Fig. 2.6 Measurement of the diameters of pristine MWCNTs using the software SemAfore 4.0 [6]

Table 2.3 Tensile properties at RT and 77 K of the epoxy matrix and epoxy nanocomposites [6]

	Tensile strength (MPa)		Tensile modulus (GPa)		Failure strain (%)	
CNT (wt.%)	RT	77 K	RT	77 K	RT	77 K
0	73.4 ± 1.3	92.7 ± 1.1	2.67 ± 0.08	4.58 ± 0.02	4.62 ± 0.31	2.50 ± 0.04
0.02	73.1 ± 1.8	94.8 ± 0.5	2.72 ± 0.05	4.80 ± 0.05	4.71 ± 0.12	2.67 ± 0.03
0.05	72.2 ± 1.6	102.5 ± 2.5	2.77 ± 0.09	4.92 ± 0.08	4.82 ± 0.18	2.82 ± 0.05
0.2	73.1 ± 0.6	116.4 ± 3.4	2.83 ± 0.08	5.02 ± 0.06	4.94 ± 0.13	3.08 ± 0.07
0.5	74.4 ± 1.2	119.4 ± 2.0	3.09 ± 0.02	5.26 ± 0.07	5.04 ± 0.07	3.20 ± 0.05
1	71.5 ± 1.5	112.8 ± 1.5	3.11 ± 0.01	5.49 ± 0.08	4.51 ± 0.19	2.64 ± 0.06
2	68.8 ± 2.3	81.4 ± 3.3	3.13 ± 0.06	5.53 ± 0.04	4.15 ± 0.14	2.20 ± 0.04

bonding between the matrix and the CNTs, the surface of CNTs would be attached with a large amount of epoxy matrix as shown in Fig. 2.7b for the 77 K case [6]. The CNT-epoxy interfacial strength must exceed the yield strength of the epoxy matrix so that the epoxy matrix could be broken and attached on the surface of CNTs. At the relatively high CNT content (2 wt.%), largely aggregated CNTs lead to a slightly lower strength than that of the epoxy matrix because the agglomeration of CNTs for the high CNT content gives rise to weak CNT–polymer interactions and high stress concentrations similar to the case of clay–epoxy composites [7], resulting in the reduction of the tensile strength.

Over 100 measurements are carried out on SEM micrographs as shown in Fig. 2.7 of the fractured surfaces of the samples using the software SemAfore 4.0 to obtain the average CNT diameters [6]. The average diameters of CNTs are estimated to be 93 ± 23 and 180 ± 61 nm, respectively, for the cases of RT and 77 K, confirming that the surfaces of CNTs are quite smooth for the RT case due to relatively weak CNT-epoxy interfacial bonding while are quite rough with matrix attachment for the 77 K case due to strong CNT-epoxy interfacial bonding.

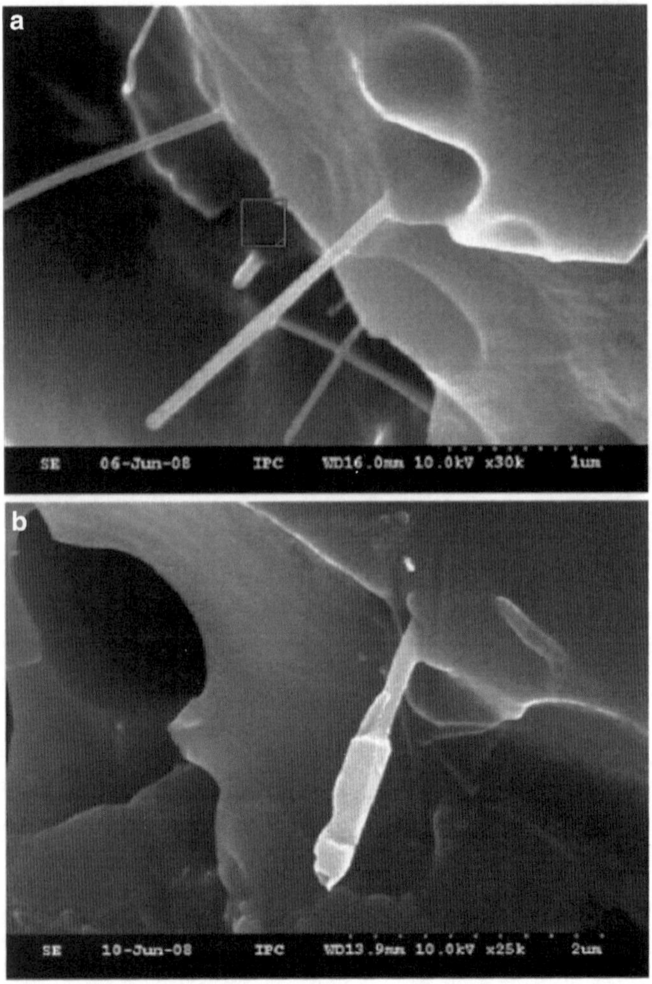

Fig. 2.7 SEM images showing CNT pullout from the tensile fracture surfaces of MWCNT/epoxy composites at (**a**) RT and (**b**) 77 K [6]

Similarly, a strong interfacial adhesion between carbon fibers and polyamide matrix is observed at 77 K [8]. A strong CNT-epoxy interfacial bonding would increase the stress transfer from the epoxy matrix to the CNTs and hence enhance the CNT reinforcing efficiency. Consequently, the strength of the epoxy composites is significantly enhanced at 77 K and the cryogenic strength is much higher than at RT.

Table 2.3 also exhibits that Young's modulus of CNT/epoxy composites at both RT and 77 K increases consistently with increasing the CNT content. Moreover, the cryogenic Young's modulus is higher than the RT modulus with the same composition. This is due to the fact that on the one hand, the molecules of epoxy matrix become stiff due to the restrained mobility of the molecules at 77 K; on the other hand, the CNTs would become stiffer at 77 K than at RT. Finally, in terms of the principles for particulate polymer composites [9] or short fiber-reinforced polymer composites [10, 11], the Young's modulus of the CNT/epoxy composites must be higher at 77 K than at RT.

In addition, it can be seen from Table 2.3 that the failure strain is enhanced by the addition of CNTs at appropriate contents to the epoxy matrix. The epoxy matrix is actually a blend showing a sea-island structure, which is consisted of brittle epoxy phase (DGEBF) and second diluent soft phase [12]. The second soft phase islands are uniformly distributed in the brittle DGEBF phase sea and thus good cryogenic mechanical properties have been achieved [12]. Here, MWCNTs are employed to further improve the cryogenic mechanical properties of brittle DGEBF epoxy resin.

A bisphenol-A diglycidyl ether (DGEBA, CYD-128) is used as the matrix reinforced by MMT and the MMT/epoxy composites are prepared under the ultrasonic treatment [13]. TEM morphologies of the raw MMT, organo-MMT prepared in HCl solution and MMT/epoxy nanocomposites (1 and 3 wt.%) are presented in Fig. 2.8a–d. The size of organo-MMT particles is about 20 nm thick and 200–500 nm long as shown in Fig. 2.8b, which is smaller than that of raw MMT in Fig. 2.8a, especially in thickness. Figure 2.8c exhibits that the organo-MMT has been intercalated and exfoliated in the nanocomposite with the 1 wt.% organo-MMT content. This is because netlike macromolecules form the interlayer of the organo-MMT during the curing process [14]. As the organo-MMT content in the composites is increased to 3 wt.%, the conglomerating probability of organo-MMT and the quantity of non-exfoliated organo-MMT inside the epoxy matrix increase. Thus, it can be seen in Fig. 2.8d that serious aggregations occur at the 3 wt.% organo-MMT content.

Figure 2.9 shows the tensile strength of the MMT/epoxy samples at RT and 77 K as a function of organo-MMT weight content from zero to 3 wt.% and the error bars represent the standard deviation. It can be seen that the organo-MMT content has marginal influence on the tensile strength of MMT/epoxy composites. As the organo-MMT content increases, the tensile strength at both RT and 77 K reaches the maximum at the 0.1 wt.% organo-MMT content. The intercalated and exfoliated organo-MMT layers shown in Fig. 2.8c serve as the short fasciculus inside the composite. The addition of short fasciculus to the epoxy matrix can reinforce the epoxy matrix and contribute to the increase of the tensile strength of the MMT/epoxy composite at the 1 wt.% organo-MMT content. The tensile strength at 77 K is generally higher than that at RT at the same content of organo-MMT due to the two similar major reasons as mentioned for MWCNT/epoxy nanocomposites [6]. The strength of the epoxy matrix becomes higher at 77 K and the interfacial bonding becomes stronger at 77 K than at RT, leading to the higher composite strength.

The tensile modulus of the epoxy composites at RT and 77 K is shown in Fig. 2.10 as a function of organo-MMT content. The tensile modulus increases almost linearly as the organo-MMT content increases. This is mainly because organo-MMT has a higher modulus than epoxy matrix. Moreover, Fig. 2.10

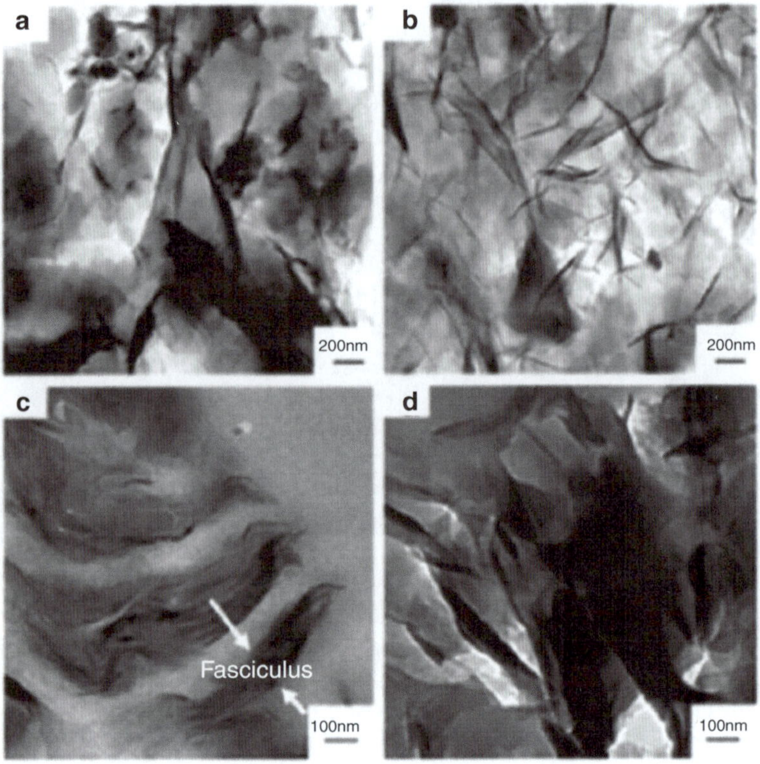

Fig. 2.8 TEM micrographs of (**a**) raw MMT, (**b**) organo-MMT, (**c**) 1 wt.% MMT/epoxy composite, and (**d**) 3 wt.% MMT/epoxy composite [13]

Fig. 2.9 Effect of organo-MMT content on tensile strength of MMT/epoxy composites at RT and 77 K [13]

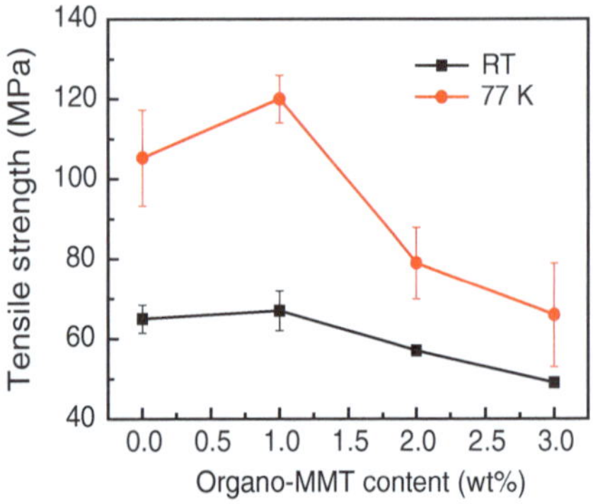

displays that the modulus at 77 K is much higher than that at RT at the same organo-MMT content. On the one hand, the modulus of the epoxy matrix at 77 K is higher than that at RT content due to the tight arrangement of polymer molecules. On the other hand, the modulus of MMT would also be higher at 77 K than at room

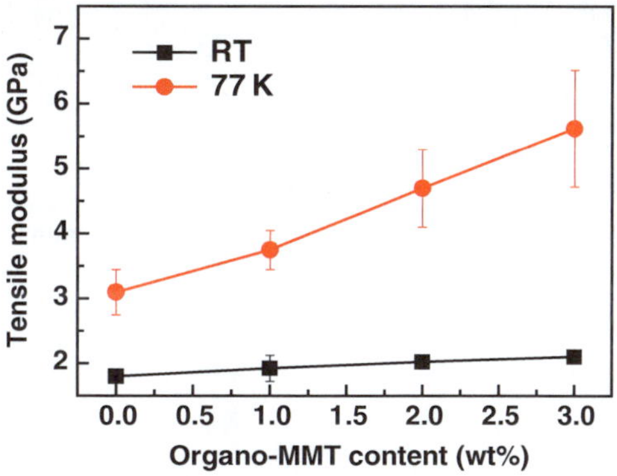

Fig. 2.10 Effect of organo-MMT content on tensile modulus of MMT/epoxy composites at RT and 77 K [13]

Table 2.4 Material properties of the polycarbonate and the MWCNT/PC composite [15, 16]

Material	Temperature (K)	Yield strength (MPa)	Tensile modulus (GPa)	Poisson's ratio
Polycarbonate	RT	34.0	2.35	0.39
	77 K	92.7	4.50	0.38
Composite	RT	36.8	2.56	0.39
	77 K	104	4.65	0.38

temperature. These two major contributions lead to the higher modulus of the composites at 77 K than at RT.

The tensile properties including yield strength, Young's modulus and Poisson's ratio at RT and 77 K of polycarbonate (PC) and 2.5 wt.% MWCNT/PC composite are given in Table 2.4 [15, 16]. It is clear that yield strength and Young's modulus at 77 K are much higher than those at RT. The Poisson's ratio is almost independent of temperature.

A recent development in polymer nanocomposites is the use of CNTs as a smart nano-filler, in which the CNTs serve as strain or damage sensors [17]. The principle is that the deformation, damage initiation, and evolution of CNT-based composites can result in obvious electrical resistance change. The electrical and mechanical responses of MWCNT/PC composites are recently investigated under tensile loading at 77 K and also RT for the purpose of comparison [15]. The load P and the normalized resistance change $\Delta R/R_0$ are plotted in Fig. 2.11 as a function of displacement δ for the 2.5 wt.% MWCNT/PC composite [15].

It can be seen from Fig. 2.11a that the resistance change is very small in the beginning. The composite has an initial resistance R_0 of 11 kΩ at RT. This small change in resistance is due to the specimen deformation [17]. But significant resistance change takes place around 0.8 mm displacement, which corresponds to the initiation of nonlinear load-displacement response. In addition, during loading of the composite specimen the resistance response and crack length show similar trends. This observation demonstrates that crack propagation results in the electrical resistance change and the resistance monitoring enables the detection of crack initiation and propagation in real-time mode.

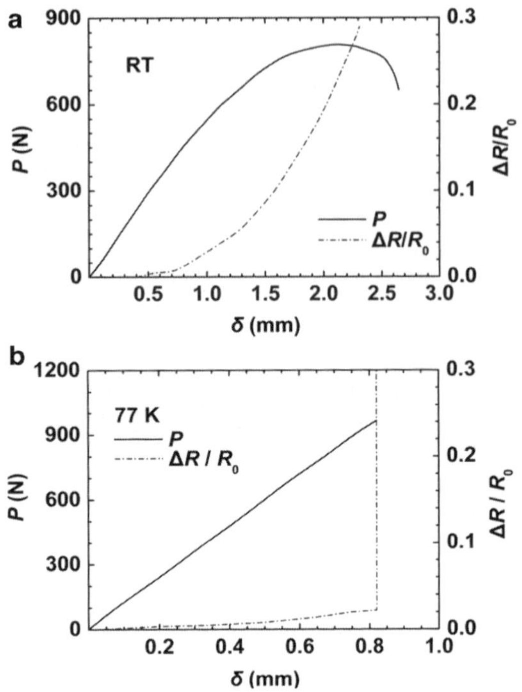

Fig. 2.11 Load and normalized resistance change vs. displacement for the MWCNT/PC composite at (**a**) RT and (**b**) 77 K [15]

Figure 2.11b shows the load P and the normalized resistance change $\Delta R/R_0$ as a function of displacement δ for the 2.5 wt.% MWCNT/PC composite at 77 K. The initial resistance R_0 at 77 K is 27 kΩ. The load-displacement curve at 77 K exhibits a linear relationship. The resistance-displacement curve displays a small resistance change, similar to the corresponding response at RT. When the maximum load is reached, the sharp increase in resistance takes place. Thus, it can be considered that the crack catastrophically propagates through the specimen at 77 K when the maximum load is reached.

2.2 Cryogenic Shear Strength

Adhesives are required to have good bond strength at cryogenic temperatures for cryogenic engineering applications. Thus, it is of significance to study formulations of adhesive resins with high shear strength at cryogenic temperatures [18, 19].

The cured samples are cycled between liquid nitrogen and ambient condition until failure. The shear strengths of some formulated epoxy resins are listed in Table 2.5. It reveals that the matrix cured by hardener HK-021 has a higher shear strength than that of the NA hardener. Moreover, after blending with multifunctional epoxies, the lap shear strength decreases. When the ratio of multifunctional epoxy is great in the formulation such as formulation 5 mixed with 100-phr

Table 2.5 Lap shear strength of some formulated adhesive resins [18][a]

			Shear strength (MPa)			
		Parts by	Initial		After 15 cycles	
Formulation no.	Composition	weight (phr)	RT	77 K	RT	77 K
1	DGEBA	100	6.6	7.1	6.4	6.8
	NA	80				
	A-178™	1				
2	DGEBA	100	8.2	9.7	8.3	9.6
	HK-021	80				
	A-178™	1				
3	DGEBA	100	5.0	4.5	4.9	4.5
	TGBA	25				
	NA	130				
	A-178™	1				
4	DGEBA	100	5.0	6.3	5.0	6.1
	TGDMDA	25				
	NA	150				
	A-178™	1				
5	DGEBA	100	5.2	5.2	5.2	5.2
	TGDMDA	100				
	NA	170				
	A-178™	1				

[a] *DGEBA* diglycidyl ether of bisphenol A; *NA* (*hardener*) nadictetrahydric-methylanhydride; *HK-021* anhydride hardener based on NA but modified by diol molecule; *A-178TM* silane coupling agent; *TGBA* (*multifunctional epoxy*) triglycidyl benzylamine; *TGDMDA* (*multifunctional epoxy*) tetraglycidyl dibenzmethyldiamine

TGDMDA, its lap shear strength decreases remarkably at RT and even more at 77 K, compared to formulation 1 with 0 phr of TGDMDA and the same hardener.

Hu and Huang also studied the lap shear strength of modified epoxy adhesives by mixing with flexible polyether and blending with multifunctional epoxy [19]. The influences of flexible polyether chain amount in the network of the adhesives and synergetic effect of mixed resins with different functionalities on the shear strength of epoxy adhesives at 77 K and RT are investigated. The lap shear strength increases at RT and 77 K with increasing polyether chain content in the network of the epoxy resin with unique functionality. When the adhesive is a mixture of epoxy resins with multifunctional epoxy, a maximum or a minimum lap shear strength at 77 K can be obtained at certain resin ratio, which depends on that whether the polyether chain is introduced into the network.

2.3 Cryogenic Impact Strength and Fracture Toughness

The fracture resistance of polymers will generally be lowered when temperature decreases from RT to cryogenic temperatures. The brittleness of polymers limits their uses in cryogenic environments. Thus, the cryogenic fracture resistance such

Table 2.6 Fracture resistance of neat resin polymers determined from fracture testing [20]

Code	Temperature (K)	K_c (MPa m$^{1/2}$)	K_{Ic} (MPa m$^{1/2}$)	G_c (kJ/m^2)	G_{Ic} (kJ/m^2)
PETI-5	79	6.01	6.01	3.15	3.15
	RT	3.68	3.68	2.98	2.98
	432	3.92	NA	3.60	NA
5050 (plaque 1)	83	7.20	6.49	5.46	4.36
	RT	5.28	5.28	5.89	5.89
	434	4.72	NA	4.20	NA
5050 (plaque 2)	84	7.01	7.01	5.11	5.11
	RT	4.84	4.84	4.97	4.97
	433	4.56	NA	3.90	NA
LCR	84	1.88	1.83	1.38	1.25
	RT	1.45	1.45	1.94	1.94
	433	0.29	NA	1.08	NA

as cryogenic impact strength or cryogenic fracture toughness is a very important parameter for polymers used in extremely low temperature environments. Sometimes brittle polymers such as epoxy resins, etc., are often modified using various modifiers.

Fracture toughness of three polymers (LaRC PETI-5, LaRC 5050, and LCR) manufactured by NASA LaCR is studied [20], where PETI-5 is a phenylethynyl-terminated imide oligomer (Mn ~ 5,500 g/mol), LaRC 5050 a thermoplastic aromatic polyimide endcapped with phthalic anhydride of a molecular weight (Mn ~ 12,000 g/mol) and a 50:50 ratio of diamine monomers in the backbone making it more flexible while LCR a liquid crystal oligomer with changes from a thermoplastic to a thermoset upon curing. Fracture toughness results for these polymers are shown in Table 2.6. It can be seen from Table 2.6 that the stress intensity factor (*K*) and fracture energy (*G*) are dependent on testing temperature. *K* shows a higher value at cryogenic temperature than at room temperature or high temperature for all samples. *G* displays different trends for different samples. *G* exhibits a higher value at cryogenic temperature than at RT but a lower value than at high temperature for PETI-5. For 5050 (plaque 1) and LCR materials, *G* shows a higher value at cryogenic temperature than at high temperature but a lower value than at RT. On the other hand, for 5050 (plaque 2), *G* displays the highest value at cryogenic temperature compared to that at RT and high temperature. It should be noted that at high temperature no tests yield valid K_{IC} and G_{IC} results according to the ASTM D 638-01 standard.

The impact strength of epoxy resins modified by poly(ethersulfone) (PES) is studied as a function of PES content [21]. Figure 2.12 shows the dependence of the impact strength at RT and 77 K on the PES content in the cured epoxy resins. The error bars denote the standard deviation. As expected, the impact strength at RT significantly increases with increasing the PES content in the modified epoxy system. An addition of 5 phr PES relative to the epoxy results in about 49 % increase in the impact strength at RT compared to the neat epoxy. This is because that the immiscible PES-rich domains in the modified epoxy system would increase the

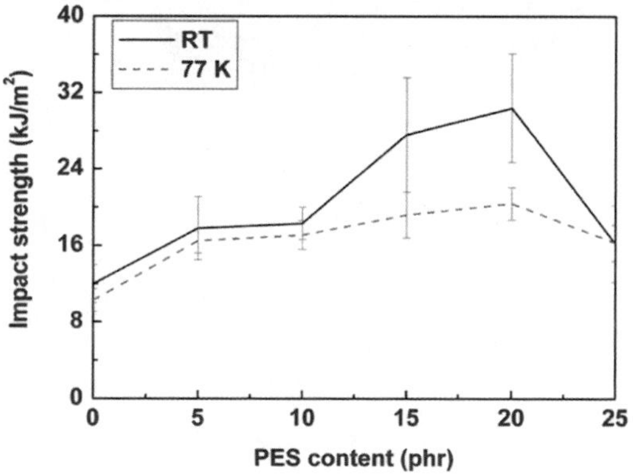

Fig. 2.12 Impact strength of the unmodified and modified epoxy resin at RT and 77 K [21]

impact strength through crack trapping and bridging of the dispersed thermoplastics [22, 23]. And the impact strength shows small variation with increasing the PES content from 5 to 10 phr and from 15 to 20 phr because of the similar morphological features developed after curing [21]. A rapid increase in the impact strength can be observed when the PES content increases from 10 to 15 phr (see Fig. 2.12). The impact strength of the PES-modified epoxy is up to the maximum value for the 20 phr case, where a co-continuous phase structure is formed for dispersed PES phase and epoxy main phase [21]. A large decrease in the impact strength is observed when the PES content is further increased to 25 phr, where the PES rich domains form large agglomerates, which act as defects and then initiate failure.

A DGEBA epoxy resin cured by methyltetrahydrophthalic anhydride is modified by a hyperbranched polymer (polyester Boltorn H30), which is a third-generation hyperbranched polyester with 32 primary hydroxyl groups on the shell and it is an amorphous solid at RT. Effects of the H30 content on the impact strength at RT and 77 K of the cured epoxies are shown in Table 2.7 [24]. It exhibits that the RT and cryogenic impact strength increases up to the maximum by adding the 10 wt.% H30 and decreases afterwards with further increasing the H30 content. In general, it is expected that the impact strength of epoxy resins at 77 K should be lower than that at RT since epoxy resins become normally more brittle at low temperatures. Here it is interesting to note that the impact strength at 77 K of the modified epoxy systems is comparable to that at RT. Furthermore, the maximum improvements in the impact strength at RT and 77 K are, respectively, 20.3 and 26.3 % over those of the pure epoxy resin by the introduction of 10 wt.% H30.

Charpy impact testing of the DGEBF epoxy matrix cured with DETD and the MWCNT/epoxy nanocomposites is conducted at both RT and 77 K and the results are shown in Table 2.8. It can be seen that the introduction of MWCNTs at appropriate contents into epoxy resin can effectively enhance the impact strength of the epoxy resin. When the CNT content is equal to 0.5 wt.%, the impact strength is largely enhanced by 76.7 % and 51.4 % at RT and 77 K, respectively. This can be explained in terms of the synthetic sequence. Since MWCNTs are first mixed with DGEBF and the resulting mixture is then blended with the DETD with an amine

Table 2.7 Impact strength at RT and 77 K of unmodified and H30-modified epoxy resins [24]

Boltorn H30 content (wt.%)	Impact strength (kJ/m^2)	
	RT	77 K
0	25.9 ± 3.2	25.4 ± 2.7
5	27.1 ± 3.3	29.1 ± 3.3
10	31.2 ± 2.7	32.0 ± 1.0
15	24.9 ± 1.6	27.3 ± 1.4

Table 2.8 Impact strength at RT and 77 K of the epoxy matrix and epoxy nanocomposites [6]

MWCNT content (wt.%)	Impact strength (kJ/m^2)	
	RT	77 K
0	37.8 ± 3.5	24.0 ± 2.3
0.02	38.7 ± 2.5	30.9 ± 2.3
0.05	39.6 ± 3.2	31.5 ± 2.7
0.2	41.4 ± 1.6	32.4 ± 4.3
0.5	49.5 ± 4.5	35.6 ± 1.9
1	42.9 ± 0.9	26.3 ± 3.7
2	40.0 ± 1.4	22.9 ± 1.6

weight equivalence of 44.5 g/mol (DETD) as the curing agent and a reactive aliphatic diluent (DILUENT) as a modifier. Because the brittle DGEBF phase can be effectively toughened by introduction of CNTs, the impact strength can be significantly enhanced by the addition of CNTs at appropriate contents when no aggregates of CNTs occur. Also, Table 2.8 shows that the impact strength is larger at RT than at 77 K with the same composition. This is because the molecular mobility of the epoxy matrix would be lowered when the temperature is down to 77 K from RT. When rapid impact loading is applied to the epoxy samples, it would be difficult to yield plastic deformation and hence relatively low impact energies are required to break the samples at 77 K.

The cryogenic critical stress intensity factor (K_{IC}) for the MWCNT/PES-modified DGEBF epoxy composites is investigated as a function of CNT content [25]. Figure 2.13 shows that the fracture toughness increases with increasing the MWCNT content up to 0.5 wt.%. The maximum K_{IC} reaches 2.02 MPa m$^{1/2}$ with an improvement of 13.5 % at 0.5 wt.% MWCNT content from 1.78 MPa m$^{1/2}$ of the PES-modified epoxy matrix. Although the K_{IC} at 77 K of MWCNT/PES-modified epoxy composites is decreased by the further increase of MWCNT content to 1.0 and 2.5 wt.%, it is still higher than that of the epoxy matrix.

2.4 Cryogenic Thermal Properties

In cryogenic engineering technologies, the thermal properties are very important parameters for polymers. Thermal conductivity plays a significant role in the processing of cooling and heating of cryogenic devices and systems. Values for thermal expansion are required for evaluating thermally induced stresses when different materials with different thermal expansion behaviors are combined.

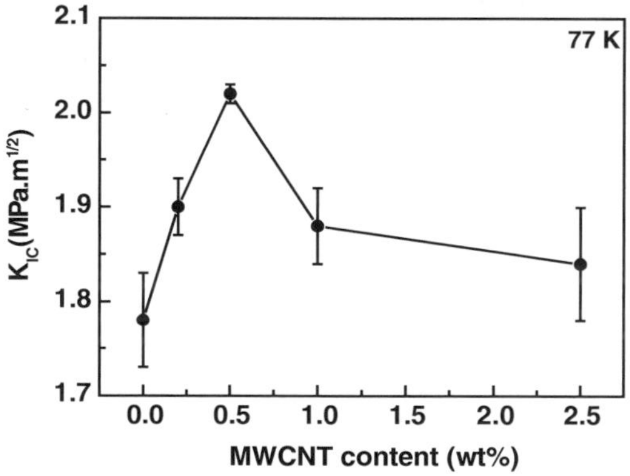

Fig. 2.13 Fracture toughness K_{IC} at 77 K of MWCNT/PES-modified epoxy composites [25]

Generally speaking, the thermal conductivity of polymeric materials is low when compared to other materials such as metals and ceramics. The thermal conductivity of nonconductive polymeric materials is due to the propagation of the thermal oscillation of atoms and molecules as a lattice wave (sound wave) in the material [3, 26]. Some methods have been taken to account for the temperature dependence of thermal conductivity for polymers [26–29]. The thermal conductivity is shown in Fig. 2.14 as a function of temperature for poly(methy1 methacrylate) [30]. And this thermal conductivity behavior is regarded to be probably common for amorphous polymers. In the region A of very low temperature, thermal conductivity decreases roughly in terms of T^2. In the region B of around 10 K, thermal conductivity is weakly temperature-dependent (namely a plateau region). In the region C of relatively high temperature, thermal conductivity decreases slowly with decreasing temperature.

For crystalline polymers at cryogenic temperatures, the thermal conductivity depends strongly on both crystallinity and temperature [30]. For highly crystalline polymers such as high density polyethylene, the thermal conductivity first increases with increasing temperature and then reaches a maximum near 100 K. For polymers with low crystallinity, the thermal conductivity increases slowly up to the glass transition temperature. At cryogenic temperatures, a phonon scattering mechanism arising from an acoustic mismatch at the interface between the amorphous and crystalline phase is very important. For amorphous polymers e.g., the case of poly(ethylene terephthalate), its cryogenic thermal conductivity at 1.5 K is about ten times higher than that of a partially crystalline sample [30].

Results for coefficient of thermal expansion (CTE) of seven epoxy resins are measured and presented in Fig. 2.15 [31]. Epoxy resins A, B, and C are mold resins used for cryogenic components (A: novolac, B: cycloaliphatic, and C: bisphenol A). Epoxy resins D and E are room temperature-cured epoxy resins with low thermal expansions. Resin D contains 42 wt.% calcium silicate fillers, and resin E contains 46 wt.% alumina fillers. Epoxy resins F and G are heat-resistant epoxy resins. The pure epoxy resins (A–C) exhibit higher CTE than the epoxy composites (D and E) containing inorganic particles. Moreover, the CTE for all resins decreases with decreasing the temperature.

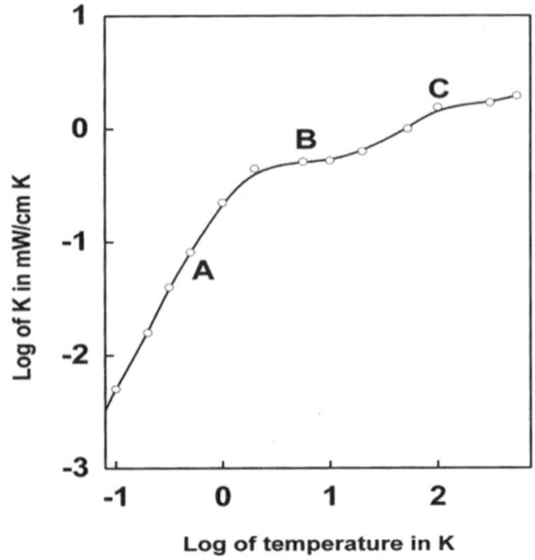

Fig. 2.14 Temperature dependence of thermal conductivity for atactic poly (methy1 methacrylate) [30]

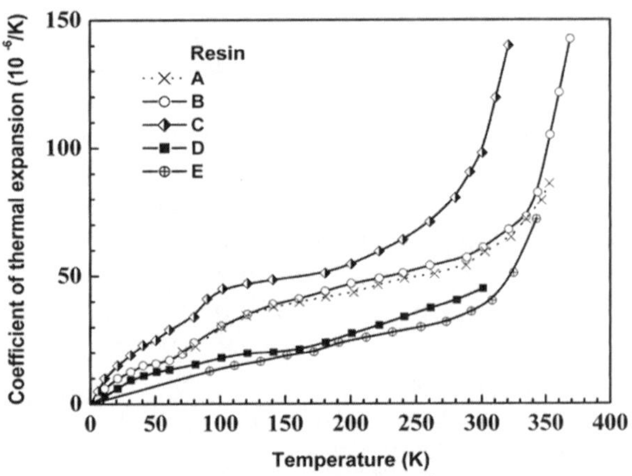

Fig. 2.15 Relationship between temperature and coefficient of thermal expansion (CTE) for the resins A–E [31]

The reduction in the CTE of polymers is desired for cryogenic applications since the CTE of polymers is normally higher than those for metals and ceramics, etc., used in cryogenic environments. The addition of silica nanoparticles can lead to the decrease in the CTE of epoxy resins. Figure 2.16 shows the average CTE from RT to 77 K of the silica/epoxy nanocomposites as a function of silica content [32]. It can be seen that the introduction of silica nanoparticles dramatically reduces the CTE. This is because silica nanoparticles have much lower CTE than epoxy resins do.

The $\Delta L/L_0$ values of silica/PI composite films with different SiO_2 weight fractions (0, 1, 3, 5, 8, 10, and 15 %) are shown in Fig. 2.17 as a function of temperature [33]. The data are fitted by least-square method and seven $\Delta L/L_0 \sim T$

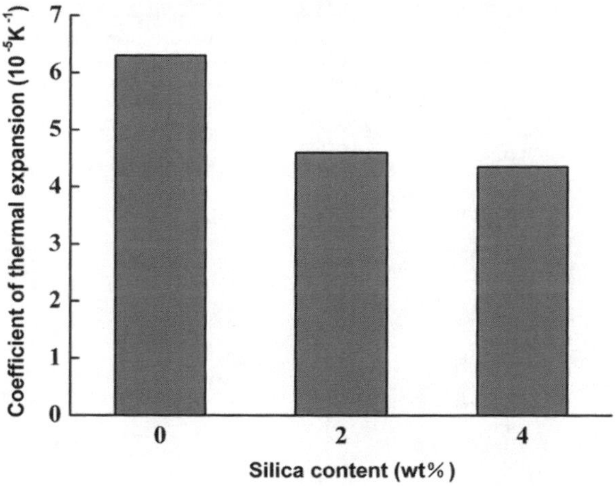

Fig. 2.16 The effect of silica nanoparticle content on the average CTE from RT to 77 K [32]

curves are obtained. The differential CTE of SiO_2/PI composite films can be obtained from Fig. 2.17, and the values are shown in Fig. 2.18 and partly shown in Table 2.9.

It can be seen from Fig. 2.18 and Table 2.9 that there is an obvious effect of SiO_2 content on the CTE of SiO_2/PI composite films. The differential composite CTE decreases with decreasing testing temperature. Moreover, the CTE increases with increasing the SiO_2 content. For example, the CTE at RT decreases from 49.9×10^{-6}/°C for the pure PI film to 25.4×10^{-6}/°C for the composite film with 8 wt.% silica. Also, when the SiO_2 content reaches to 15 wt.%, the differential CTE corresponds to the lowest values. The morphology study [34, 35] shows that the particles are well dispersed in the polyimide matrix and the mean particle size decreases dramatically as the silica content decreases. Combination of the morphological observation and the results for the CTE indicates that the silica particle size has a much less effect on the CTE but the particle content shows a relatively significant effect on the CTE.

2.5 Cryogenic Creep and Sliding Performance

Cryogenic creep fracture of the polymers is the probable cause of the delayed coil quenching in superconductive technology. Research on friction property of polymers and composites at low temperatures is attributed to the application of superconducting magnetic winding systems.

The creep strains, etc., at 77 K in epoxy resins B, E, and G under a compressive stress of −100 MPa are shown in Fig. 2.19 [31]. It is shown that epoxy resins exhibit large creep deformations even under elastic deformation at a cryogenic temperature. The cryogenic creep strains are in typical stage-I logarithmic creep, in which the strain hardening mechanism is dominant. The creep strains are almost

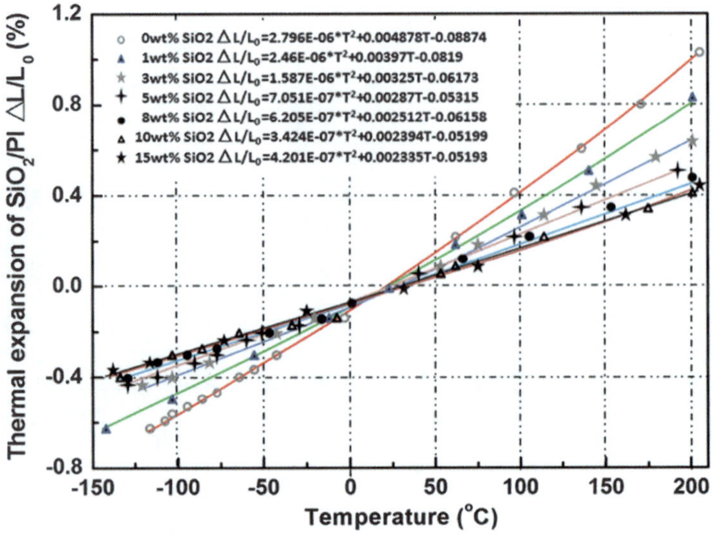

Fig. 2.17 Thermal expansion of silica/polyimide (PI) composite films [33]

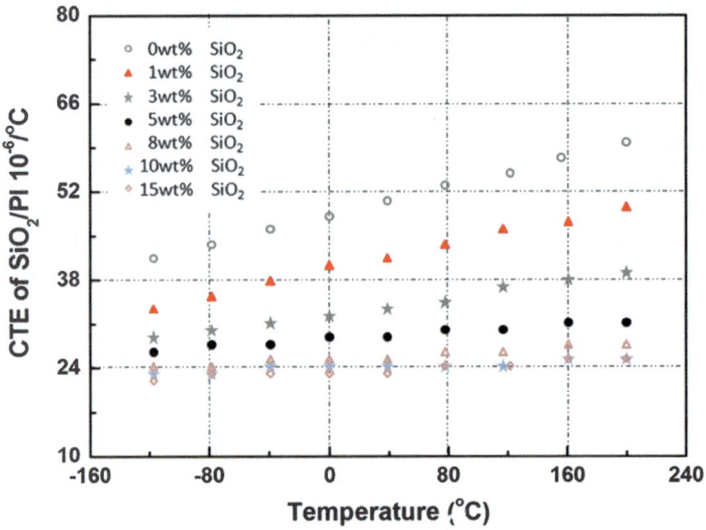

Fig. 2.18 Differential CTE of silica/PI composite films [33]

proportional to the logarithmic time after about 10^3 s. Extrapolation displays that the creep strains at 3×10^8 s can be evaluated as twice as those at 10^5 s. This type of cryogenic creep deformation may cause a delayed resin fracture at an operating temperature. Figure 2.19 also shows the results for the recovery strains after the unloading. The recovery strains are similar to or larger than the creep strains up to

Table 2.9 Differential CTE of SiO_2/PI films calculated from Fig. 2.17 at three typical temperatures for the 0–15 wt.% silica content [33]

	CTE (10^{-6}/°C)						
T(°C)	0 wt.%	1 wt.%	3 wt.%	5 wt.%	8 wt.%	10 wt.%	15 wt.%
−100	43.2	34.8	29.3	27.2	23.9	23.3	22.5
20	49.9	40.7	33.1	29.0	25.4	24.1	23.5
200	60.0	49.5	38.9	31.7	27.6	25.3	25.0

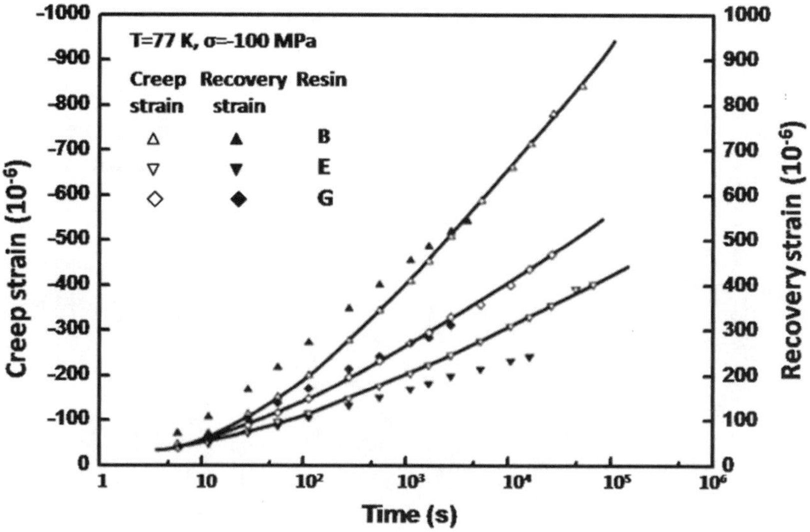

Fig. 2.19 Creep and recovery strains under −100 MPa at 77 K in resins B, E, and G mentioned in Fig. 2.14 [31]

10^3 s. When the load is applied again after recovery, creep deformation will occur, similar to that at the first loading.

Details and frictional property of some polymer composites are shown in Table 2.10 and Fig. 2.20 [36]. The average diameter of short carbon fibers is approximately 14.5 μm, with an average fiber length of about 90 μm. The particle size of polytetrafluoroethylene (PTFE) is around 4 μm, whereas that of the graphite flakes is around 20 μm. TiO_2 (Kronos 2310) with an average diameter of 300 nm is applied as an additional nano-size filler. Figure 2.20 gives the results for the friction coefficient of some composites based on different polymers as a function of temperature. It is observed that the friction coefficient of all composites is reduced with lowering the testing temperature. There are two possible reasons for the reduction in frictional coefficient: (1) the increase of stiffness of the polymer composites at low temperatures; and (2) the liquid lubricating effect of the cryogenic media (either liquid nitrogen or liquid hydrogen) between the two contact partners.

Table 2.10 Details of composites based on various polymer matrices [36]

Matrix	Short carbon fiber (vol.%)	Graphite (vol)	PTFE (vol.%)	TiO_2 (300 nm) (vol.%)
Epoxy	15	15	0	5
Polyetheretherketone (PEEK)	15	5	5	0
Polyamide 6,6 (PA 6,6)	15	5	0	5
Polyetherimide (PEI)	5	15	0	5

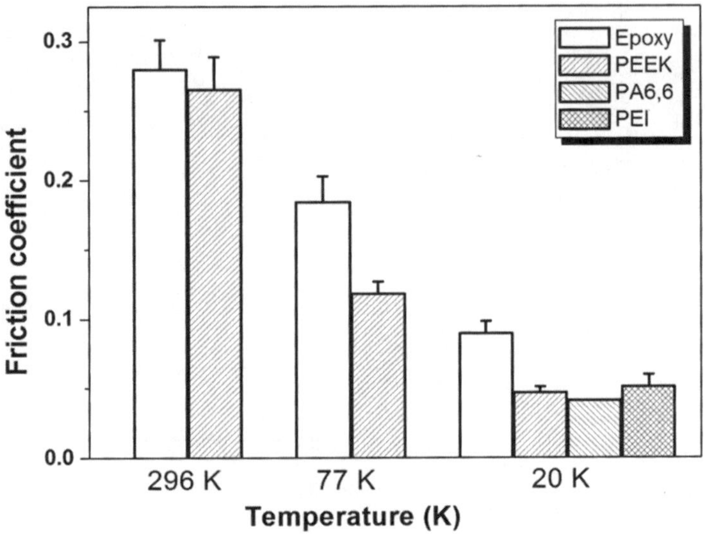

Fig. 2.20 Friction coefficient of different matrix composites in liquid hydrogen (20 K) and liquid nitrogen (77 K) [36]

2.6 Cryogenic Dielectric Properties

It is scarcely seen materials specially designed for cryogenic dielectric applications utilizing high-temperature superconductors (HTSs). So, electrical insulation materials used in conventional high voltage and nonconducting structural materials are adapted as cryogenic dielectric materials for HTS power applications.

The comparison of dielectric breakdown data for polyamide 6,6 (PA 6,6) and polypropylene-laminated paper (PPLP) is shown in Fig. 2.21, where E_{bd} is the dielectric breakdown field. It shows that PA 6,6 has significantly (57 %) higher breakdown strength than PPLP [37]. This means that high voltage equipment with thinner insulation can be made using PA 6,6 instead of PPLP.

TiO_2 nanoparticles with a diameter smaller than 5 nm are synthesized in situ to fabricate a cryogenic resin nanocomposite. The nanocomposite with 2.5 wt.% TiO_2 shows a uniform distribution of the particles as clusters in the epoxy matrix.

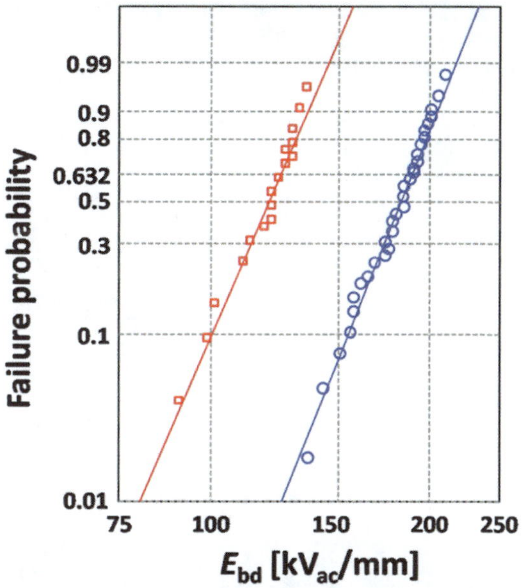

Fig. 2.21 Comparison of polyamide 6,6 (PA 6,6) dielectric breakdown data with polypropylene-laminated paper (PPLP). The data for PA 6,6 and PPLP are presented with *circles* (*open circle*) and (*open square*) symbols, respectively [37]. The measurements are performed in liquid nitrogen open bath at 77 K

The dielectric breakdown data of the nanocomposite show significant improvement over the pristine epoxy, as shown in Fig. 2.22. The data are presented in terms of Weibull statistics [39, 40]

$$P(\alpha,\beta,E_{\mathrm{bd}}) = 1 - \exp[-(E_{\mathrm{bd}}/\alpha)^{\beta}] \tag{2.1}$$

where E_{bd} is the dielectric breakdown field, α the characteristic breakdown strength at which the failure probability $P = 0.6321$, and β the slope which corresponds to inverse scattering in the breakdown values. The parameters α and β are estimated, respectively, as 93.4 kV/mm and 5.98 for the pristine epoxy resin, whereas the nanocomposite shows 109.9 kV/mm and 9.48 for α and β. The improvement in the characteristic breakdown strength is approximately equal to 20 % due to 60 % improvement in the slope. Finally, a possible dielectric design parameter can be adopted by using (2.1) as the 1 % failure probability field, which is 56 % higher in the nanocomposite (67.7 kV/mm) than in the unfilled epoxy resin (43.3 kV/mm). This improvement means 0.64 times less insulation thickness for a given voltage level.

2.7 Influence of Water Absorption and Cryogenic Aging on Cryogenic Properties

Polymers may absorb some amount of water and serve under cryogenic thermal cycling. Effects of water absorption and cryogenic aging on cryogenic properties are briefly summarized below.

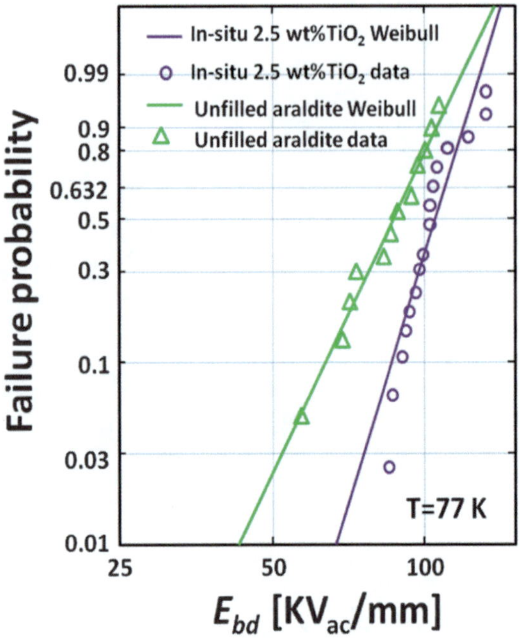

Fig. 2.22 The dielectric breakdown failure probability data measured with 500 V/s at a power frequency of 60 Hz at 77 K [38]. The *solid lines* correspond to estimates from a Weibull statistical analysis

The results for the loss factor (tanδ), the shear modulus (G), and the CTE (α) are presented in Figs. 2.23 and 2.24, respectively, for semicrystalline polyetheretherketone (PEEK) and polyethersulphone (PES), which are in the dry and nearly water-saturated state. It can be seen from Fig. 2.23 that the shear modulus of PEEK is increased below the water peak at around 170 K, the damping peak at very low temperatures is suppressed by water absorption, and nearly no influence on the CTE is observed below 320 K.

The influence of water absorption on the cryogenic properties of PES is similar. One reason might be that water is attached to a different polar group (sulphone group). As is seen from Fig. 2.24, the existing damping peak at 175 K is increased only slightly while the shear modulus is increased remarkably below 170 K. This is also true for the Young's modulus shown in Table 2.11, in which the results for the Young's modulus, E, are given. A slight reduction of E_{wet} at RT and a slight increase of E_{wet} at 77 K can be observed. However, the shear modulus is much less influenced by water above 200 K. Moreover, there is only a slight influence of water on the thermal expansion of PES. Below around 180 K, the thermal expansion is reduced by water while above 180 K, the CTE is increased by water.

The situation is somewhat different for polyamides, which are nonaromatic and have a low main glass transition temperature. No damping peaks exist at very low temperatures and no influence of water is observed below 140 K at low water concentrations as seen in Fig. 2.25 for PA 12 [41]. A small increase in shear modulus due to water is observed in the temperature region starting at around 140 K. Below that temperature no shear modulus increase is observed. A shear modulus decrease by water (softening effect), however, starts above 280 K.

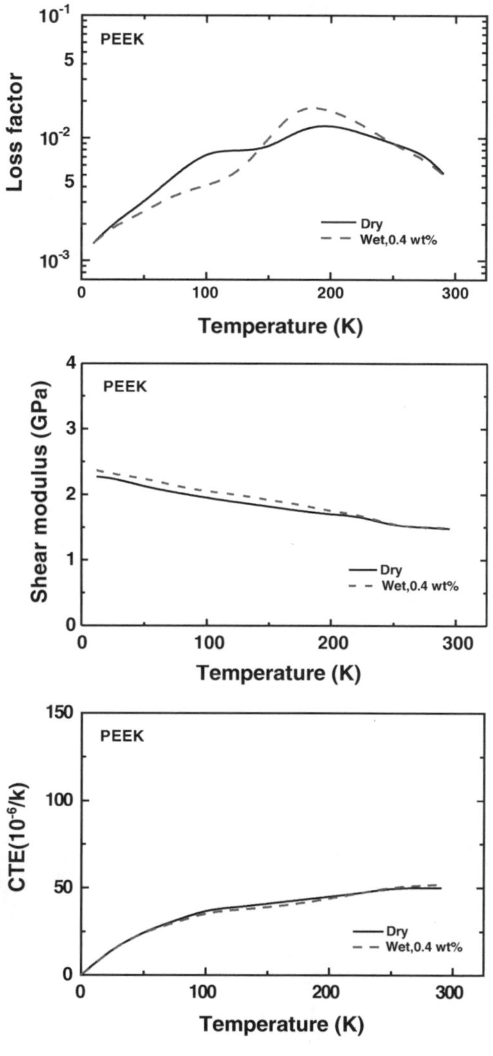

Fig. 2.23 Loss factor, shear modulus, and thermal expansion of dry and wet PEEK [41]

This may be the influence of the main glass transition whose tail can be seen from the damping spectra above 300 K. In Table 2.11, the Young's modulus of PA 12 is given for the dry and wet samples. The Young's modulus at RT is reduced by water while that at 77 K is almost not influenced by water. In addition, at a temperature below around 80 K, the effect of water on the thermal expansion of PA 12 can be neglected while above around 80 K the thermal expansion is reduced by water.

Two grades of solvent-cast polyarylate films manufactured by ISONOVA, namely 15F and 25F with thicknesses of 120 and 100 μm, respectively, are selected for investigation of the cryogenic aging treatment on their properties [42]. Electrical and mechanical properties of some polymers are studied by considering

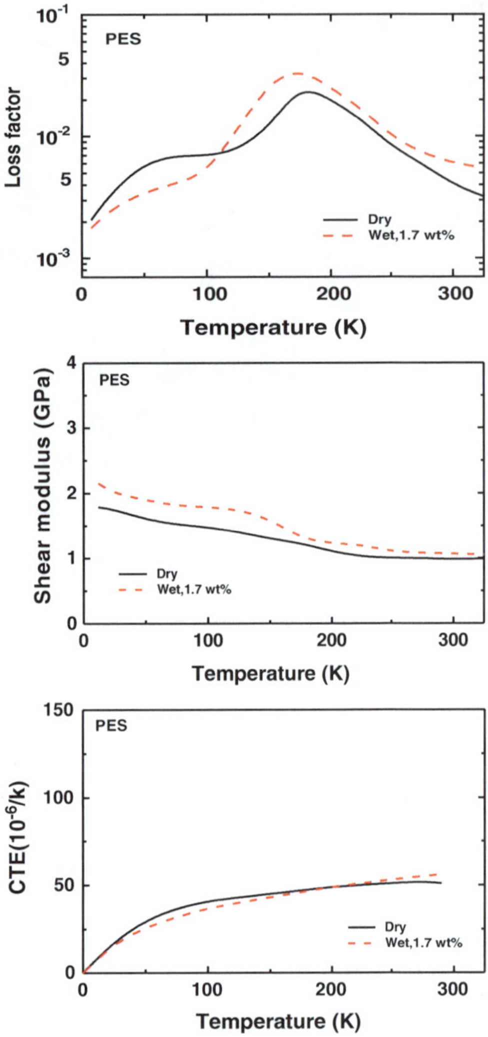

Fig. 2.24 Loss factor, shear modulus, and thermal expansion of dry and wet PES [41]

Table 2.11 Elastic modulus of dry and wet PES and PA12 samples [41]

	PES		PA 12	
Testing temperature	E_{dry}	E_{wet} (1.4 wt.% water)	E_{dry}	E_{wet} (0.4 wt.% water)
RT (GPa)	2.83	2.70	2.05	1.83
77 K (GPa)	4.58	5.66	6.65	6.70

cryogenic aging treatment like soaking in liquid nitrogen for 24 h [42]. Table 2.12 shows the volume and surface resistance of the polymer films measured by applying a d.c. potential of 500 V and a circular size sample of 6 cm in diameter. The cryogenic aging seems to have no significant influence on resistance in the

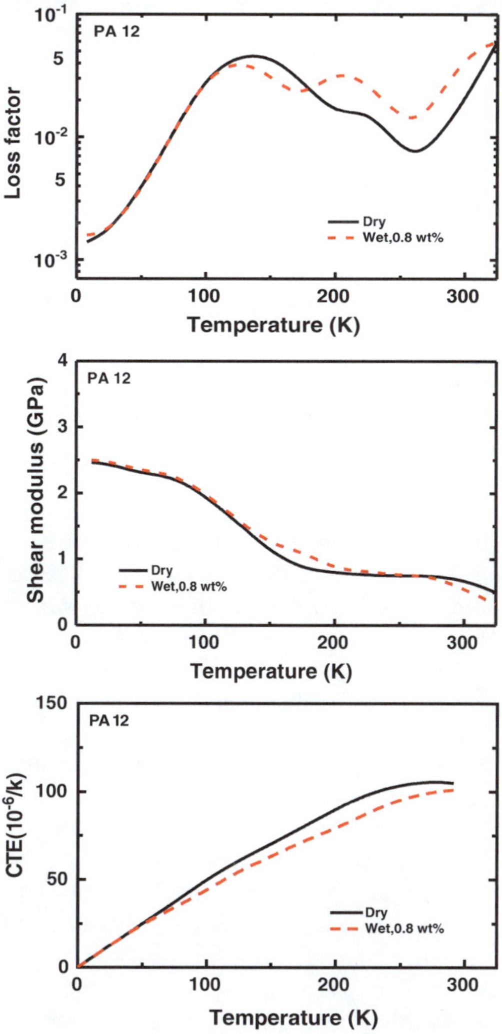

Fig. 2.25 Loss factor, shear modulus, and thermal expansion of dry and wet PA 12 [41]

experimental error range. The only appreciable change is observed in the increase of surface resistance of aged 25F film. This behavior is regarded to result from surface preparation and contamination.

Table 2.12 also shows the results for the mechanical properties of control and aged samples. The tensile strength and elongation at break does not change much with cryogenic aging. The Young's modulus appears to exhibit a slight increase for both aged films. In summary, it seems that the cryogenic aging does not have a significant influence on the tensile properties.

Table 2.12 Volume and surface resistivity, and mechanical properties of polyarylate films [42]

	Grade 15F		Grade 25F	
Property	Control	Aged	Control	Aged
Volume resistance (Ω cm)	1.1×10^{16}	5.5×10^{15}	6.2×10^{15}	7.6×10^{15}
Surface resistance (Ω)	9.2×10^{13}	1.3×10^{14}	6.3×10^{13}	1.1×10^{17}
Tensile strength (MPa)	81	80	74	76
Young's modulus (GPa)	0.80	0.91	1.04	1.06
Elongation at break (%)	82	76	47	48

2.8 Concluding Remarks and Outlooks

With the new developments of space, superconducting, electronic and defense technologies as well as large cryogenic engineering projects such as International Thermonuclear Experimental Reactor (ITER), etc., polymers are increasingly applied as electrical insulators, thermal insulators, vacuum sealants, and matrix materials for composites, etc., used in cryogenic environments. And the polymers need to withstand exerted load by applied magnetic field, etc., severe cryogenic environment and thermal cycling. The cryogenic mechanical and physical properties of polymers are thus critical parameters for determining the applicability of polymers for use in the severe cryogenic environments. The cryogenic properties studied previously include tensile strength, modulus, elongation at break, fracture resistance, thermal conductivity, thermal expansion, shear strength, and dielectric breakdown strength, etc. Some attempt is also made to examine the effects of water absorption and cryogenic aging, etc., on the cryogenic properties of polymers. The previous knowledge on cryogenic properties of polymers provides useful information and strong evidence that can tell us what polymers may be employed in cryogenic engineering applications.

So far there are only a limited number of polymers successfully used in cryogenic engineering areas possibly because the cryogenic mechanical and physical properties of most polymers are low and cannot meet the requirements by the cryogenic engineering applications. Modification of polymers is possibly a good means to enhance their cryogenic properties. However, relatively little work has been reported on modification of polymers using modifiers to enhance the cryogenic properties of polymers so that the modified polymers may meet the high requirements by the cryogenic engineering applications. Moreover, very little work is made on hybridization of polymers using multifunctional fillers to get multifunctional composites for cryogenic applications. Therefore, future research work should be extensively directed towards enhancements of cryogenic properties of polymers by modifying methods so that modified polymers will receive greater success in their applications in cryogenic engineering areas. Multifunctional composites are also an important direction for future work in this subject so that polymers can be successfully applied in multifunctional devices such as cryogenic sensors and actuators, etc., used in cryogenic environments.

Acknowledgements This work was financially supported by the National Natural Science Foundation of China (Nos. 10672161, 50573090, 10972216, 51073169, and 11002142), the National Basic Research Program of China (No. 2010CB934500), Key Research Program of Beijing City Science and Technology Committee (No H020420020230), and the Overseas Outstanding Scholar Foundation of the Chinese Academy of Sciences (Nos. 2005-1-3 and 2005-2-1).

References

1. Nobelen M, Hayes BS, Seferis JC (2003) Cryogenic microcracking of rubber toughened composites. Polym Compos 24(6):723–730. doi:10.1002/pc.10066
2. Ueki T, Nishijima S, Izumi Y (2005) Designing of epoxy resin systems for cryogenic use. Cryogenics 45(2):141–148. doi:10.1016/j.cryogenics.2004.07.002
3. Yano O, Yamaoka H (1995) Cryogenic properties of polymers. Prog Polym Sci 20 (4):585–613. doi:10.1016/0079-6700(95)00003-X
4. Nishino T, Okamoto T, Sakurai H (2011) Cryogenic mechanical behavior of poly(trimethylene terephthalate). Macromolecules 44(7):2106–2111. doi:10.1021/ma200111v
5. Zhang YH, Wu JT, Fu SY, Yang SY, Li Y, Fan L, Li RKY, Li LF, Yan Q (2004) Studies on characterization and cryogenic mechanical properties of polyimide-layered silicate nanocomposite films. Polymer 45(22):7579–7587. doi:10.1016/j.polymer.2004.08.032
6. Chen ZK, Yang JP, Ni QQ, Fu SY, Huang YG (2009) Reinforcement of epoxy resins with multi-walled carbon nanotubes for enhancing cryogenic mechanical properties. Polymer 50 (19):4753–4759. doi:10.1016/j.polymer.2009.08.001
7. Basara G, Yilmazer U, Bayram G (2005) Synthesis and characterization of epoxy based nanocomposites. J Appl Polym Sci 98(3):1081–1086. doi:10.1002/app. 22242
8. Rosso P, Friedrich K, Wollny A (2002) Evaluation of the adhesion quality between differently treated carbon fibers and an in-situ polymerized polyamide 12 system. J Macromol Sci B41 (4–6):745–759. doi:10.1081/MB-120013062
9. Fu SY, Feng XQ, Lauke B, Mai YW (2008) Effects of particle size, particle/matrix interface adhesion and particle loading on mechanical properties of particulate-polymer composites. Compos Part B Eng 39(6):933–961. doi:10.1016/j.compositesb.2008.01.002
10. Fu SY, Lauke B (1998) An analytical characterization of the anisotropy of the elastic modulus of misaligned short-fiber-reinforced polymers. Compos Sci Technol 58(12):1961–1972. doi:10.1016/S0266-3538(98)00033-5
11. Fu SY, Lauke B (1998) The elastic modulus of misaligned short-fiber-reinforced polymers. Compos Sci Technol 58(3–4):389–400. doi:10.1016/S0266-3538(97)00129-2
12. Chen ZK, Yang G, Yang JP, Fu SY, Ye L, Huang YG (2009) Simultaneously increasing cryogenic strength, ductility and impact resistance of epoxy resins modified by n-butyl glycidyl ether. Polymer 50(5):1316–1323. doi:10.1016/j.polymer.2008.12.048
13. Yang JP, Yang G, Xu GS, Fu SY (2007) Cryogenic mechanical behaviors of MMT/epoxy nanocomposites. Compos Sci Technol 67(14):2934–2940. doi:10.1016/j.compscitech.2007.05.012
14. Deng YM, Gu AJ, Fang ZP (2004) The effect of morphology on the optical properties of transparent epoxy/montmorillonite composites. Polym Int 53(1):85–91. doi:10.1002/pi.1410
15. Shindo Y, Kuronuma Y, Takeda T, Narita F, Fu SY (2012) Electrical resistance change and crack behavior in carbon nanotube/polymer composites under tensile loading. Compos Part B Eng 43(1):39–43. doi:10.1016/j.compositesb.2011.04.028
16. Takeda T, Shindo Y, Narita F, Mito Y (2009) Tensile characterization of carbon nanotube-reinforced polymer composites at cryogenic temperatures: experiments and multiscale simulations. Mater Trans 50(3):436–445. doi:10.2320/matertrans.MBW200817

17. Thostenson ET, Chou TW (2006) Carbon nanotube networks: sensing of distributed strain and damage for life prediction and self healing. Adv Mater 18(21):2837–2841. doi:10.1002/adma.200600977
18. Chen Q, Gao BJ, Chen JL (2003) Properties of impregnation resin added to multifunctional epoxy at cryogenic temperature. J Appl Polym Sci 89(5):1385–1389. doi:10.1002/app. 12431
19. Hu XL, Huang PC (2005) Influence of polyether chain and synergetic effect of mixed resins with different functionality on adhesion properties of epoxy adhesives. Int J Adhes Adhes 25 (4):96–300. doi:10.1016/j.ijadhddh.2004.08.001
20. Pavlick MM, Johnson WS, Jensen B, Weiser E (2009) Evaluation of mechanical properties of advanced polymers for composite cryotank applications. Compos Part A Appl Sci 40 (4):359–367. doi:10.1016/j.compositesa.2008.12.009
21. Yang G, Zheng B, Yang JP, Xu GS, Fu SY (2008) Preparation and cryogenic mechanical properties of epoxy resins modified by poly(ethersulfone). J Polym Sci A Polym Chem 46 (2):612–624. doi:10.1002/pola.22409d
22. McGarry FJ (1970) Building design with fibre reinforced materials. Proc R Soc Lond Ser A 319(1536):59. doi:10.1098/rspa.1970.0165
23. Kinloch AJ, Young RJ (1989) Fracture behaviour of polymers. Applied Science, New York
24. Yang JP, Chen ZK, Yang G, Fu SY, Ye L (2008) Simultaneous improvements in the cryogenic tensile strength, ductility and impact strength of epoxy resins by a hyperbranched polymer. Polymer 49(13–14):3168–3175. doi:10.1016/j.polymer.2008.05.008
25. Yang JP, Chen ZK, Feng QP, Deng YH, Liu Y, Ni QQ, Fu SY (2012) Cryogenic mechanical behaviors of carbon nanotube reinforced composites based on modified epoxy by poly (ethersulfone). Compos Part B Eng 43(1):22–26. doi:10.1016/j.compositesb.2011.04.037
26. Morgan GJ, Smith D (1974) Thermal conduction in glasses and polymers at low temperatures. J Phys C 7(4):649–664. doi:10.1088/0022-3719/7/4/004
27. Gush HP (1991) A compact low thermal-conductivity support for cryogenic use. Rev Sci Instrum 62(4):1106. doi:10.1063/1.1142017
28. Greig D (1988) Low temperature thermal-conductivity of polymers. Cryogenics 28 (4):243–247. doi:10.1016/0011-2275(88)90008-2
29. Nittke A, Scherl M, Esquinazi P, Lorenz W, Li JY, Pobell F (1995) Low-temperature heat release, sound-velocity and attenuation, specific-heat and thermal conductivity in polymers. J Low Temp Phys 98(5–6):517–547. doi:10.1007/BF00752280
30. Choy CL (1977) Thermal-conductivity of polymers. Polymer 8(10):984–1004. doi:10.1016/0032-3861(77)90002-7
31. Usami S, Ejima H, Suzuki T, Asano K (1999) Cryogenic small-flaw strength and creep deformation of epoxy resins. Cryogenics 39(9):729–738. doi:10.1016/S0011-2275(99)00084-3
32. Huang CJ, Fu SY, Zhang YH, Lauke B, Li LF, Ye L (2005) Cryogenic properties of SiO_2/epoxy nanocomposites. Cryogenics 45(6):450–454. doi:10.1016/j.cryogenics.2005.03.003
33. Chen XG, Guo JD, Zheng B, Li YQ, Fu SY, He GH (2007) Investigation of thermal expansion of PI/SiO_2 composite films by CCD imaging technique from -120 to 200 °C. Compos Sci Technol 67(14):3006–3013. doi:10.1016/j.compscitech.2007.05.029
34. Li Y, Fu SY, Li YQ, Pan QY, Xu GS, Yu CY (2007) Improvements in transmittance, mechanical properties and thermal stability of silica-polyimide composite films by a novel sol-gel route. Compos Sci Technol 67(11–12):2408–2416. doi:10.1016/j.compscitech.2007.01.003
35. Zhang YH, Li Y, Fu SY, Xin JH, Daoud WA, Li LF (2005) Synthesis and cryogenic properties of polyimide–silica hybrid films by sol–gel process. Polymer 46(19):8373–8380. doi:10.1016/j.polymer.2005.07.012
36. Zhang Z, Klein P, Theiler G, Huebner W (2004) Sliding performance of polymer composites in liquid hydrogen and liquid nitrogen. J Mater Sci 39(9):2989–2995. doi:10.1023/B:JMSC.0000025824.18291.f0
37. Tuncer E, Polizos G, Sauers I, James DR, Ellis AR, Messman JM, Aytug T (2009) Polyamide 66 as a cryogenic dielectric. Cryogenics 49(9):463–468. doi:10.1016/j.cryogenics.2009.06.008
38. Polizos G, Tuncer E, Sauers I, More KL (2010) Properties of a nanodielectric cryogenic resin. Appl Phys Lett 96(15):152903. doi:10.1063/1.3394011

39. Weibull W (1951) A statistical distribution of function of wide applicability. J Appl Mech 18 (3):293–297
40. Rowland SM, Hill RM, Dissado LA (1986) Censored Weibull statistics in the dielectric-breakdown of the oxide-films. J Phys C Solid State Phys 19(31):6263–6285. doi:10.1088/0022-3719/19/31/020
41. Baschek G, Hartwig G, Zahradnik F (1999) Effect of water absorption in polymers at low and high temperatures. Polymer 40(12):3433–3441. doi:10.1016/S0032-3861(98)00560-6
42. Patterson RL, Hammoud A, Fialla P (2002) Preliminary evaluation of polyarylate dielectric films for cryogenic applications. J Mater Sci Mater Electron 13(6):363–365. doi:10.1023/A:1015656700768

Chapter 3
Friction and Wear of Polymer Materials at Cryogenic Temperatures

Géraldine Theiler and Thomas Gradt

Contents

Abstract Polymers are extensively used for sliding systems in cryogenic applications because of their favourable friction and wear behaviour in the absence of external lubrication. Since important new technologies are based on applications under extreme conditions, such as at low temperatures, new requirements on material properties, in particular regarding their operability and reliability, must be met. Up to now, most tribological investigations have been carried out in inert cryogens or cryogenic gas (He, N_2). Few experiments have been performed in vacuum environment at cryogenic temperatures. Rarely were testing in reactive media, such as LH_2 or LOX. Due to the wide range of operating conditions in cryogenic applications, it is difficult to state general rules. Therefore, this chapter

G. Theiler (✉) • T. Gradt
Federal Institute for Materials Research and Testing, BAM 6.3, Macrotribology and Wear Protection, 12200 Berlin, Germany
e-mail: geraldine.theiler@bam.de

S. Kalia and S.-Y. Fu (eds.), *Polymers at Cryogenic Temperatures*,
DOI 10.1007/978-3-642-35335-2_3,

tends to give an overview on theories and experimental studies on polymer tribology at cryogenic temperatures.

Keywords Cryogenic fluids • Cryotribology • Friction • Polymers

3.1 Introduction

The cryogenic properties of polymers are drawing attention with increased new technologies. For tribological applications, materials have to be resistant to extreme conditions of simultaneous influences of low temperatures, mechanical deformation at the surface, special environments and the mutual interactions of these influences. In the case of tribological applications in special reactive environments like hydrogen and oxygen, possible chemical reactions due to mechanical activation are also to be expected. This involves new requirements on material properties, especially regarding their operability and reliability. Polymers are used extensively in cryogenic applications because of their favourable friction and wear behaviour in the absence of external lubrication. Polymers like Polytetrafluoroethylene (PTFE) are particularly advantageous because of their ability to act as solid lubricants. The use of polymer materials for tribological stressed systems at cryogenic temperatures covers a wide range of applications such as:

- Storage and distribution of refrigerants: pumps, valves, bearing, seals
- Space technology including turbo pump bearings in rocket engines, gyroscopes, moving parts in satellites
- Superconductivity: fusion technology, cryogenic electronic and magnet systems

In any of these applications, stringent operation conditions lead to a continuous material development and testing.

Ultra-high vacuum, a wide temperature range, presence of atomic oxygen and limited applicability of liquid lubricants, make open space an extremely hostile environment for tribologically stressed components in satellites [1, 2]. The main polymer composites used are glass-fibre reinforced PTFE and MoS_2-filled Polyimide (PI). This is why extensive works have been carried out with PTFE, PI and also Polychlortrifluorethylene (PCTFE) materials. These have to withstand low temperatures but also ultra-high vacuum.

One example of applied superconductivity is controlled nuclear fusion where large-scale superconducting magnets are necessary for plasma confinement. The Tokamak and the stellarator configuration employ a variety of superconducting magnets that are operated at cryogenic temperatures. Both types need support elements which link cryogenic devices to room temperature (RT) environment. In particular the stellarator type, with its complicated magnet configuration, also requires support elements between the coils. These components are operated in high vacuum at temperatures of 4–5 K. Further requirements are high mechanical strength, high thermal insulating power and the ability to allow relative movement without creating mechanical disturbances such as stick-slip. The problem of relative motion between surfaces in the magnet system is also present inside the system in LHe due to various forces including winding pretension, differences in material thermal contraction and the large electromagnetic forces. Small mechanical disturbances resulting in frictional heating can be sufficient to exceed the wire's critical temperature, resulting inevitably to quenching.

Another specific area of interest is hydrogen technology, which is considered to be the new generation of environmentally compatible energy carrier instead of gasoline in the transportation industry. It may also serve as an energy storage medium, which is a prerequisite for an energy supply based on renewable sources. Because of the low density of hydrogen gas at ambient temperature, liquefaction (boiling temperature: 20 K) for transportation of large quantities of hydrogen is state of the art. Insulation and sealing of components, permeation through the materials and in particular hydrogen-induced embrittlement are of increased importance. Tribological experimental data of materials at low temperatures and in hydrogen environment are hardly available due to cost and safety reasons. Therefore, it is necessary to fill the lack of information since conventional materials cannot automatically be employed [3].

Due to the wide range of operating conditions in cryogenic applications (inert or reactive environments, sliding velocities, contact pressures, etc.) it is difficult to fit experiments with theories. This chapter gives an overview of the theories and experimental studies that have been conducted on polymer tribology at cryogenic temperatures, in particular those carried out at the Federal Institute for Materials Research and Testing, BAM. Because of the temperature-dependent characteristics of polymer materials, thermal and mechanical properties relevant of tribological applications will be initially briefly summarized.

3.2 State of the Art

3.2.1 Low-Temperature Properties of Polymer Materials

It is well known that the properties of polymers strongly depend on the temperature. As seen in previous chapters, polymeric chains get frozen at low temperatures. Below the glass transition, other relaxations can occur, which are related to movements of molecular segments and side chains. Thermal vibrations and deformations are controlled mainly by the weak interchain binding forces (van der Waals) [4]. These restrictions of movements can seriously affect the properties of polymers and thereby their tribological properties, for example by hindering the formation of thin shear layer. Some of the main low temperature properties of polymer materials, relevant in tribology, are mechanical deformation, thermal conductivity and thermal expansion (contraction).

According to Hartwig [4], the thermal conductivity of polymers at cryogenic temperatures depends weakly on the composition. For semicrystalline polymers, it depends mainly on the crystalline content and on the crystallite size. At very low temperatures and not too high stress levels the deformation behaviour is determined

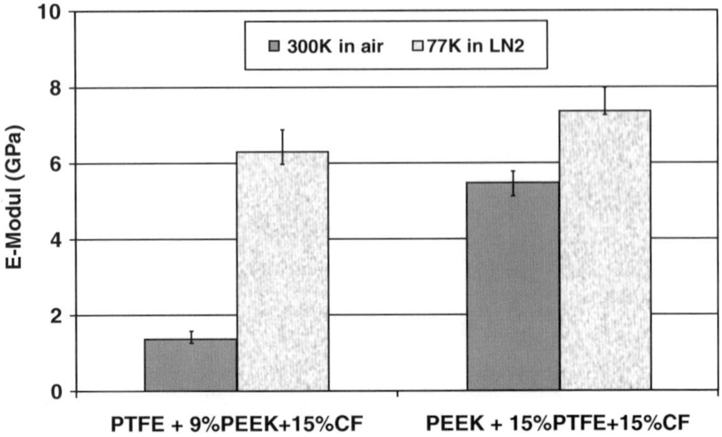

Fig. 3.1 Young's modulus of PTFE and PEEK composite at 300 and 77 K [7]

by van der Waals binding potential. Time-dependent viscoelastic effects are negligible and most polymers behave elastically. Mechanical properties of unfilled polymers have been extensively studied at low temperatures [4–6], and show increased moduli with decreasing temperature down to LHe. It has been observed that the properties of PTFE composites are more affected at low temperatures than the ones of Polyetheretherketone (PEEK) composites that already have a high Young's modulus at ambient temperature (Fig. 3.1) [7].

Problems arise for most designs of cryogenic support structures from the thermal expansion, especially if different materials are involved. In composite materials, different thermal coefficients of expansion between matrix and fibres can lead to mechanical stresses at the interface of components. Thermal cycle experiments between RT in air and LN_2 and between RT and LHe showed significant changes at the surface of polymer composite including cracks and extraction of particles (Fig. 3.2) [8].

3.2.2 Cryogenic Environments

Cryogenic engineering is concerned with technologies for the temperature range below 120 K (−153 °C). These applications may run in liquid cryogens, cryogenic gases or in vacuum at low temperatures. Most of the tribological studies were performed in LN_2 as it is an inert and low cost cryogen. Tests in LHe became more and more important as applications of superconductivity increased. LH_2 and LOX were used up to now mainly in aerospace propulsion. Therefore, only few laboratories developed test equipment adapted to these environments. With the increasing interest in hydrogen as an energy carrier, more facilities are now available for material testing in hydrogen environment but only few in LH_2 at 20 K. Tribological experiments in vacuum are usually restricted to temperatures above 77 K, rarely down to 4 K.

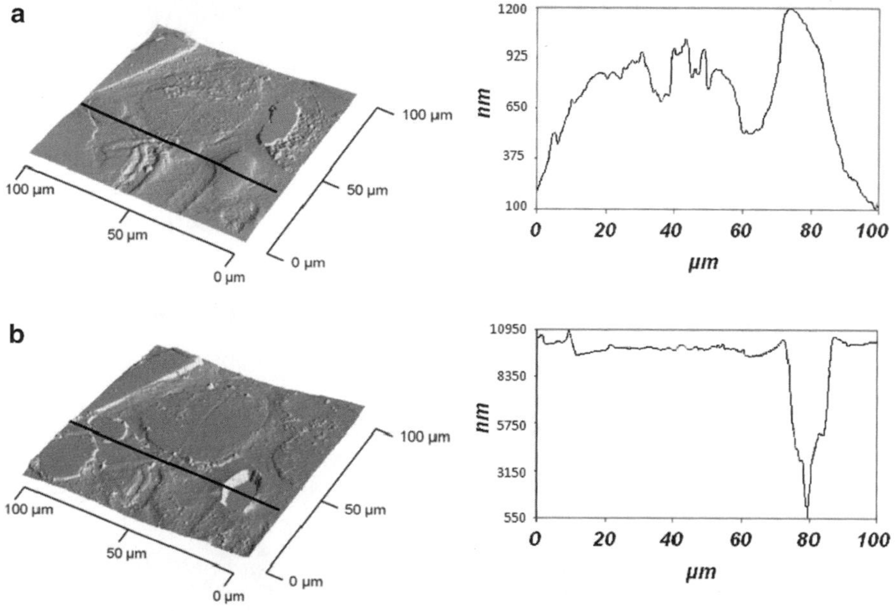

Fig. 3.2 AFM images and linescans of bronze/CF-filled PTFE composite (**a**) before and (**b**) after thermal cycles in LN_2

Table 3.1 Properties of cryogenic fluids

Property		LHe	LH_2	LN_2	LOX	CH_4
Boiling temperature T_b,1 bar	K	4.2	20.4	77.3	90	111.7
Heat of evaporation at T_b	kJ/kg	20.9	446	199.1	213	510,3
Density at T_b	Kg/m^3	124.8	70.8	804.2	1,141	422.6
Specific heat at T_b	kJ/kgK	4.4	9.3	2	1.7	
Thermal conductivity at T_b	10^{-3} W/mK	27.2	119	139.8	1.52	
Viscosity at T_b	10^{-7} kg/ms	35	133	1,650	2,000	

The tribological behaviour of a friction couple is substantially dependent not only on the temperature but also on the environment [3]. This is mainly due to different thermal and chemical properties of the environment. Table 3.1 summarizes some properties of five cryogenic liquids.

3.2.3 Friction Models

As described in [9], friction is essentially an energy dissipation process which is determined by heat generation and dissipation at the friction contact. It has been characterized by the two following models:

- The adhesion model of friction, which applies to the immediate interfacial zone and corresponds to high rates of energy dissipation.
- The deformation model of friction, which involves deformation and ploughing within a larger volume of material and hence lower rates of energy dissipation.

Thereby the friction can be described by $F = F_a + F_d$, where F_a and F_d are the adhesion and deformation components of friction respectively.

In general, the adhesion component of friction (F_a) is determined by the formation and rupture of interfacial adhesion bonds. The attractive interaction forces include primary chemical bonds, i.e. metallic, covalent and ionic, as well as secondary van der Waals bonds [10]. The adhesion between polymeric solids is mainly determined by van der Waals forces.

F_a introduces strain due to the interfacial stresses generated by the adhesive forces operating between the two solids. It can be expressed as

$$F_a = \tau_s A_r, \tag{3.1}$$

where τ_s is the shear strength of the contacts and A_r is the real contact surface.

Analogously, the adhesive friction coefficient can be expressed as

$$\mu_a = \frac{\tau_s A_r}{F_N}. \tag{3.2}$$

The determining factors for the adhesive friction coefficient are the adhesive interaction and the real contact surface, and depend therefore of both the shear strength and the real contact area.

In a simple way, the deformation component of friction has been described [9] as

$$\mu_d \propto E^{-1/3} \tag{3.3}$$

and is therefore inversely proportional to the Young's modulus (E).

The relationship between deformation and adhesion components of friction has been discussed among others by Bartenev and Lavrentev [11, 12]. They concluded that the relationship between F_a and F_d is determined by the type and nature of polymer, the surface roughness and the testing conditions.

3.2.4 *Cryotribology*

Low temperatures imply lower thermal conductivity of sliding material and better cooling efficiency of the surrounding medium. This obviously influences a change in the friction behaviour at cryogenic temperatures.

Attempts to find an empirical relationship between the friction force and the temperature were made in the past decades. According to the molecular-kinetic theory, Bartenev [12] proposed the following relationship between the friction force and the absolute temperature for high elastic polymer, where A is the real area of contact, A_0 the elementary real area of contact of one chain of polymer network, U the activation energy of the friction process referred to a single chain and λ the mean length of a jump. The quantity T_0 is a constant with the dimension of the temperature.

$$F = \frac{2U}{\lambda A_0} \cdot A_r \left(1 - \frac{T}{T_0}\right) \tag{3.4}$$

Other authors [13, 14] proposed an adjusted Arrhenius equation to fit the friction behaviour of PTFE where C is a constant $(\mathrm{m/s})^{-1}$, V the sliding velocity (m/s), R the universal gas constant and E_a the activation energy (J/mol).

$$\mu = (CV)^n \cdot \mathrm{e}^{\left(\frac{E_a}{R}\left(\frac{1}{T} - \frac{1}{T_0}\right)\right)} \tag{3.5}$$

Equations (3.4) and (3.5), however, could not be verified at cryogenic temperatures.

At very low temperatures, two main theories have been used to describe the friction behaviour of polymeric materials:

As the Young's modulus increases at low temperatures, the deformation components of friction decreases as well as the real contact area. Furthermore, since the adhesion component of the friction force F_a is related to the real contact area A_r according to (3.1), the corresponding adhesive friction should decrease with decreasing A_r. Consequently, the deformation and the adhesion component of friction should decrease at low temperature, accordingly to the two models of friction described above.

On the other hand, according to (3.1), F_a can also increase at low temperature due to an increase of the shear strength [15]. This theory was supported in particular by Gardos, who suggested that friction of PTFE should be higher at low temperatures due to large "lumpy" transfer film particles adhering to the counterface [16].

Therefore, the adhesive component of friction is the combined effect of a decrease of the real contact area and an increase in shear strength at low temperatures.

It should be added that the relationships between the elastic modulus and the temperature, as well as, between shear strength and the temperature are not always proportional due to glass transition of polymer [15]. If the main glass transition of polymer appears above 150 K, some small mechanical losses have been found at about 50 K for certain polymers [17], as well as tunnelling effect at very low temperatures [18]. These processes, which are polymer dependent, can also have an influence on the polymer properties, and therefore on the friction behaviour at cryogenic temperatures. As reported in Ludema and Tabor [19], the relationship between friction and temperature (down to −100 °C) for various polymers against steel depends on each polymer and does not follow a linear trend. Therefore, according to the statements above, it would not be possible to give a general rule for all polymers.

Another important factor that influences the frictional behaviour of polymers is obviously the environmental medium. That is why the authors will tend to review experiments according to the environment.

3.3 Review of Experiments

3.3.1 Experiments in LN_2

3.3.1.1 Influence of the Composition

Low temperature investigations of polymers have been mainly performed in liquid cryogens. Due to their exceptional lubricating properties PTFE and PTFE composites have been studied since many decades [20] and successfully utilized in space applications. Since then, in addition to PTFE which still is of major interest, a variety of polymers like Ultra High Molecular Weight Polyethylene (UHMWPE), PCTFE, Polyamide (PA), PI, PEEK have been tested against steel in LN_2 [3, 21–24]. Most experiments showed lower friction and wear in LN_2 compared to RT, attributed to the decreasing deformation.

While the friction and wear values generally decrease in cryogens and in particularly in LN_2, the mechanisms were found to be dependent on the polymer materials. Whereas the wear mechanism of PI was found to be mainly adhesive in LN_2 contrarily to ambient air [23], strong abrasion has been observed with PA6 in LN_2 due to hydrogen bonds that become very strong at low temperatures [24]. Concerning the wear mechanism of PTFE composites in LN_2 it was found to be strongly dependent on the sliding velocity. At very low velocities (0.006 m/s) the wear is characterized by highly localized deformation and scoring with evidence of surface fatigue [21]. At higher velocities, PTFE enables some transfer to the steel counterface due to the heat generated during sliding at 0.2 m/s and even more at 1 m/s [3, 24]. The influence of the velocities will be reported in more details later on.

The influence of the polymer composition has been further investigated within PTFE and PEEK matrix against steel. Particularly in LN_2, which is an effective coolant with a relatively high viscosity, lower friction values were observed compared to RT independently of the material composition (Fig. 3.3). Whereas the composition of the materials plays an important role at RT, the type of fillers (a) as well as the percentage of fillers and fibres (b and c) do not influence the friction coefficient in LN_2 as much [3, 7, 25].

Regarding the wear behaviour, the influence of the filler and matrix were more important (Fig. 3.4). Better wear resistance was found with bronze-filled PTFE composite compared to PEEK-filled PTFE, which was associated partly to a better thermal conductivity of the bronze-filled composite [7, 25]. The performance of PEEK composite, however, is far the best in LN_2 with the lowest specific wear rate. It has been shown that at low temperatures, the transfer of material onto the disc doesn't build a homogenous film as it does at RT as shown by the optical

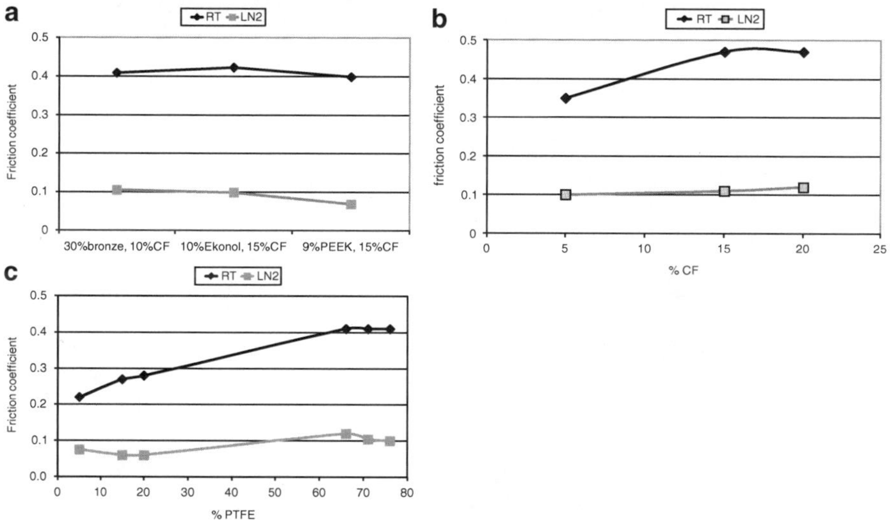

Fig. 3.3 Friction coefficient at RT and in LN_2 ($v = 0.2$ m/s). (**a**) PTFE and PEEK composites filled with 15 % CF; pure PEEK without CF. (**b**) PTFE composites filled with 20 % Ekonol®. (**c**) PTFE composites filled with different types of fillers

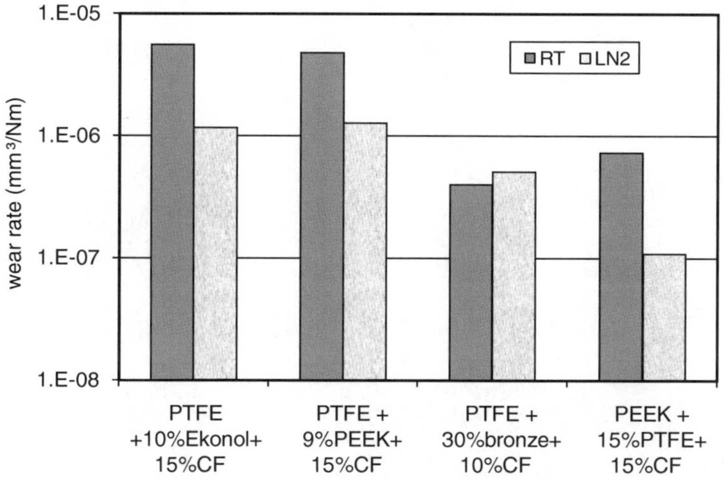

Fig. 3.4 Wear rate of PTFE and PEEK composites at 300 K in air and 77 K in LN_2

microscopy images (Fig. 3.5). The destruction of the PTFE band structure is hindered by the frozen state of the molecules, which also prevents the composite material from wearing out. Friction depends no longer on the formation and adhesion of the transfer film, but more on the real surface area in contact between pin and disc [7].

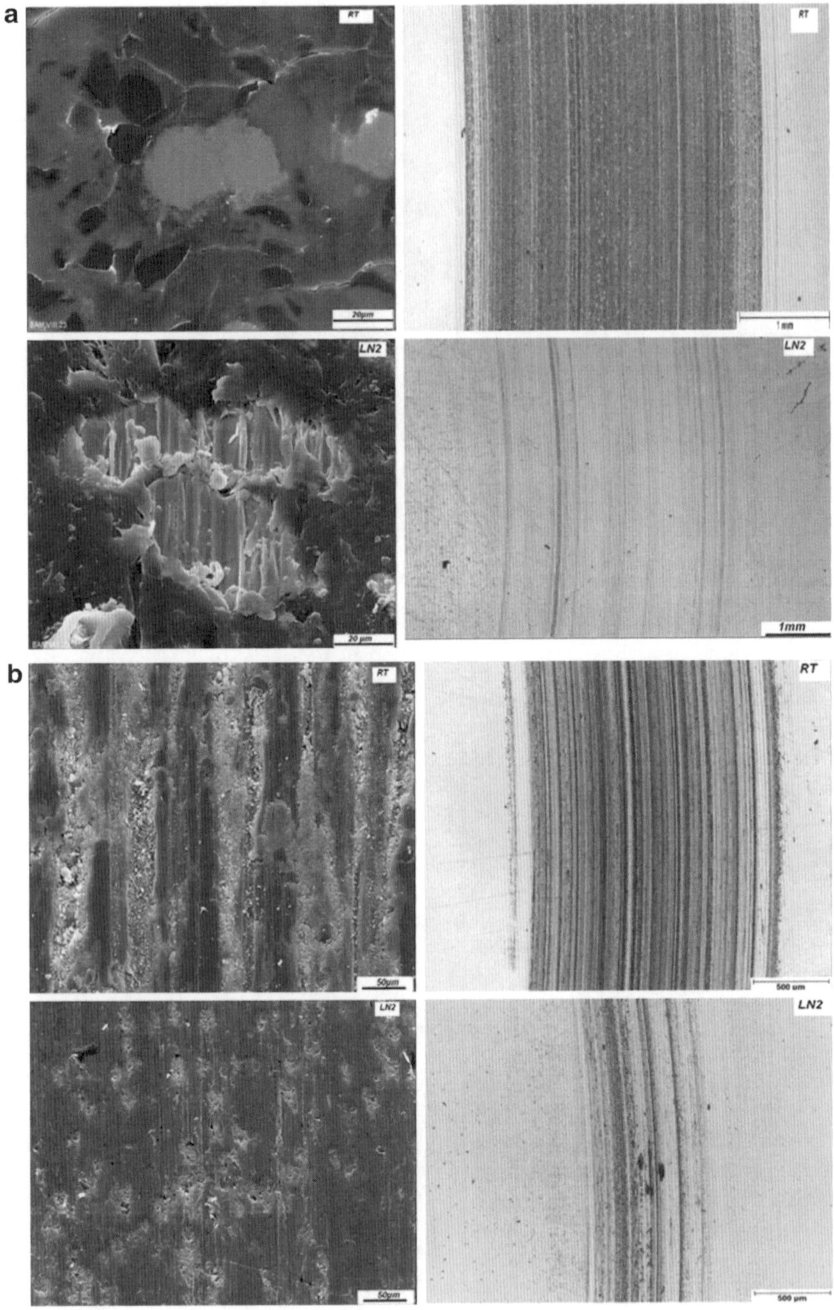

Fig. 3.5 SEM images of polymer composite and counterface after tests at RT and in LN_2 at $v = 0.2$ m/s. (**a**) PTFE composite against steel. (**b**) PEEK composite against steel

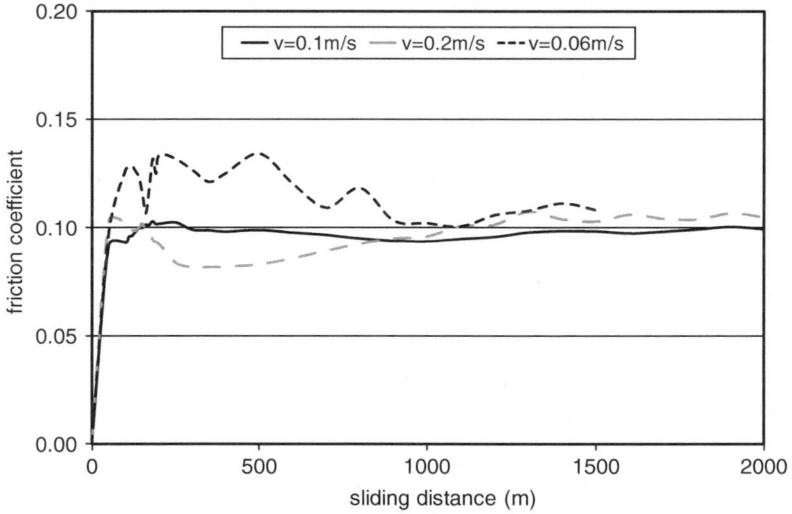

Fig. 3.6 Friction coefficient of PTFE filled with 30 % bronze and 10 % CF in LN_2 [25]

3.3.1.2 Influence of the Velocities

The influence of the velocities has been investigated in particular in LN_2. As mentioned in Michael et al. [22] and Rabinowcz [26], high speeds and low temperatures should prevent excessive junction growth, which produces a small contact area and low friction values, whereas, at low speeds and high temperatures, the strain-rate-dependent junction shear stress drops markedly, resulting in similarly low friction values. In [27], Michael showed however that at low temperature (77 K), the friction coefficient is independent of the speed.

The influence of velocity was also investigated with PI against steel in LN_2 [23]. Friction is nearly constant from 0.05 until 4 m/s. Below 2 m/s, the friction coefficient is lower in LN_2 than in ambient air and adhesion is predominant. Above 2 m/s, deformation regains importance. Similarly, the friction coefficient of PTFE composites was constant at $v = 1$, 0.2 and 0.06 m/s (Fig. 3.6) in LN_2 [25]. Therefore an optimal cooling seems to be reached in LN_2 until 1 m/s. It was also suggested that LN_2 has an influence in removing wear particles from the surface, probably due to its relatively higher viscosity.

3.3.2 Experiments in Helium Environment

As seen above, friction and wear of polymer composites decrease in LN_2 compared to RT. However, this decrease is not proportional to the temperature of the cryogens but depends on the temperature at the friction contact. Experiments in cryogens

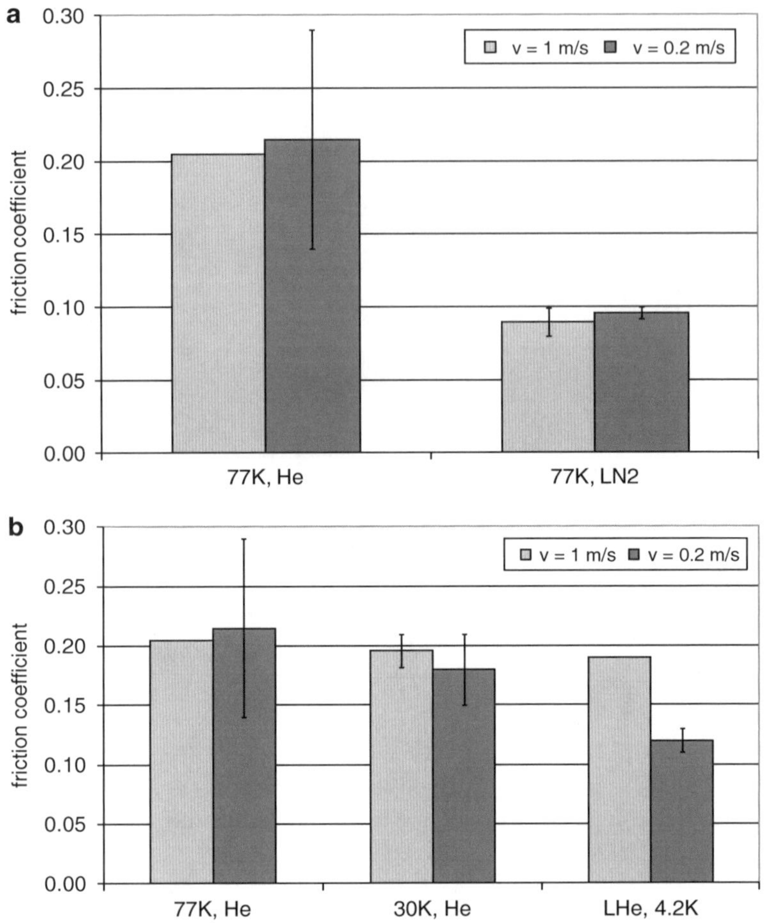

Fig. 3.7 Friction coefficient of PTFE filled with 30 % bronze and 10 % CF at $v = 1$ and 0.2 m/s. (**a**) at $T = 77$ K in He-gas and in LN_2 (**b**) in He-gas at $T = 77$, 30 and 4.2 K

showed higher friction in LHe than in LN_2 for several polymers [2, 7, 22, 24]. The thermal properties of the cryogenic medium, particularly the heat of evaporation and thermal conductivity, have a significant influence on the tribological performance of the polymer composites, especially since the thermal conductivity of the polymer decreases.

As explained in Friedrich et al. [7] and Theiler [25], experiments at $T = 77$ K indicate a much more favourable tribological behaviour in LN_2 than in helium gas due to a better cooling efficiency of the cryogenic fluid compared to the gas (Fig. 3.7a). In the same study, the friction coefficient of PTFE composite against steel were measured in helium environment at $T = 77$ K, $T = 30$ K and $T = 4.2$ K (Fig. 3.7b). Contrary to the experiments in LN_2, the influence of the velocities is significant in helium environment. At 0.2 m/s, the coefficient of friction decreases with the temperature whereas at higher speed friction becomes independent of the temperature. Due to the lower frictional heat at low sliding speed, the effect of low temperatures on the tribological behaviour of these composites was more clearly detected under these conditions, with a change in wear mechanism from mainly adhesive to more abrasive.

3.3.3 Stick-Slip Behaviour

Kensley [28] studied several metal-insulator combinations and found that LHe promotes stable sliding, whereas a stable sliding regime in LN_2 was much more difficult to achieve. He suggested that the existence of a critical temperature at which a transition from stable to unstable behaviour occurs at ~40 K.

Friction of polymer against polymer generally has a higher adhesion than against metal. Iwasbuchi investigated Polyethylene reinforced plastic vs. itself or vs. epoxy in air at RT, in LN_2 and in LHe [29]. Friction was the lowest at 77 K and the highest at 4 K.

On-going experiments carried out at BAM at low temperatures showed that the friction behaviour at the first pass is quite dependent on the polymer investigated. Similarly to Iwabuchi et al. [29], friction is lower at 77 K and higher at 4 K for Polyethylene terephthalate (PET) against PET (Fig. 3.8a). However, the friction behaviour of PTFE composite against PET is characterized by a lower friction at RT, high static friction in LN_2 and stick-slip at 4 K (Fig. 3.8b). In these experiments, the friction behaviour is similar in LN_2 for the two different friction couples, but polymer chemical structure has an influence in LHe.

3.3.4 Experiments in LOX and LH_2

In reactive LH_2, the first major works were performed by NASA Lewis Research Center [30–32]. Wisander studied PTFE laminate against stainless steel. The wear tracks were deeper and rougher in LH_2 than in LN_2. In a further work, he investigated several polymer composites in LH_2 for applications in turbo pumps, i.e. at high velocities (11.6 m/s) [32]. Promising results were obtained with graphite-filled PA, graphite and glass-fibre-filled PCTFE and PTFE-filled polyparaphenyl (PPP).

Experiments in LOX were conducted by Bozet who studied the influence of the environment on the wear behaviour of graphite-filled PI [33]. Compared to LN_2, the wear increased in LOX due to chemical process, in particular to the oxidation of the surface of the PI resulting in embrittlement and removal by abrasive of the mating material surface.

Theiler et al. studied the influence of liquid and gaseous hydrogen on the tribological behaviour of PTFE and PEEK composites. Compared with air, nitrogen

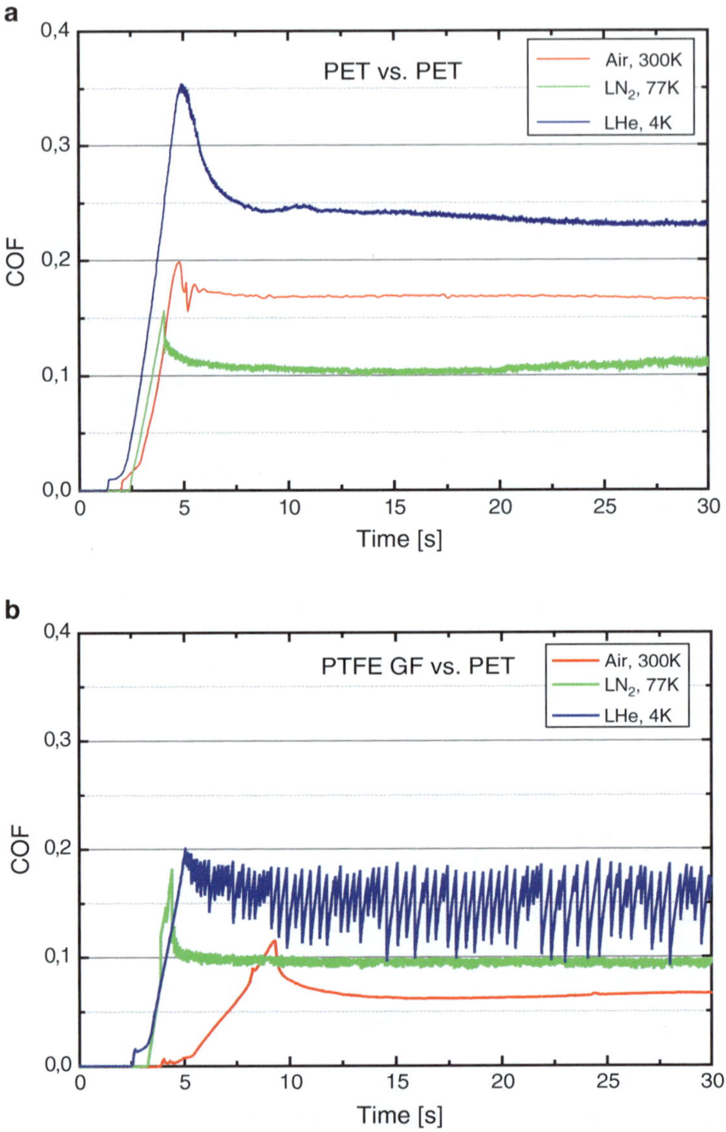

Fig. 3.8 Friction behaviour of polymer against polymer at 5 MPa during the first few seconds. (**a**) PET vs. PET. (**b**) PTFE composite vs. PET

or helium environments, the adhesion of the transfer is hindered in hydrogen (Fig. 3.9). Chemical analyses showed that the reduction properties of hydrogen have an influence on the triboreaction at the contact surface, enhancing the formation of iron fluorides to the detriment of peroxides that have a good adhesion to metal surfaces [25].

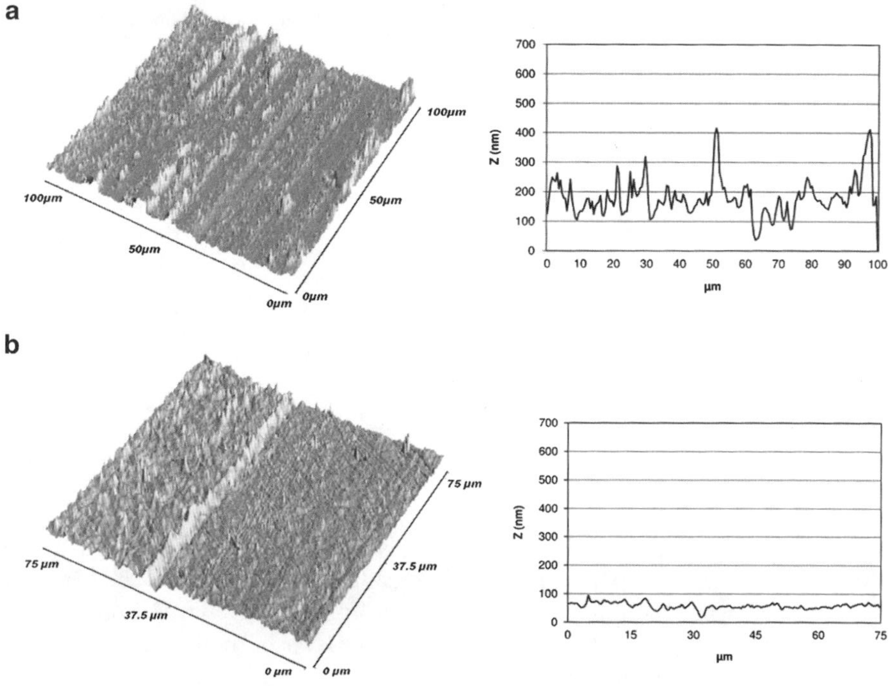

Fig. 3.9 AFM images of the surface of the discs after tests against bronze/CF-filled PTFE at $v = 0.2$ m/s in (**a**) LN_2 and (**b**) LH_2

Further friction experiments were conducted with several composites in LH_2. As seen in Fig. 3.10 the friction of most polymer composites is lower in LH_2 compared to ambient air, particularly the ones containing graphite as solid lubricant [34].

3.3.5 Experiments in Vacuum

Testing in vacuum is certainly the best way to study the influence of the temperature on the friction behaviour without having any effect of the cryogenic medium. According to the authors, only few researchers performed experimentations at cryogenic temperatures.

Bartenev and Elkin [11, 12] studied the friction force for viscoelastic polymer against steel between −200 °C (73 K) and 160 °C (473 K). At low temperatures, friction decreases due to a decrease in contact area. At higher temperatures, friction decreases with temperature according to the molecular-kinetic theory. The temperature dependence of the friction force was characterized by two maxima, a principal and a low-temperature maximum. The low-temperature maximum observed in the glassy state was associated with the mechanical loss maximum which could be related to microdeformation of the asperities as the polymer slides over the rigid

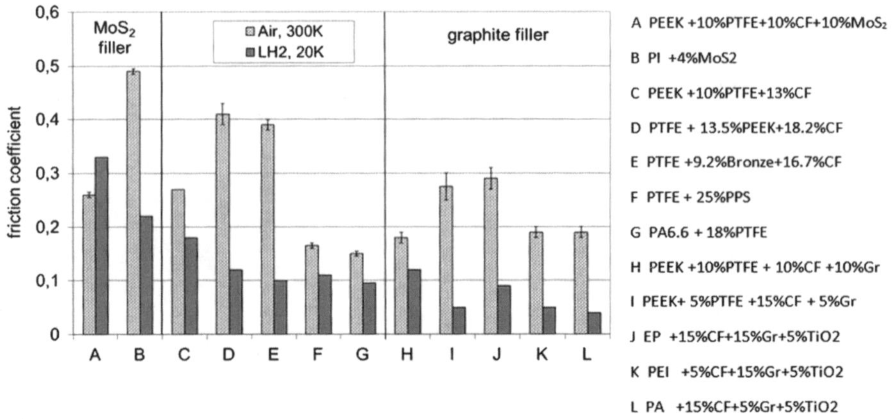

Fig. 3.10 Friction coefficient of various polymer composites in LH_2 (20 K) and in air (300 K) [34]

surface. Bartenev suggested however that this low-temperature maximum could also occur as a result of two competing factors. One is the growth of the forces of adhesion and hence the force of friction as the temperature falls. The other is the reduction in the actual area of contact and hence the force of friction upon transition from forced-elastic to purely elastic contact.

Similar observations were made by Martin who conducted few experiments in ultra-high vacuum [35]. Polyethylene film against metal was investigated as a function of temperature in the range 123–393 K. The maximal friction is reached at ambient temperature (300 K). Decreasing the temperature decreases the friction coefficient due to higher hardness. On the other hand, increasing the temperature above 300 K also decreases friction, which was also attributed to increased molecular flow.

More recently Burton [36] investigated the friction coefficient of PTFE coating against steel in vacuum using a sliding block cryotribometer down to 4 K. Results showed very little dependence of the temperature between 4 and 200 K. Contrary to his expectation, he concluded that at sufficiently low temperatures, the rate of thermally activated processes becomes vanishingly small.

3.4 Conclusion

Sliding systems in cryogenic environments are present in a wide variety of technical applications ranging from turbo pumps in rocket engines to supporting and insulating elements in applied superconductivity.

Many polymer materials maintain their favourable tribological properties in dry sliding also at low temperatures. Therefore, they are extensively used in cryogenic engineering where conventional lubrication by oil or grease is not possible. Compared to RT, many polymers show lower friction and reduced wear under cryogenic conditions. Because of their chemical inertness, polymers are also appropriate for operation in reactive media, such as hydrogen or oxygen.

As stick-slip is a major problem in the operation safety of superconducting magnets, knowledge about the frictional behaviour of polymers used for electrical insulation or as spacers must be gained.

Although friction and wear have been well studied since many decades, the tribological behaviour of polymer materials doesn't systematically follow theories at very low temperatures. Due to the wide range of operating conditions in cryogenic applications, it is difficult to state general rules. Therefore, experimental studies should be performed when new materials or requirements are demanded.

References

1. Gradt Th, Theiler G (2009) Influence of solid lubricant fillers on the tribological behaviour of PEEK composites in vacuum, Proc. 13th European Space Mechanisms and Tribology Symposium, 23.–25. Sept. 2009, Vienna, Austria, ISBN 978-92-9221-234-6, ISSN 1609-042X
2. Theiler G, Gradt T (2010) Friction and wear of PEEK composites in vacuum environment. Wear 269:278–284
3. Theiler G, Hübner W, Gradt T, Klein P, Friedrich K (2002) Tribological behaviour of PTFE-composites against steel at cryogenic temperature. Tribol Int 35:449–458
4. Hartwig G (1994) Polymer properties at room and cryogenic temperatures. Plenum Press, New York
5. Ahlborn K (1991) Cryogenic mechanical response of carbon fibre reinforced plastics with thermoplastic matrices to quasi-static loads. Cryogenics 31(4):257–260
6. Reed RP, Clark AF (eds.) (1983) Materials at low temperatures. American Society for Metals, Metals Park, Ohio
7. Friedrich K, Theiler G, Klein P (2010) Polymer composites for tribological applications in a range between liquid helium and room temperature. In: Sinha SK, Brisco BJ (eds) Polymer tribology. Imperial College Press, London
8. Theiler G, Meine K, Hübner W, Klein P, Friedrich K (2002) Thermal shock cycles experiments between room and cryogenic temperatures on polymer composites. In: Conference Proceeding 19th ICEC (International Cryogenic Engineering Conference), Grenoble, France
9. Briscoe BJ (1986) Interfacial friction of polymer composites. General fundamental principles. In: Friedrich K (ed) Friction and wear of polymer composites. Elsevier, Amsterdam
10. Czichos H (1986) Introduction to friction and wear. In: Friedrich K (ed) Friction and wear of polymer composites, Composites Materials Series, vol 1. Elsevier, Amsterdam, pp 1–23
11. Bartenev GM, Lavrentev VV (1981) Friction and wear of polymers, vol 6, Tribology series. Elsevier, Amsterdam
12. Barthenev, Elkin (1965) Friction properties of high elastic materials. Wear 8:8
13. Blanchet TA, Kennedy FE (1992) Sliding wear mechanism of polytetrafluoroethylene (PTFE) and PTFE composites. Wear 153:229
14. Cook NK, Burris DL, Dickrell PL, Sawyer WG (2005) Cryogenic friction behaviour of PTFE based solid lubricant composites. Tribol Lett 20(2):109–113
15. Yamaguchi Y (1990) Tribology of plastic materials, vol 16, Tribology series. Elsevier, Amsterdam
16. Gardos MN (1986) Self-lubricating composites for extreme environmental conditions. In: Friedrich K (ed) Friction and wear of polymer composites, Composites materials series, vol 1. Elsevier, Amsterdam, pp 397–447
17. Yano O, Yamaoka H (1995) Cryogenic properties of polymers. Prog Polym Sci 20:585–613

18. Slutsker A, Hilyarov V, Polikarpov Yu, Karov D (2010) Possible manifestations of the quantum effect (Tunneling) in elementary events in the fracture kinetics of polymers. In: "Physics of the Solid State" Physics of the Solid State, vol 52, no 8, Pleiades Publishing, pp. 1637–1644
19. Ludema KC, Tabor D (1966) The friction and visco-elastic properties of polymeric solids. Wear 9:329–348
20. Wisander D (1959) Wear and friction of filled PTFE compositions in liquid nitrogen. ASLE Trans 2(1):58–66
21. Glaeser WA, Kissel JW, Snediker DK (1974) Wear mechanisms of polymers at cryogenic temperatures. Polym Sci Technol 5B:651–662
22. Michael PC, Rabinovich E, Iwasa Y (1991) Friction and wear of polymeric materials at 293, 77 and 4.2K. Cryogenics 31:695–704
23. Bozet JL (1993) Type of wear for the pair Ti6Al4V/PCTE in ambient air and in liquid nitrogen. Wear 162–164:1025–1028
24. Hübner W, Gradt T, Schneider T, Börner H (1998) Tribological behaviour of materials at cryogenic temperatures. Wear 216:150–159
25. Theiler G (2005) PTFE- and PEEK composites for tribological applications at low temperatures and in hydrogen. BAM Dissertation-Band 14, Publisher : BAM, Ed. NW-Verlag, Bremerhaven
26. Rabinowcz E (1965) Friction and wear of materials. Wiley, New York
27. Michael PC, Iwasa Y, Rabinowicz E (1994) Reassessment of cryotribology theory. Wear 174:163–168
28. Kensley RS, Maeda H, Iwasa Y (1981) Transient slip behaviour of metal/insulator Pairs at 4.2K. Cryogenics 21:479–489
29. Iwabuchi A, Arai H, Yoshino Y, Shimizu T, Sugimoto M, Yoshida K, Kashima T, Inui H (1995) Frictional properties of ceramics, MoS_2 coated films and polyethylene fibre reinforced plastics at 4.2K in liquid helium. Cryogenics 35(1):35–40
30. Brewe DE, Scibbe HW, Anderson WJ (1966) Film transfer studies of seven ball bearing retainer materials in 33K hydrogen gas at 0.8 million DN value. NASA technical note D-3730
31. Wisander DW, Johnson RL (1968) Wear rate and friction coefficient in liquid nitrogen and hydrogen of steel sliding on polymer laminates. NASA technical note D-4463
32. Wisander DW, Johnson RL (1969) Friction and wear of nine selected polymers with various fillers in liquid hydrogen. NASA technical note D-5073
33. Bozet JL (2001) Modelling of friction and wear from designing cryogenic valves. Tribol Int 34:207–215
34. Theiler G, Gradt Th (2007) Polymer composites for tribological application in hydrogen environment. In: Conference Proceeding Second International Conference on Hydrogen Safety, Saint Sebastian, Spain
35. Martin J-M, Liang H, Le Mogne T, Malroux M (2003) Low-temperature friction in the XPS analytical ultrahigh vacuum tribotester. Tribol Lette 14(1):25–31
36. Burton JC, Tabork P, Rutledge JE (2006) Temperature dependence of friction under cryogenic conditions in vacuum. Tribol Lett 23(2):131–137

Chapter 4
Mechanical Behavior of Polymer Composites at Cryogenic Temperatures

Sanghamitra Sethi and Bankim Chandra Ray

Contents

S. Sethi • B.C. Ray (✉)
Department of Metallurgical and Materials Engineering, National Institute of Technology, Rourkela 769 008, India
e-mail: drbcray@gmail.com

S. Kalia and S.-Y. Fu (eds.), *Polymers at Cryogenic Temperatures*,
DOI 10.1007/978-3-642-35335-2_4, © Springer-Verlag Berlin Heidelberg 2013

Abstract High specific strength, stiffness, excellent environmental fatigue resistance and low weight remain the winning alliance that impels fibrous composite materials into new arenas, but other properties are also equally important. Fibrous reinforced plastics (FRPs) offer good vibrational damping and a low coefficient of thermal expansion, characteristics that can be engineered for specialized applications. Commercial composites are used in large markets such as automotive components, boats, consumer goods, and corrosion-resistant industrial parts. Advanced composites, initially developed for military aerospace applications, offer performance superior to that of conventional structural metals and now find applications in satellites, aircraft, and sporting goods and in the energy sector in oil and gas exploration and wind turbine constructions. Cryogenic applications of polymeric fiber composites are mainly in superconductivity, space technology, and handling of liquefied gases. By contrast, because of the heterogeneous nature and anisotropic behavior of FRPs, a structural designer faces challenges in predicting the integrity and durability of FRP laminates during service periods. Polymer composites soften, creep, and distort when heated to high temperatures (>100 °C), accompanied by collapse of free volume as the molecular adjustments take place. This can result in buckling and failure of load-bearing composites structures. Severe environmental exposure affects the physical and mechanical properties of polymeric composite materials, resulting in an undesirable degradation and damage.

Cryogenic fuel tanks are the most common structural application of FRP at low temperatures. Expose to cryogenic temperatures can cause microcracks as well as delamination in the composites due to thermal residual stresses. These microcracks provide a pathway for the ingress of moisture or corrosive chemicals and are a possible pathway for loss of cryogenic fluids in the tanks. Matrix resins at low temperatures are brittle and do not allow relaxation of residual stresses or stress concentration to take place. At low temperatures, polymers are well below their glass transition temperature and show little viscoelastic behavior. Molecular motion of segments or side groups is still possible, but the degrees of freedom decrease with decreasing temperature. This motion influences the damping behavior of the polymers under cyclic mechanical load. If the temperature-dependent relaxation time of molecular motion is equal to the time of external deformation, maximum power dissipation occurs. Simultaneously, a change in the shear modulus is observed. The goal of this chapter is to extensively study the in-plane mechanical properties of FRP composites at cryogenic temperatures. The composites considered include carbon, glass, and Kevlar fiber-reinforced polymers with different resin matrices.

Keywords Polymer matrix composite • Cryogenic temperature • Failure and fracture • Damage and degradation • Mechanical properties • Scanning electron microscope • Interphase • Atomic force microscope • Alternating differential scanning calorimeter

Abbreviations

AFM	Atomic force microscope
CFRP	Carbon fiber-reinforced plastic
FRP	Fibrous reinforced plastic
FTIR	Fourier transform infrared spectroscopy
GFRP	Glass fiber-reinforced plastic
ILSS	Interlaminar shear strength
IPN	Interpenetrating network
PMC	Polymer matrix composites
SEM	Scanning electron microscope
T_g	Glass transition temperature
TMDSC	Temperature-modulated differential scanning calorimetry

4.1 Introduction

4.1.1 Status and Strength

During the last few decades, composite materials have become an interesting topic to engineers and manufacturers. The first appearance of composites dates back to 1500 BC when Mesopotamians and Egyptian settlers made their strong and durable homes with a mixture of mud and straw. Straw acts as reinforcement and fulfills all the optimum properties. The major categories of composites are metal matrix composite (MMC), ceramic matrix composite (CMC), and polymer matrix composite (PMC). The material applications and degradation characteristics are very specific to the composite material. This chapter focuses mostly on PMCs, with special emphasis on performance at low temperatures.

The term "composite" covers a very broad class of materials made of several (at least two) components. In contrast to alloys, the composite constituents are generally distinct at the macroscopic level. However, composite materials are often engineered for specific requirements. Commercial composites are used in large markets such as automotive components, boats, consumer goods, and corrosion-resistance industrial parts. Advanced composites, initially developed for military aerospace applications, offer performance superior to that of conventional structural metals and now find applications in satellites, aircraft, and sporting goods and in the energy sector in oil and gas exploration and wind turbine construction. Aircraft doors, fairings, and interior panels are perfect for FRP technology. Other potential applications could include the horizontal tail leading edge, pylon stub fairing, and engine cowl panel troop door deflector. The large Airbus A380 has 11 times greater length and over 1,000 times the weight of the Wright's 1903 Flyer, this super-jumbo will be the first airliner to have a carbon fiber composite (CFRP) centre-wing box. In addition, the rear pressure bulkhead, crucial for passenger safety, will be of CFRP, as will the entire rear fuselage aft of the bulkhead. This is one of the largest carbon fiber parts yet produced by Airbus. The composite A500 carbon Aero from Adam Aircraft flew successfully in mid-2003. Composite materials are being used in several crucial parts of helicopters such as the Boeing-Sikorsky Comanche helicopter and Bell-Boeing V22 Osprey. The parts include exhaust doors, shrouds for transmission, and tail rotor blades. The composite deck is expected to last for more than 50 years with less maintenance. Currently, a car manufacturer is using composite materials in car metal bumpers, radiator end tanks, door handles, and front and rear ends [1, 2]. The world's oldest composite industry was founded in Russia. Looking to the future, Russians will be fighting against the constraints of gravity with composite applications and there will be some exploring opportunities available. Russia has a very comprehensive and extensive rail and road network, with major cities having underground transport systems. As more and more cargo is exported and more people travel by rail, refurbishment of rolling stocks is strongly needed. The temperature in winter is very cold and as a result the coaches and wagons face severe corrosion problems. Composite materials seem to be the best choice for refurbishment of old systems. A recent example for this is the Bridge-in-a-Backpack for the 2014 Winter Olympics in Russia. This is an innovative inflatable composite-concrete arch bridge developed to reduce construction time and cost, increase in lifespan, reduce maintenance costs, and reduce the carbon footprint of bridge construction.

Polymer matrix is mainly used for the hull bottom of V24 racing powerboats. Here, the hull bottom is designed to both minimize the wetted area as the boat rises in the water at speed and to help aerate the water's surface layer for providing a more forgiving ride. Windmill blades made up of FRP are exposed to a large number of sustained and occasional environmental loads such as bending, vibration, UV, cold and heat, moisture, and also impact loading. Such loads combined with large dimensions perfectly illustrate the challenge of durability assessment of composite materials.

At cryogenic temperatures (i.e., below $-150\ ^{\circ}C$), fibrous polymeric composites undergo inelastic deformation and damage when subjected to mechanical loading. Despite such a restriction, the field of polymer composites still remains broad. The nature and geometry of the material constituents define the response of the composite to external mechanical and environmental loads to a large extent. At low temperatures, the material might have some nucleating cracks and delaminated areas, possibly by misfit strains at the interface region due to differences in the coefficient of thermal expansion (CTE). Also, matrix resin is brittle at low temperatures unless modified by a modifier agent to increase its toughness value and retaining its strain capabilities. However, at low temperatures the performance of fiber has not been well evaluated and characterized. Unless and until a complete and comprehensive study of the constituents (fiber, matrix resin, and interphase) of FRP composites has been made and assessed, the full potential of these material may not be fully utilized for cryogenic applications.

4.1.2 Prospects and Problems

Structures like the "Benghazi Dome", an advanced roof in Dubai Airport, and the roof of London's Covent Garden Flower Market show the long-term durability and integrity of composite materials. A church spire made up of GFRP-reinforced materials stands near Birmingham, UK, erected in 1961 and still in service. There are many instances where composites have been added to pilot or demonstration projects but where better care should have been taken in their design and implementation. Boeing has awarded a project for developing damage tolerant design. A marine engineer will be designing a propeller in which the plastics are strong enough to serve the requirement. Many ships have propellers made of metal, which are corroded easily in sea water and set up galvanic action. To overcome these problems, plastics can play a vital role. Resin Systems (Edmonton, Canada) is launching a range of composite hockey stick shafts that are urethane-based composite resins with fiber replacement technology. This type of hockey stick has better cut resistance and impact resistance as compared to other competitive products. Most polymer matrices are characterized by the presence of amorphous phases. Due to this non-equilibrium state, polymer composites are particularly sensitive to environmental factors such as temperature, time, and exposure to liquids, gases, electrical fields, and radiation. Static and dynamic mechanical loads can interact with the environmental parameters and accelerate the degradation process. Defects along the matrix and reinforcement interface further amplify the action of environmental factors. In this chapter, we investigate the effects of low temperature on the structural integrity of PMC.

Hydrogen technology is progressing rapidly. Liquid hydrogen vessels for vehicles will be made with lightweight materials of high specific strength and fatigue resistance and low thermal conductivity. Wind tunnels yield much higher Reynolds numbers when operated at cryogenic temperatures. CFRPs are prime candidates due to their high specific strength and high fatigue endurance limit for this application. The maximum use of FRP composites in large structural applications requires a better understanding of interdependent thermal and mechanical responses when subjected to elevated and high temperature. The thermo-mechanical behavior of FRP composites mainly depends on the polymer resin used [3–6, 8]. For better application of fibrous composites, a comprehensive understanding of the mechanisms of aging and environmental degradation is needed. Environmental attack can degrade the fiber–matrix interface mostly by mechanochemical processes. The thermal aging behavior of epoxy resins is important because of their expanding use for structural applications, where increased temperatures are a common environmental condition. A material that exhibits ductile behavior at room temperature may become brittle at low temperatures or may soften and creep at elevated temperatures [7]. With these temperature fluctuations, changes in strength and/or stiffness are observed. The effect of the combination of thermal and mechanical stresses may cause a change in location of failure as well as a change in the threshold load factor needed to cause failure. It is very difficult to generalize the effect of changes in temperature. Matrix-dominated properties such as compressive, flexural, shear, and transverse properties are mostly affected by the temperature of the matrix. Thermal conditioning imparts better adhesion and, thus, an improved interlaminar shear strength (ILSS) value, especially for lower conditioning times . Considering the dependency on temperature for the physical properties of many polymeric materials, various relaxation effects are extremely important. Thermal degradation of resin involves chemical reaction and physical changes. Chemical reaction is represented by oxidation, further crosslinking and reaction of unreacted monomers, whereas physical change is represented by the viscoelastic behavior. The viscoelastic yield behavior generally depends upon the temperature and loading rate. At the macromolecular scale, chain scission and crosslinking affect the polymer network and, thus, alter the mechanical properties of the oxidized layer [8]. At the macroscopic level, the hindered shrinkage of the oxidized layer induces a stress gradient that is susceptible to the initiation and propagation of cracks [1]. Thermal conditioning behaviors of glass/epoxy and carbon/epoxy composites are in huge demand because of their expanding use for structural applications where increased temperatures are a common environmental condition. The fracture processes of polymers are strongly influenced by deformation or yielding processes, which depend on temperature and time. At very low temperatures, no yielding is possible and fracture is brittle. At high temperatures, two characteristic phenomena occur after the yield point: strain softening and strain hardening [8]. The specimens tested at a lower temperature are characterized by a greater level of microcracking and delamination. These effects are believed to be due to higher thermal residual stresses. Due to the heterogeneous nature of the fiber and matrix, there are large thermal expansion and contraction mismatches, which generates thermal stresses. This weakens the fiber–matrix interface and, as a result, their ILSS values deteriorate progressively.

4.2 Architecture and Character of Synthetic Fibers and Chemistry of Matrix Resins

A large choices of synthetic fibers is currently available for tailoring the FRPs to meet the needs of a wide spectrum of applications, from casual to supercritical range. Glass, carbon, and Kevlar fibers are the major classes of reinforcement for advanced composites. Glass is an amorphous material (i.e., non-equilibrium) in nature. Glass has been used for various purposes for thousands of years. There are different grades of glass fibers: E-glass (E for electrical) is based on $CaO–Al_2O_3–SiO_2$. Its excellent process ability enables the drawing of long fibers at relatively low cost. S-glass on the other hand is based on $SiO_2–Al_2O_3–Mg$ and exhibits a higher stiffness and strength. Unfortunately, the higher temperature resistance of S-glass also translates into more difficult fiber manufacturing processes, yielding higher fabrication costs. Carbon fibers are generally used in more advanced applications. Carbon fibers can be developed from different precursors such as rayon, cellulose, and polyacrylonitrile (PAN). Manufacturing processes are generally complex and details kept proprietary. The price of carbon fibers still remains high and their use is generally confined to key components where the high stiffness-to-weight ratio enables significant cost reductions. For calculation purposes, it might also be meaningful to classify composites with respect to their reinforcement geometries. Multiple reinforcement geometries exist, including continuous fiber reinforcement, fabrics, short (mat) reinforcement, particulate reinforcement, and (occasionally) 3D reinforcement. Depending on the application, such reinforcements can be included in various composite structures such as single layers, laminated structures, or sandwich composites.

Aramid fiber is used to make a variety of clothing, accessories, and safety equipment. It is lightweight and extraordinarily strong, with five times the strength of steel on an equal-weight basis. Best known for its use in ballistic- and stab-resistant body armor, Kevlar brand aramid fiber has shown its own heroism in helping to save the lives of thousands of people around the world.

4.2.1 Glass Fiber

Glass fibers are widely used in all types of polymer composite. It is used in aircrafts, civil infrastructures, and sports goods [9–12]. A well-known application is the hull of yachts. This is ideal use of fabric with polyester resin and glass fiber in a marine environment, giving freedom from corrosion. The nonmagnetic nature of glass fiber is an excellent choice for minesweeper hulls, instead of magnetic steel with its disadvantages. In Germany during World War I, glass fibers (replacing asbestos) were produced by touching a heat-softened glass rod to a spinning wheel [13].

4.2.1.1 Structure and Properties of Glass Fiber

After World War II, glass fibers entered the market in a greater volume for reinforcement of polymers. Various types of glass fiber exist, but the chemical composition is mostly based on silica (SiO_2) with addition of oxides of alumina, sodium, magnesium, and calcium. Fibers can be produced in either a continuous filament or staple form. The molten glass is extruded through a hole of 1 or 2 mm diameter in a spinner jet. The fiber diameter usually lies between 5 and 15 μm and is greatly influenced by the winding speed and viscosity of the molten glass.

The composition of E-glass fiber is: SiO_2 52.4%, Al_2O_3 14.8%, Fe_2O_3 14.4%, CaO 17.5%, MgO 4.5% and B_2O_3 8.0% by weight [10].

The important features of GFRP composite are good corrosion resistance, lightweight, high toughness, high absolute and specific strength values, and process and shape freedom. From a micromechanics point of view, crucial points are a glass fiber that transfers the load from fiber to fiber properly and the selection of matrix resin [14].

In response to the numerous and diverse range of GFRP, thermosetting polymer resins are more reliable. Generally, high strength curing systems are more durable and environmentally friendly.

4.2.1.2 Silane Treatment of Glass Fibers

Over the last three decades, glass fibers have played a dominant role in various composite applications. This is partly due to the development of silane coupling agents that can act as a protective layer for glass fibers and as a coupling agent to endorse adhesion with the polymer matrix resin. The polyhedron network structure of sodium silicate glass is schematically shown in Fig. 4.1. Each polyhedron can be seen to be a combination of oxygen atoms around a silicon atom, bonded together by covalent bonds [15]. Here, the sodium ions are linked with charged oxygen atoms by ionic bonds but are not linked to the network. This 3D network of glass exhibits isotropic properties.

In the case of carbon and Kevlar, the network shows anisotropic properties. The unique characteristic of glass fiber is that the elastic modulus along the fiber axis is same as that in the transverse direction [7]. The physical and mechanical properties of glass fiber are: diameter 5–25 μm, density 2.54 kg/m^3, elastic modulus at 25°C 7,700 kg/mm^2, tensile strength 2.4 GPa, Young's modulus 72.4 GPa, and CTE 5×10^{-6} °C [10].

To explain the interfacial bonding mechanisms of silane coupling agents with the fiber surface, several theories have been proposed. Chemical bonding is widely spread [16, 17]. Silane mostly contains organofunctional groups for different resin chemistries. The structure of organosilanes is X3–Si–R. This acts as a siloxane bridge between the resin and the glass fiber by forming a chemical bond. This multifunctional molecule reacts at one end with the glass fiber surface and the other end with polymer surface. The R group generally reacts with the matrix resin and the X group, which can hydrolyze to form a silanol group in an aqueous solution, reacts with the hydroxyl group of the glass surface.

The thickness and orientation of the fiber are determined by conditions of deposition, topography of glass fiber surface, concentration of solution, and treatment duration [18].

The microstructure of the coupling agent depends upon various factors. The coupling agent–resin matrix interface is a diffusion boundary where intermixing takes place due to penetration of the resin into the chemisorbed silanes layers and the migration of the physisorbed silane molecules into the matrix phase [19].

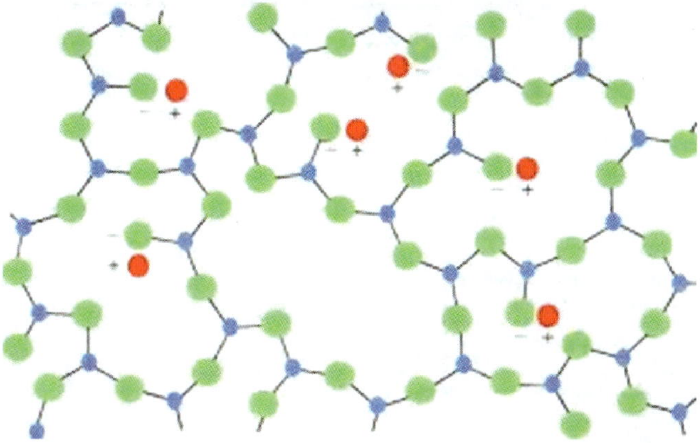

Fig. 4.1 Two-dimensional polyhedron network structure of sodium silicate glass. http://www.tech.plym.ac.uk/sme/mats324/mats324c1%2ofibers.htm

The interphase material tends to have a lower glass transition temperature (T_g), higher modulus and tensile strength, and lower fracture toughness than the bulk matrix [20].

Apart from mechanical properties, the characteristic of silane treatment is its ability to provide the glass fiber with a water-resistant bond. The degradation of untreated glass fiber–resin matrix interface is more affected by water. The water penetrates into the interface region by diffusion along the fibers and is eventually absorbed by the glass fiber. The water molecules form weak hydrogen bonds with the oxide groups of the glass fiber. This hydrolyzes the existing bond and thus weakens the adhesion level, which tends to degrade the mechanical properties [20].

The glass fiber shows four different types of fracture surfaces. These regions include a smooth mirror region, a mist region with some radial ridges, a hackle region consisting of big ridges, and branching of crack propagation.

Silane agents are intended to act as a protective coating for the glass fiber surface and as a coupling agent to promote adhesion with the polymer matrix. The silane agents are applied to the glass fiber surface as a size along with other components. Use of a silane coupling agent for improving the bond quality is common and a wide variety of organofunctional silanes have been developed. In chemical bonding theory, the bifunctional silane molecules act as a link between the resin and the glass by forming a chemical bond with the surface of the glass through a siloxane bridge, while its functional groups bond to the polymer resin. The general chemical formula is shown as XaSi-R, i.e., multifunctional molecules that react at one end with the glass fiber surface and at the other end with the polymer phase. R is a group that can react with the resin, and X is a group that can hydrolyze to form a silanol group in aqueous solution and thus react with a hydroxyl group of the glass surface (as shown in Fig. 4.2). The R group may be vinyl, aminopropyl, methacryloxypropyl, etc.; the X group may be chloro, methoxy, ethoxy, etc.

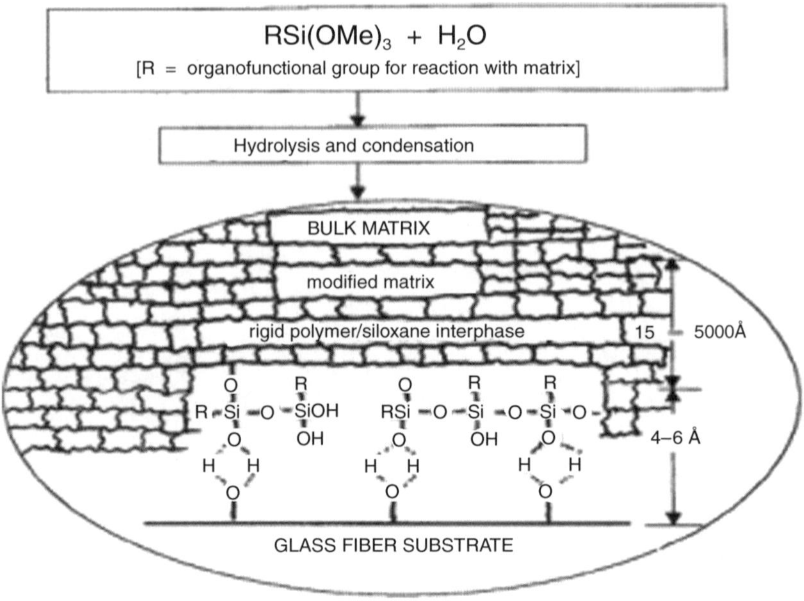

Fig. 4.2 Functions of coupling agents at the interface region between glass fibers and bulk epoxy region

The subject of silane chemistry and its interaction with both glass surface and polymer resin have been studied extensively. The following points should be noted:

1. The chemical reactivity of the organofunctional group of the silane to form covalent bonds with the polymer matrix
2. The primary or secondary chemical bond formation at the glass interface
3. The ability of the polymer matrix to diffuse into the siloxane interphase to form a rigid, tough, water-resistant interpenetrating polymer network (IPN) as a transition zone between the bulk matrix and the glass reinforcement [18]

An IPN is layer of silane agent that usually does not produce an optimum mechanical strength, and there must be other important mechanisms taking place at the interface region.

The presence of a stiff fiber mitigates the effect of a soft interphase, increasing the effective modulus of the interphase beyond that of the bulk matrix in close vicinity of the fiber [21, 22]. As stated above, FTIR and other spectroscopic techniques have been used to quantify the rates of interdiffusion and chemical reaction between silanes and polymeric matrices (as shown in Fig. 4.3). Generally, it may be concluded that the formation of IPNs contributes to an increase in interfacial adhesion and that the conditions under which the silanes are applied has a significant effect on the mechanical properties of the resulting composites [23, 24].

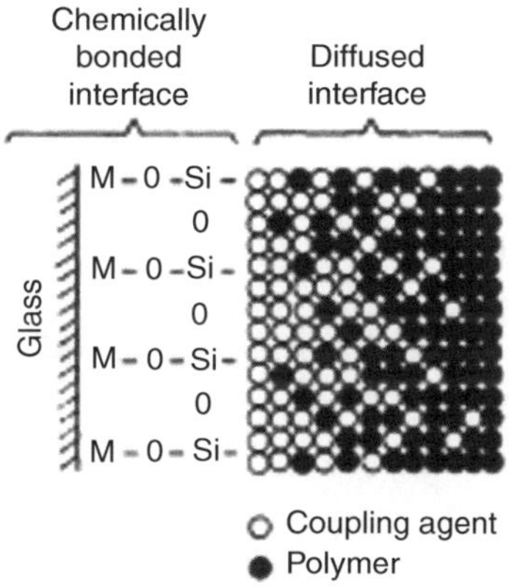

Fig. 4.3 Chemical reaction between silanes and polymeric matrices

4.2.2 Carbon Fiber

Carbon fibers make up a significant volume fraction of modern structural airframes. Embedded into polymer matrices, they provide significant strength and stiffness gains by unit weight compared with competing structural materials [25]. In the last few decades, vast improvements in the carbon fibers have been obtained. In large part, these improvements were accelerated in the field of aerospace, space, and various industries in response to higher strain-to-failure fibers. Higher strain, higher modulus carbon fibers merge with the resin matrix to provide optimum properties (i.e., load-bearing capacity) in various applications [26].

This evolving carbon fiber is manufactured from organic fibers (precursors) by treating with heat to obtain an ordered carbon structure. The precursors are mainly PAN and pitch [7]. Scanning tunneling microscopy (STM) reveals a systematic change in the surface profile as a function of oxidative fiber surface treatment. The topography of PAN-based carbon fibers consists of graphitic platelets of dimensions $30 \times 20 \times 100$ nm. The adhesion between fiber and matrix is a function of surface treatment. If the amount of oxidation is high, then there is decline in interfacial bond strength [27].

PAN-based fibers can be divided into different categories: (1) high strength, (2) high modulus, and (3) ultrahigh modulus [1]. PAN fiber is a long-chain linear polymer composed of a carbon backbone with attached carbonitrile groups. By treatment of this fiber in an oxidizing atmosphere at temperatures of 200–300 °C, we obtain carbon fibers (shown in Fig. 4.4) [28]. The properties of PAN-based carbon fiber are: tensile strength 2,300 MPa, tensile modulus 145 GPa, compression strength 1,600 MPa, compression modulus 135 GPa, and short beam shear stress 120 MPa [1].

The strength of a carbon fiber depends on the type of precursor fiber used. Its heat treatment temperature along with presence of flaws or defects has a great impact on the durability of the fiber [29]. With increasing the heat treatment temperature to about 1,000 °C, the strength is increased [30, 31].

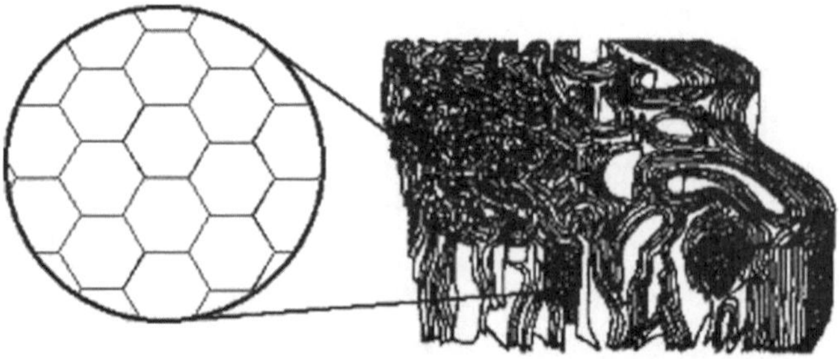

Fig. 4.4 Drawing of a 3D model of a carbon fiber [62]

4.2.2.1 Surface Treatments of Carbon Fibers

CFRPs have poor shear strength. In particular, the lack of shear strength of polymers reinforced with high modulus fibers is generally attributed to a lack of bonding at the fiber–matrix interface [32]. The mechanisms of bonding between carbon fibers and polymer matrices are as complex as those of glass fibers, and there are more complications associated with the carbon fiber surface because it is highly active and readily absorbs gases. A schematic model for the chemical reaction between oxidized fibers and epoxy resin has been presented as shown in Fig. 4.4 and the change in chemistry at the carbon fiber–epoxy interface by sizing well illustrated [20].

Chemical bonding due to the presence of functional groups changes the energetics of the carbon fiber surface considerably [33]. However, there is a limit to the improvement in bond strength by surface treatment alone. A decrease in bond strength and other deteriorating effects are expected to occur if the surface treatment is excessive, leading to severe damage of the fiber.

4.2.2.2 Effect of Surface Treatment on Composite Properties

The ILSS, flexural strength, and tensile strength are increased as the principal effects of fiber surface treatment on composite properties. The enhancement of these strength properties depends on the fiber elastic modulus, the degree of surface treatment, and the types of resin and curing agent used. The largest improvement in ILSS is obtained for high modulus fibers. Enhancement of the interfacial bond strength changes the failure mode from "interface-dominated" to "matrix-dominated," which is mainly responsible for the increase in interlaminar fracture toughness [34].

The longitudinal tensile strength of the composite is found to be increased with interfacial bond shear strength. When the interface bond strength is very high, the failure location changes from the interface to the surrounding matrix and the composite become brittle. With increasing interface bond strength, the compressive strength is also enhanced to a greater extent than the tensile strength.

4.2.3 *Aramid Fiber*

Aramid fiber is a generic term for aromatic polyamide fibers with 85% of amide linkage linked directly to two aromatic rings [35]. At present and in the foreseeable future, the principle aramid fiber for composite reinforcement of high performance composites is Kevlar. It is available in four categories: Kevlar, Kevlar 29, Kevlar 49, and Kevlar 149. More recently, the next-generation fiber DuPont Kevlar AP has been developed for advanced performance of composite materials. It increases the design flexibility and durability of various structures. DuPont Kevlar K29 AP and Kevlar K49 AP are the future prospects for commercially available fibers. Due to their high strength, stiffness, and low density, aramid fibers have been widely used in aerospace composites, aircraft, and for military applications. In addition, the longitudinal failure mode of these fibers makes them applicable for the ballistic resistance of composite materials.

An increasing use of Kevlar fibers is in low thermal expansion composite circuit boards, which leads to negative CTE. The strong covalent bonds in the fiber axis direction provide high longitudinal strength, whereas the weak hydrogen bonds in the transverse direction result in low transverse strength [36].

4.2.3.1 Surface Chemistry

Aramid fibers exhibit skin and core structures with distinct physical properties. The surface consists of an oxidized hydrocarbon with approximately twice the oxygen content and half of the nitrogen content of the bulk. Figure 4.5 presents a schematic diagram of Kevlar 49, which exhibits higher surface oxygen content than Kevlar 29. With the help of X-ray photoelectron spectroscopy (XPS), it is observed that the oxide primarily consists of hydroxyl and ether linkages, with a similar amount of carboxyl and carbonyl groups [37, 38]. The thickness of the oxide layer is of the order of 20 Å. From a morphological point of view, the fiber consists of extended chains, which are highly oriented in the fiber direction. Hydrogen-bonded sheets of these chains exist in the fiber direction [39].

Properties of Kevlar fiber at 23 °C are: tensile strength 3,024 MPa, tensile elongation 2.48%, tensile modulus 121.9 GPa, and tensile strain 2.5% [1].

When Kevlar 49 fibers are placed under compression along the longitudinal axis, they buckle and form kink bands. Both the weak lateral H-bonding of Kevlar 49 fibers and the presence of residual impurities in the interfibrillar regions contribute to the buckling susceptibility of the fiber.

Kevlar 49 epoxy composites fail in tension by axially splitting 20–50 times the diameter of the fiber along their length [1]. Depending on temperature, Kevlar fiber loses its mechanical properties with increasing temperature, but at low temperature there is essentially no loss in mechanical properties [1]. At 200 °C, there is a 20% loss in strength modulus and 25–40% loss in tensile strength [1, 40–42].

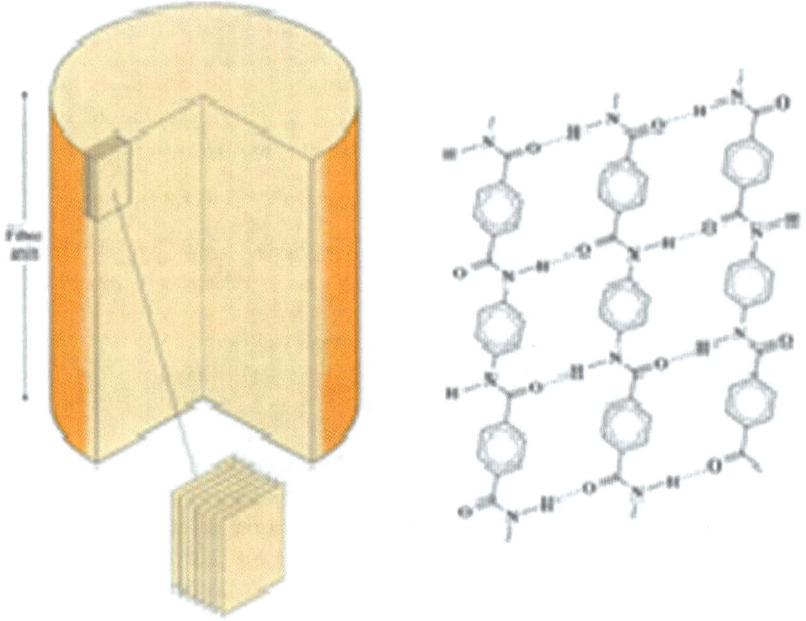

Fig. 4.5 Scheme of Kevlar 49 (aramid fiber)

4.2.3.2 Environmental Liability

Kevlar 49 is hygroscopic in nature and can sorb up to 3.5–4.5 wt% in ambient temperature at 100% relative humidity [43]. Moisture sorption from humid air occurs rapidly. Equilibrium moisture content is attained in 16–36 h [44]. At ambient temperature, no swelling is observe in Kevlar fiber when moisture is absorbed over a long duration of time. However, above 100 °C the superheated steam is the prime reason for permanent structural damage, which leads to 10 wt% moisture gain [42]. The unique chemistry and morphology of Kevlar polyaramid is also manifested in the behavior of its composites. It is assumed that the adhesion strength between Kevlar fiber and epoxy is excellent due to intermolecular forces and that covalent bonds exist between the Kevlar fiber and the matrix region.

4.2.4 Unsaturated Polyesters

Unsaturated polyesters are very unique polymers for polymer fiber reinforced composites. A good balance between mechanical properties and chemical resistance is required at room and moderate temperatures. Unsaturated polyesters show high shrinkage on curing and absorb water under certain conditions. These polymers are made by a reaction of a glycol such as propylene glycol with an unsaturated acid and crosslinking the resulting polymer with unsaturated monomer. This polyester alkyd is blended with styrene monomer in the ratio of about two parts alkyd and one part styrene. The poor impact resistance, limited shelf life and high curing shrinkage make the polymer utilized in high performance applications [45]. The notable areas of application for fibrous polyesters include appliances, business equipment, sporting goods, construction, and transportation. In contrast, polyester resins undergo net shrinkage after long exposure to water [46]. Three processes (hydrolysis, hydrophilization, and migration of low molecular weight electrolytes) predominate at different stages of water sorption behavior in composites. It was observed that composites are more susceptible to damage and/or debonding by thermal fatigue compared to epoxy resin matrix composites [3–6, 47].

4.2.5 Vinyl Esters

Vinyl esters are a mixture of unsaturated carboxylic acids and epoxide resin. The C=C double bond is formed only at the end of a vinyl ester molecule and therefore crosslinking can take place. Figure 4.6 presents the preparation of vinyl ester resin from a mixture of unsaturated carboxylic acids and epoxide resin [48]. Due to less crosslinkages, it is more flexible and has higher toughness than polyester resin. The volume shrinkage (5–10%) of vinyl ester resins, although lower than that of polyester resins, is higher than that of the epoxy resin. It can be used in polymer composites as pipes, storage tanks, and scrubbers.

Over the last few decades, high quality fiberglass vinyl ester swimming pools have come on the market, first manufactured in the USA. However, vinyl ester is more sensitive to moisture, temperature, and mix ratio (with catalyst). Due to the low ester content and low vinyl functionality, it has greater resistance to hydrolysis and is therefore preferred in applications requiring chemical resistance [49].

4.2.6 Epoxy Resins

Epoxy resins are the most common material for high performance advanced fibrous polymeric composites, but they are inherently brittle because of their high degree of crosslinking. The densely crosslinked structures are the basis of superior mechanical properties such as high modulus, high fracture strength, and solvent resistance [50]. However, these materials are irreversibly damaged by high stress due to the formation and propagation of cracks. At cryogenic temperatures, the mechanical properties of polymer composite are influenced by matrix failure or propagation of microcracks. This is nucleated by the residual stress and strain in the matrix resin system [51]. At low temperatures, the matrix becomes stiffer and stronger but less ductile (brittleness is high). In alliance with very low or negative CTE fibers, more and more residual stresses are generated on the matrix region. Because of this, few free stresses and

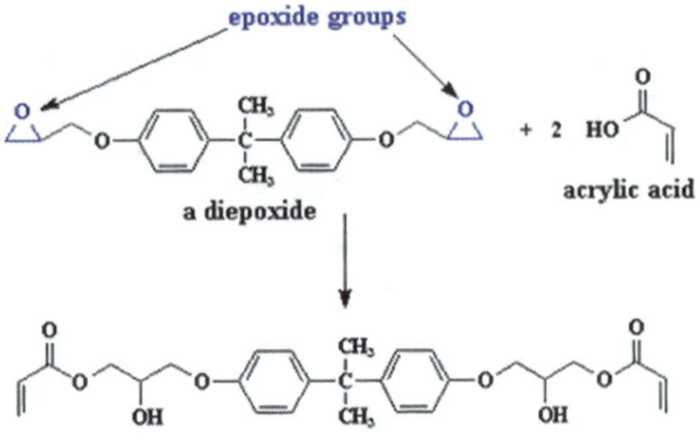

Fig. 4.6 Preparation of vinyl ester resin

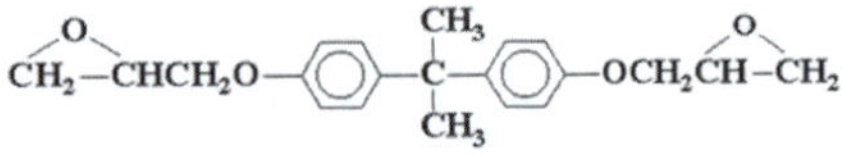

Fig. 4.7 Chemical structure of epoxy resin

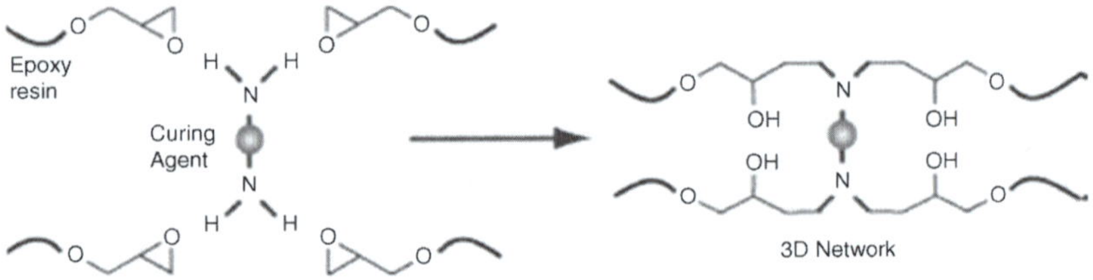

Fig. 4.8 Curing of epoxy resin

strains are accessible for the external loading of composite material at cryogenic temperatures. The low temperature shear strength of epoxy resins lies between 150 and 200 MPa at a temperature of about 77 K [52]. The most important epoxy resins are characterized by the existence of the epoxy group, which is a three-membered ring with two carbons and an oxygen atom (shown in Fig. 4.7).

Diglycidyl ethers of bisphenol are the most widely used epoxy resins. Epoxy resins contain a three-membered ring with one oxygen and two carbons. This epoxy group is helpful for the adhesion of reinforcement surface, which increases the strength and durability of the composite material. Epoxy resins have enormous applications as adhesives, bonding agents, and sealants. These resins easily react with epoxies and have good thermal stability and greater crosslinking density. Figure 4.8 shows the curing reaction of epoxy resin in a 3D network.

Usually, epoxy resins are used in conjunction with a curing agent to reduce curing time and to activate desirable properties. Aerospace applications require highly toughened epoxy resin to impart damage tolerance and necessary interlaminar fracture toughness. Epoxy resin is more expensive than polyester resin but it is not as suitable for elevated temperature use.

4.2.7 Low Temperature Physics of Reinforcement Fibers, Matrix Resins, and the Interphase Region

Composites have the advantage of the stiffness and strength of high performance fibers by dispersing them in a matrix, which acts as a binder and also transfers forces to the fibers across the fiber–matrix interface. Severe environmental exposure affects the physical and mechanical properties of polymeric composite materials, resulting in an undesirable degradation. As the material is intrinsically heterogeneous, damage and degradation will therefore often be strongly influenced by local processes. The anisotropy contributes to more complexity in the assessment of the damage mechanisms and in their impact on the composite responses. Molecular motion of the polymer matrix at low temperatures will be abysmal; molecular motion will be small and re-orientation modes will be relatively simple. Many of the amorphous polymers exhibit cryogenic relaxation peaks that are determined by intermolecular forces [53]. At low temperatures, the intrinsic conductivity tensor of carbon fibers becomes more isotropic. This is due to the fact that only long-wave phonons are activated, which propagate in a similar mode as in transverse and longitudinal directions of graphitized areas. In the case of Kevlar fiber, the mean transverse stiffness is similar to that of isotropic polymer matrix resin at low temperatures.

In the interface region at low temperatures, the glass fiber contracts less than the polymer matrix and, thus, the interfacial bond strength increases. For Kevlar fibers, the transverse contraction of hydrogen-bonded molecules is more than that of the isotropic polymer matrix resin. Thus, the interfacial bond strength and adhesion decreases. However, the interfacial adhesion of carbon/epoxy composite shows very tiny signs of temperature dependence [52].

4.3 Molecular Motion and Relaxation Behavior of Matrix Resins

The chemistry of polymer matrix resin has a great impact on understanding the potential of composite materials. The polymer matrix most used in advanced PMC is thermosetting polymer matrix. Figure 4.9 shows the interface region between fiber and matrix resin, with distinct layers. This region is the site of synergy in composite materials and its influence on the overall mechanical properties is significant. Although fiber serves as the load-carrying member, the matrix binds the fibers together and therefore the mechanical properties are decisively dependent on the resin matrix. Curing of amorphous resin leads to crosslinking, chain branching, and chain extensions. It impacts high stiffness, high strength, and good thermal and oxidative stability. As the curing proceeds, the glass transition temperature (T_g) of the growing chains increases as the network molecular weight increases. The low temperature curing resins have T_g as high as 120 °C. However, 120 °C curing glass/epoxy composites have higher ILSS values as well as high flexural strength [54].

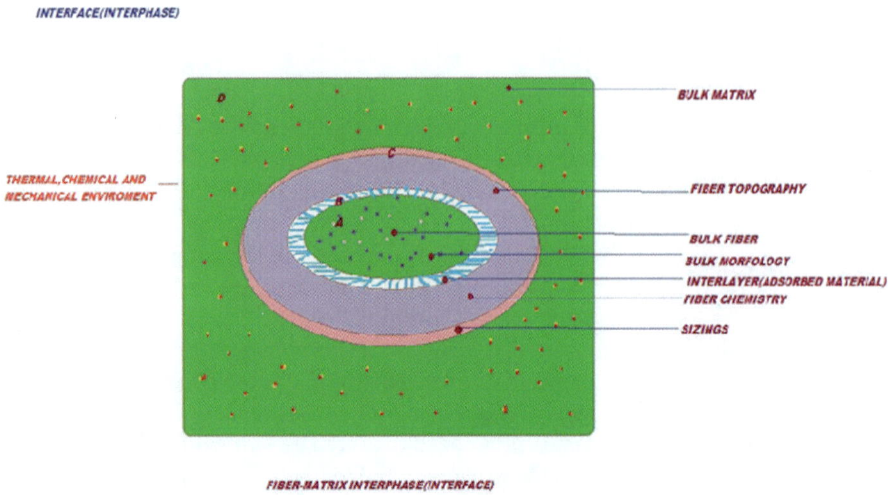

Fig. 4.9 Representation of interphase region between fiber and matrix

The T_g is measured by differential scanning calorimetry (DSC) in modulated mode. It covers two time scales of measurement: fast (fixed by the modulation time period) and slow (fixed by the average, underlying heating or cooling rate of the polymer matrix resin) [55].

At low temperatures, several factors affect the polymer matrix and can be modified to achieve its optimum properties. The main factor is its crosslinkage density, which influences the mechanical properties. Water penetration in composite materials of different interfacial strengths has been studied and explained as the result of two parallel processes, i.e., diffusion through the polymer matrix and diffusion through a network of microchannels formed along the imperfectly bonded polymer–fiber interface. Such channels along the exterior of the loose fiber are presumed to contribute mesoscopic "free volume" to the composite, which is conducive to diffusion. An analytical method based on an adaptation of the free volume theory approach has been proposed for estimating the incipient looseness and any further interfacial relaxation caused by water absorption in composite material.

Upon a gradual temperature decrease, polymers (especially those with irregular structure or composed of monomers fairly different in their structure and volume) form a glassy state (i.e., undergo the glass transition). The T_g is a function of molecular architecture, monomer units, presence of impurities or low molecular weight species, and the rate of temperature change. Normally, there is some free space between polymer monomers even in melt. In general, one can distinguish between the total volume and the occupied volume. The difference between these two is considered to be a free volume. The amount of the free volume changes with temperature.

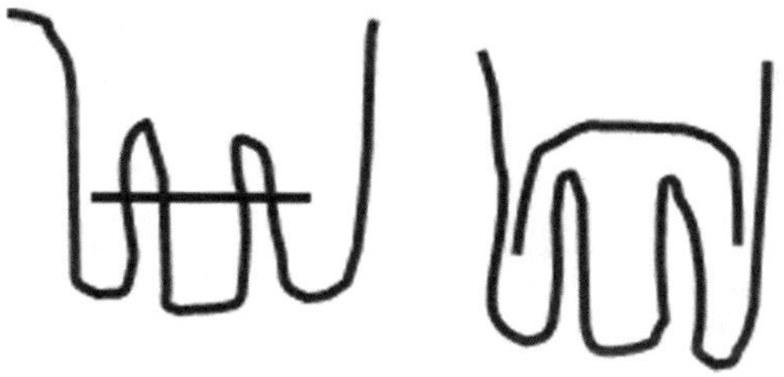

Fig. 4.10 Movement of amorphous polymer molecules: *left*, tunneling; *right*, potential barrier jumping

4.3.1 Relaxation Behavior and T_g

The shift of molecular chain and group with the unfreezing of molecular mobility upon warming is the cause of appearance of the T_g. At low and medium temperatures, a limited mobility of molecular parts becomes active in a weak glass transition. Amorphous polymers obey two types of position changes: tunneling and potential barrier jumping.

These position changes can have a drastic effect on the mechanical properties of the polymer matrix, which influences the properties of the composite material. Figure 4.10 presents tunneling and potential movement of polymer molecules. The tunneling and potential barrier jumping of molecules associated with time is known as relaxation time. The mechanical behavior of the polymer matrix solely depends upon the relaxation time of the matrix. In the absence of force, it follows the state of thermal equilibrium position. When force is applied, it changes from their original position with the relaxation time. In contrast to the applied load, relaxation time also depends upon the temperature [56]. At low temperatures the relaxation time is longer, whereas at very low temperatures the relaxation time is shorter because tunneling transition plays the dominant role. At elevated temperatures, the potential barrier jumping plays the major role, which is associated with a longer relaxation time.

4.3.2 Effect of T_g on Mechanical Action

T_g is used for evaluating the flexibility of a polymer molecule and the type of response the polymeric material would exhibit to mechanical stress. Above their T_g, polymers will exhibit a delayed elastic response (viscoelasticity) and dimensional stability. Chain flexibility, intermolecular forces, chain geometry, molecular weight, and stiffness of the crosslinking chains are important factors for T_g but are not important for free volume [57]. However, a change in T_g value is directly related to the molecular relaxation processes of the polymer matrix [32, 58]. General common-sense indicates that the higher the T_g, the better are the mechanical properties. Above the T_g value, the polymer matrix resin tends to thermo-oxidative degradation. Fibers are stable below T_g but the mechanical properties of the interface region can undergo degradation, which ultimately affects the durability of the composite material [59].

The interphase material has a lower T_g, higher modulus and tensile strength, and lower fracture toughness than the bulk matrix. Thermal conditioning is used to induce further polymerization, with the development of a penetrating and/or semipenetrating network at the fiber–matrix interface. The response of the fiber–matrix interface within the composite plays an important role in determining the gross mechanical performance, because it transmits the load from the matrix to the fibers, which contributes the greater portion of the composite strength. The better the interfacial bond, the better will be the ILSS, and resistance to delamination, fatigue and corrosion.

4.3.3 Viscoelastic (Elastomer) Behavior at Low Temperature

At low temperatures, especially below T_g, the polymer matrix resin shows tiny viscoelastic behavior. The degree of freedom decreases with decrease in temperature [60]. Due to the inclement climate of Russia, low-temperature studies of fibrous composites have been vitally important for the successful development of industries and transportation. The matrix resin undergoes a low rate of deformation and has low rigidity, a large difference between shear modulus and bulk modulus, and significant losses in mechanical energy in dynamic deformation. These unique properties are realized in the temperature range close to room temperature. The low-temperature resistance of the polymer gives the ability to preserve this set of unique mechanical properties at decreased temperatures. The rates of formation of rubber-like deformation components are 100 times lower, even at high temperatures, and drop sharply as temperature decreases.

Environmental attack can degrade the fiber–matrix interface mostly by mechanochemical processes. The diffusion coefficient of a polymer chain has a strong power-law dependence on the degree of polymerization. Polymers above their T_g will exhibit a delayed elastic response (viscoelasticity) and dimensional stability. Chain flexibility, intermolecular forces, chain geometry, molecular weight, and stiffness of the cross linking chains are important factors for T_g.

4.4 Mechanical Properties of FRP Composites at Low Temperature

The mechanical properties of composites are reflection of the properties of individual constituents without rule of mixture, but the temperature-induced integrity of each phase of the composites should be critically considered for the evaluation and prediction of properties at different temperatures.

Temperature has an important role in changing the deformation behavior of the materials, but its dominance differs from material to material. The deformation behavior of a polymer is highly sensitive to fluctuations in temperature.

4.4.1 Deformation Characteristics of FRP at Different Temperatures

The advantages of composite materials make them an exciting option for a wide range of applications. A judicial selection of fiber reinforcement, polymer resin matrix, and the interface/interphase offers unique physical and mechanical properties. However, a proper and uniform load transfer across the interface region between fiber and matrix has a big importance in composite materials. Environment and temperature have an unfavorable effect on these valuable properties, leading to premature failure and fracture of the material. These factors have a great effect on the life span of the fiber-reinforced composites. A number of potential solutions have been proposed to conquer these limitations.

Polymeric composites are susceptible to temperature and moisture when operating in changing environmental conditions. Composites are ductile at room temperature, brittle at low temperatures, and show viscoelastic behavior at elevated temperatures. The thermal aging behavior of epoxy resin has a unique value, especially due to their expanding use for structural applications where increased temperature is a common environmental condition. However, the swing behavior of composite materials at different temperatures alters the mechanical properties, which leads to failure. It is very difficult to generalize the effect of a change in temperature. Matrix-dominated properties such as compressive, flexural, shear, and transverse properties are those most affected by the temperature dependence of the matrix [61]. Thermal conditioning often imparts better adhesion between the fiber–matrix interface [62–65]. When considering the temperature dependence of the mechanical and physical properties of polymeric materials, various relaxation behaviors are critically important [57]. A dynamic mechanical relaxation occurs in polymer molecules because of heat transfer between the intermolecular mode (strain-sensitive mode) and the intramolecular mode (strain-insensitive mode) [66]. Thermal degradation of the resin matrix involves chemical reactions (such as oxidation and crosslinking) and physical changes (viscoelastic behavior of matrix resin) [67]. The nature of the viscoelastic yield depends on the time, temperature, and loading rate. At the macromolecular level, chain scission and crosslinking affect the polymer network and thus alter the mechanical properties of the oxidized layer. The hindered shrinkage of the oxidized layer induces a stress gradient that is susceptible to the initiation and propagation of cracks [68]. Thermal conditioning of fibrous composites is important because of their wide use for structural applications where increased temperatures are a common environmental condition. The fracture processes of polymers are strongly influenced by deformation or yielding processes, which depend on temperature, time, and loading rate. At very low temperatures, no yielding is possible in the material and the fracture is brittle. At high temperatures, two characteristic phenomena occur after the yield point: strain softening and strain hardening [67]. The specimens tested at a lower temperature are characterized by a greater level of microcracking and delamination. These effects are due to higher thermal residual stresses [68]. The heterogeneous and anisotropic nature of the fiber and matrix results in thermal expansion and contraction mismatch, which generates thermal stresses.

When subjected to elevated and high temperatures, the mechanical properties of FRP composites such as the E-modulus and viscosity experience significant changes. At a certain temperature, a composite material can be considered as mixture of materials that are in glassy, leathery, rubbery, or decomposed states [69, 70]. The mechanical properties of the mixture are determined by the content and the property of each state. Thermal conditioning at above ambient temperature might possibly improve adhesion level at the interfaces. The interphase is defined as a region that is manifested as a result of bonding and reactions between the fiber and the matrix. This region is the site of synergy in composite materials and it influences the overall mechanical properties significantly [62, 64, 65]. It is generally recognized that the bond strength variation at the interface greatly affects the integrity of composite materials. The bond strength depends on the quality of interfacial adhesion. Adhesion chemistry at the interface may be influenced by post-curing phenomena and this effect is supposed to increase with more conditioning time, limited by some optimum value [71]. Thermal conditioning causes matrix shrinkage due to loss of volatile material and additional cure of matrix resin [72]. The fracture behavior of the CFRP composites at different temperatures is complicated. Various fracture modes such as fiber breakage, fiber–matrix interfacial fracture, and delamination occur between adjacent plies of the laminates. ILSS values decrease nearly linearly with increasing temperature of the material [73]. By contrast, GFRP composites lose their rigidity and strength at room temperature and high temperatures [74]. Above the T_g, polymer composites are prone to thermo-oxidative degradation. Although the fibers are stable below T_g, the matrix and especially the fiber–matrix interface can undergo degradation that affects the physical and mechanical properties of the composite structures over time. Thermal conditioning induces further polymerization with the development of a penetrating and/or semipenetrating network at the fiber–matrix interface. The response of fiber–matrix interface within the composite plays an important role in determining the gross mechanical performance because it transmits the load from the matrix to the fibers, which contribute the greater portion of the composite strength [59].

The polymer composites have a wide range of application at cryogenic temperatures in space, superconducting magnet, and electronic technologies [75]. Potholing or localized surface degradation, delamination, and microcracking are some of the more dramatic phenomena that can occur as a result of cryogenic cycling [76]. Compressive residual stresses develop due to differential thermal contraction and influence the overall thermo-mechanical properties of the composite. In some cases, the resulting stresses are sufficient to initiate plastic deformation within the matrix immediately around the fiber [77]. The thermal residual stresses can also be large enough to initiate material damage such as matrix microcracking. These microcracks can reduce both the stiffness and the strength of the material, as well as act as sites for environmental degradation and nucleation of macrocracks [78]. The propensity for microcracking and the microcrack morphology depends on the type of fiber and matrix used. At low temperatures, the crosslinked polymers show brittle behavior and do not allow relaxation of the thermal residual stresses.

The mechanical and thermal properties of several types of epoxy systems were designed on the basis of the chemical structure, network structure, and morphology required for cryogenic application. Epoxy resins have been widely used as the matrix of composites material due to their good electrical insulation, advantageous mechanical properties, and ease of fabrication [79]. Additionally, the epoxy-based composites have also been used for cryogenic applications. At low temperatures, internal stress due to thermal contraction is generated in a matrix resin. Fracture of the matrix is formed when the stress intensity factor exceeds the fracture toughness of the resin [80]. It is, therefore, the fracture toughness of the epoxy resin that improves even at cryogenic temperature. To attain the high performances of polymers, it is necessary to control the molecular structure, i.e., the chemical and network structure. To improve the fracture toughness of epoxy, a plasticizer is usually added. The morphology, such as the sea-island structure and IPN, also determine the physical properties of epoxy [81]. When the morphology is optimized, a modified toughened polymer can be obtained showing high crack resistance without degradation of other properties. Changing the types and amount of modifier or the curing process can change the morphology. It is very important to clarify how to control the molecular structure to obtain a high performance polymer effectively and economically at cryogenic temperatures [82].

4.4.2 Implications of Hygrothermal Environment on Failures

As with any engineering material whilst in service, composites can be exposed to a diverse range of environments and degrading factors, the severity of which depend upon the particular application. Glass-fiber reinforced (GRP) composites are often used in marine craft such as canoes, fishing trawlers, patrol boats, naval mine hunting ships, in non-pressure hull casing, and for the sonar domes and masts of submarines. The use of fiber-reinforced composites (FRPs) in aircraft structures is increasing as new generations of aircraft take advantage of their attractive structural properties. The weight change behavior of FRP composites in humid and thermal environments appears to be a complex phenomenon. In particular, the fiber–matrix interface is highly prone to in-service degradation [62, 64, 65].

Several models have been put forward to address the issue of the state of water molecules in epoxy resins. Zhou suggested two types of water molecules with either single or double H bonds to epoxy networks [83], whereas Marechal supposed that water molecules either establish two hydrogen bonds or do not establish any H bond with their H atoms [84]. Liang et al., investigated whether two kinds of water molecules exist: one confined to relatively free water, and the other attributed to bond water, forming a strong H bond with hydrophilic groups of epoxy networks [85]. Maxwell and Pethrick studied whether the water could exist in clusters in which it was relatively free or could be molecularly dispersed within the resin and bond to it, probably by H bonding [86]. The absorption and diffusion of water in polymeric materials such as epoxy systems is related to the free volume and the polymer–water affinity [87]. Epoxy-based composite laminates absorb moisture from the surrounding environment very quickly and attain equilibrium. The equilibrium water content depends not only on the free volume in the matrix (related to relatively free water) but also on how many H bonds are formed between water and network polar groups [88]. With PALS and calculations, Zheng and Morgan found that equilibrium water content is controlled by the free volume fraction at absolute zero (0 K), which is decided by the amount of epoxy group before curing [89]. In other words, the polar groups at the inner surface of free volume provide the bonding sites for water molecules, while the limited space of nanopores restricts the formation of free water–water H bonding [90]. Therefore, water diffuses into the composite material driven by the water gradient. Browning and colleagues reported preferential paths of diffusion due to non-uniform swelling of the resin through the thickness before equilibrium is reached [91]. Thus, it appears that a moisture gradient exists in the composites before the resin becomes saturated with moisture [92]. The effect of moisture on the matrix includes reversible processes, such as hygroelasticity and swelling. The constituent that is most sensitivity to moisture is the fiber–matrix interface and, therefore, moisture promotes damage mechanisms that are controlled by the interfacial performance [93]. Moisture wicking along the fiber–matrix interface degrades the fiber–matrix bond resulting in loss of microstructural integrity. The net effect of moisture absorption is deterioration of matrix-dominated properties such as compressive strength, ILSS, fatigue resistance, and impact tolerance.

The properties of the fiber–matrix interphase in GRP composites can play a dominant role in governing the overall composite performance. Understanding the interactions occurring at the interphase and being able to tailor them to give desired composite properties are of great importance. The sizing surface free energy and its solubility in the matrix were shown to strongly affect interphase formation, fiber–matrix adhesion, composite shear, and flexural strength.

In an outdoor environment, aircraft are subject to continuously varying temperature and relative humidity and, as a result, there will be periods of exposure that produce moisture absorption and moisture desorption in an FRP laminate [94]. In diffusion theory, the rate of absorption is mainly a function of temperature, the amount of moisture absorbed, and relative humidity [95]. However, epoxy resins will absorb moisture from their surroundings (up to 3–4% of their original weight), which causes a lowering of T_g; 1% absorbed water reduces T_g by 20 °C.

For a given brittle-fiber/brittle-matrix composite, high strength requires a strong interfacial bond, but this may lead to low fracture energy absorption. However, by proper control of the physical and mechanical properties of the fiber–matrix interface, high strength characteristics can be combined with high toughness.

In order to fully utilize the potential of such composites without introducing a reduction in strength, it is necessary to understand the failure mechanisms leading to eventual fracture. Transverse cracks are initiated at low loading levels in the first vulnerable plies and continue to accumulate under increased loading conditions up to a saturation crack density [83].

The durability of glass-fiber/polymer composites is dictated by their response to moisture. Environmental attack by moisture can degrade the strength of the glass fiber; plasticize, swell, or microcrack the resin; and degrade the fiber–matrix interface by either chemical or mechanical attack.

4.4.3 Loading Rate Sensitivity of FRP at Low Temperatures

4.4.3.1 Mechanical Behavior

In addition to strength and modulus, another important property of a composite is its resistance to fracture. A measure of fracture resistance of a material is given by the specific work of fracture, or fracture toughness, and is defined here as the energy required for creating a unit area of new transverse fracture surface of the composite [96]. When a composite is loaded, widespread microscopic damage arises throughout the body. Large damage can be sustained to a critical value at which failure occurs by the propagation of cracks. These cracks are much more complex in nature than are cracks in homogeneous materials. The failure of a composite involves the fractures of the load-bearing fibers and the matrix as well as a complex combination of cracks propagated along the interfaces. Therefore, the fracture toughness of a composite depends not only on the properties of the constituents but significantly on the efficiency of bonding across the interface. To understand the properties of the interface and its role in controlling fracture, it is essential to know the mechanisms of adhesion at the interface involved. Since the nature of bonding is dependent on the atomic arrangement and chemical properties of the fiber as well as on the molecular conformation of the matrix, it follows that the interface is specific to each fiber–matrix system. The major mechanisms according to Hull [10] include adsorption and wetting, interdiffusion, electrostatic attraction, chemical bonding, and mechanical bonding. In addition, hydrogen bonding, van der Waals forces, and other low energy bonds may also be involved.

4.4.4 Status of the Interface on Mechanical Behavior

The fiber–matrix interface of an FRP composite is highly inclined to in-service degradation. The physicochemical and mechanical aspects of the interface are important for the composite material. The adhesion and type of bonding play a great role in obtaining optimal properties of interface region. Furthermore, characterization of the macroscopic, microscopic, and nanoscopic nature of the interface is required for best use of an advanced polymer composite. The process of adhesion between the fiber and the matrix and the mechanism of subsequent processes also have some impact on absorption, wetting, and electrostatic attraction at the interface region. The nature of bonding, which is crucial for the adhesion, is dependent not only on the molecular arrangement, but also on the morphological and diffusivity properties of the fiber and the matrix.

The strength of the interface region may be due to formation of inter-diffusion bonds between the fiber and matrix. The bond strength in PMCs will depend on the amount of molecular entanglement, the number of molecules involved, and the strength of the bonding between the molecules [20]. IPN and inter-diffusion are two examples for the explanation of the adhesion between fiber and the polymer resin matrix.

We assume that the interphase region is in 3D form and acts as a softer phase than the bulk matrix. The average modulus of the interphase of a thickness 500 nm formed between a single carbon fiber and epoxy matrix is about one-quarter of that in the bulk matrix. However, the presence of stiff fiber mitigates the effect of a soft bulk matrix in close vicinity of the fiber [97].

4.4.5 Failure Revelations Through Microscopy and Microanalysis

Fractography analysis of a failed polymer composite material is currently a challenging experience. PMC material shows a number of failure modes, which are the root cause of degradation of a component. The failure may be due to environmental factors or may be in-service degradation [98–100]. Thus, there is a need for failure analysis through microscopy and microanalysis.

The currently available most powerful tools for the characterization of failure components are: Fourier transform infrared spectroscopy (FTIR), atomic force microscopy (AFM), scanning electron microscopy (SEM), and Auger electron microscopy (AES). Each technique has some complexity, advantages, and limitations. This section covers some microscopy techniques that are useful for fractography analysis.

FTIR spectroscopy has been extensively used for surface characterization of the polymer matrix surface, indication of various functional groups of the surface of the material, and the chemical modification at the interface region. In the case of GFRP, the reactivity of the silane coupling agent with the epoxy resin was found to vary with the drying conditions of the silane coupling agent [101]. Infrared spectroscopy gives information on molecular vibrations or, more precisely, on the transition between irrational and rotational energy levels in molecules. When infrared light of that frequency is incident on the molecule, energy is absorbed and the amplitude of that vibration is increased. An infrared spectrum is obtained when the frequency of molecular vibration corresponds to the infrared radiation absorbed. The normal range of an infrared spectrum used by an organic chemist for structural work extends from 4,000 to 400 cm.

FTIR images are sequentially acquired during the curing time in the region of view image, which includes the glass fiber–epoxy mixture interface. The mechanism of the curing can follow two different but simultaneous paths. Figure 4.11 shows FTIR spectra of an epoxy–amine mixture as a function of the curing process. The curing process of an epoxy system at the interface formed with a silane coated glass fiber is studied using FTIR imaging. Chemically specific images for OH and H–N–H within the system are obtained. The analysis of these images suggests that there is a variation in the chemical structure of the matrix from fiber to polymer bulk due to different conversions arising from a gradient in the initial composition [102]. It was observed that the rate of curing reaction changed depending on the distance to the glass fiber, and that this change was also associated with changes in the stoichiometry. There are certain limitations to refraction and reflection at the fiber surface in the spectroscopy that will finally affect the FTIR spectra of glass/epoxy composites. Due to this, only a small percentage of light reaches the detector. It is difficult to separate these optical effects from the sample effects. However, these optical effects can be minimized by taking FTIR spectra over the same region of the microcomposites. This is currently the most advanced tool for the characterization of the composite material at the nanometer scale. Application of AFM is growing rapidly as it is useful for observing the surface topography of the composite material. This technique can be applied to both conductive and non-conductive specimens. Depending on the situation, forces that are measured in AFM include mechanical contraction force, van der Waals forces, capillary forces, chemical bonding, electrostatic forces, and magnetic forces. As well as force, additional quantities may simultaneously be measured through the use of specialized types of probe. The CFRP composite in Fig. 4.12 shows height image scans in tapping mode after hygrothermal treatment at different time intervals. Thermal degradation of plastics involves chemical reaction and physical changes. Chemical reaction is represented by oxidation, crosslinking, and further reaction of unreacted monomers, whereas physical change is typical of viscoelastic behavior [67].

In temperature-modulated DSC (TMDSC), the rates of heating are varied by overlapping the underlying linear heating rates with a sinusoidal temperature modulation. This separates the reversible nature of thermal expansion from irreversible deformation, which arises from creep under the applied load or changes in dimensions due to relaxation of orientation. It helps in generating images based on local changes in the thermal expansion of a specimen. The DSC measurements are performed on a Mettler-Toledo 821 with intra-cooler, using the STAR software with temperature DSC (ADSC) module. The temperature calibration and the determination of the time constant of the instrument are performed using standards of In and Zn, and the heat flow calibrated using In. An underlying heating rate of 10 °C/min is used. In order to calibrate the heat flow signal, a blank run with an empty pan on the reference side and an empty pan plus a lid at the sample side is performed before the sample measurements. However, a modulated temperature program (such as those used in modulated temperature-differential scanning calorimetry, MT-DSC) is

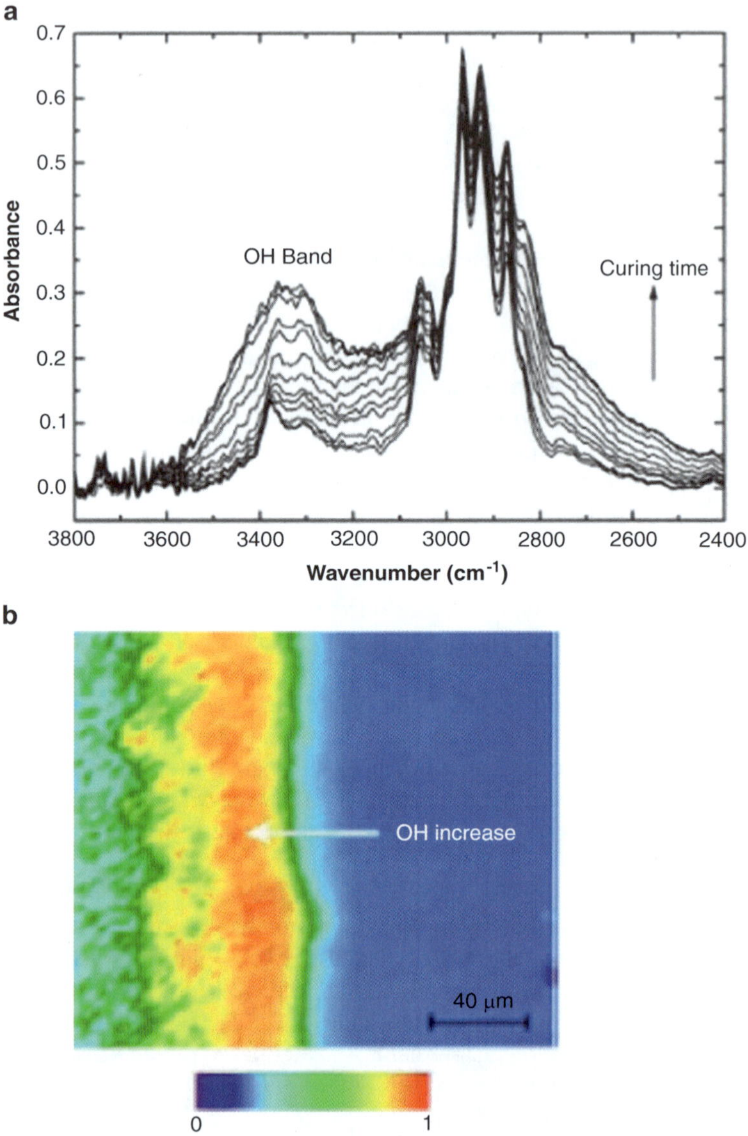

Fig. 4.11 (**a**) FTIR spectra of epoxy-amine mixtures as a function of time of the curing process. (**b**) FITR spectra showing the presence of OH signal for the poly(aminopropylsiloxane) sample

employed rather than the conventional linear temperature ramp. It is then possible to separate the temperature-dependent thermal expansion from the time- (and temperature-)dependent creep or stress relaxation behavior. Figure 4.13 shows the variation in T_g values with different time intervals of hygrothermal treatment. T_g is likely to increase with small duration of treatment time. This may be because of

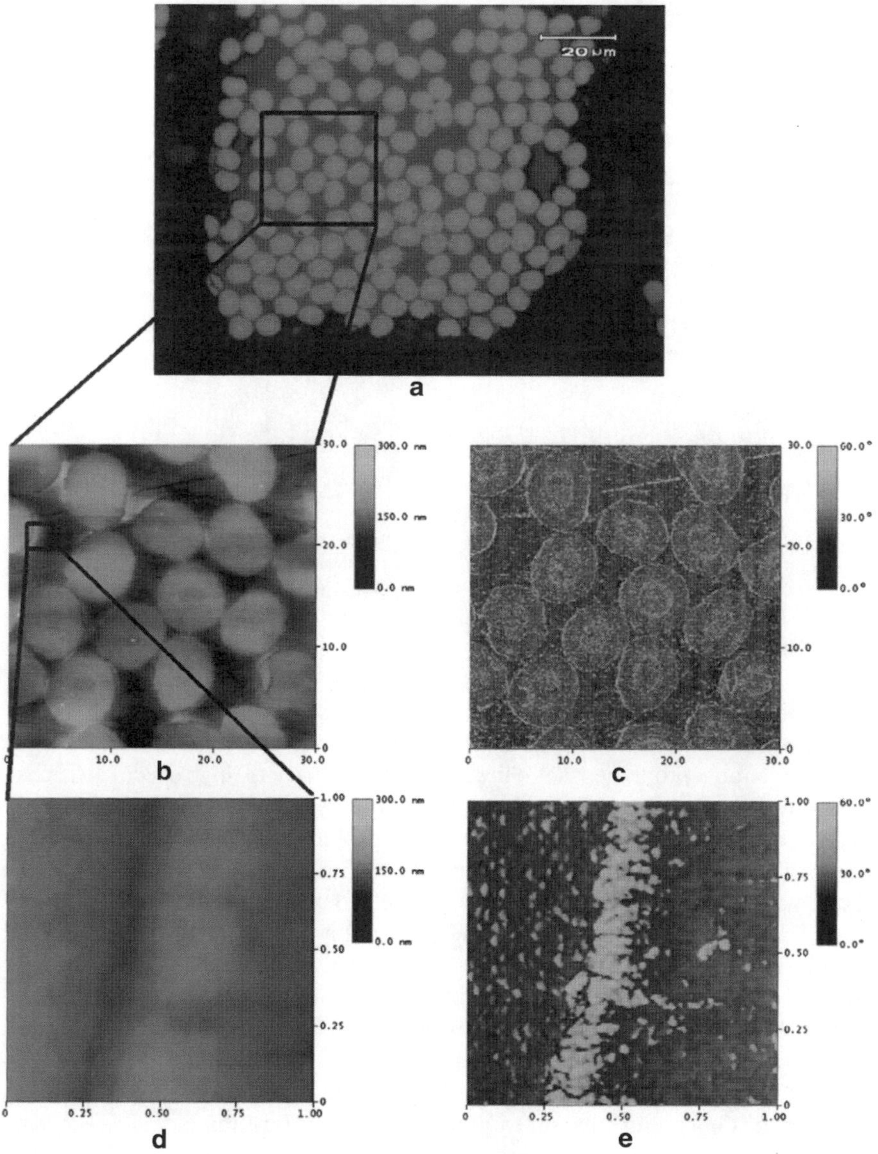

Fig. 4.12 (*continued*)

formation of double H-bonds. It appears that the change in T_g as a function of the conversion is directly related to the molecular relaxation processes [103]. During the isothermal cure of an epoxy resin, the system changes from a viscous liquid to a highly crosslinked network. The enthalpy and the volume of the system are expected to decrease continuously during the curing until the vitrification process.

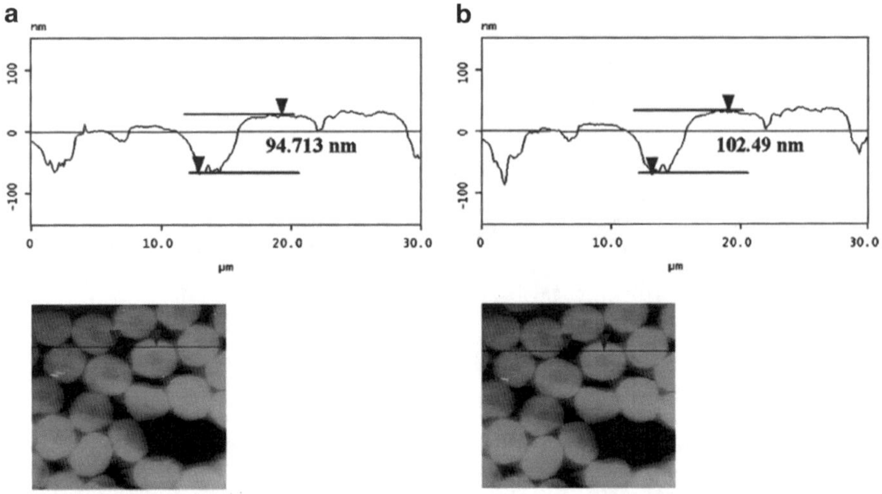

Fig. 4.12 (**a**) Optical microscope and AFM micrographs of AS4/VRM34: optical micrograph (1); height image of cross-section (2); phase image of cross-section (3); height image of interface (4); and phase image of interface (5). (**b**) AFM topography images of same cross-section area of AS4/VRM34 using the tapping mode (1) and contact mode (2). Both AFM images in different mode exhibit equivalent surface profiles

Fibrous composites are increasingly being used in many applications owing to various desirable properties including high specific strength, high specific stiffness, and controlled anisotropy. But unfortunately, polymer composites are susceptible to heat and moisture when operating in changing environmental conditions. The structural integrity and lifetime performance of fibrous polymeric composites are strongly dependent on the stability of the fiber–polymer interfacial region. Plasticization induces plastic deformation in addition to lowering the T_g, whereas swelling is related to the differential strain, which is created by the expansion force exerted by the liquid while stretching polymeric chains. It is generally believed that the polymer is likely to be tough if homogeneous yielding occurs. Even brittle crack propagation in polymers usually involves localized viscoelastic and plastic energy-dissipating processes, which take place in the vicinity of the crack tip. The propagation of the debonding cracks may involve breaking of the primary and secondary bonds between the fiber and the polymer matrix. It should be noted that multiple matrix cracking may represent a significant source of toughness.

The T_g of matrix resin is a parameter of utmost importance for the design and development of polymeric composites. The reliability and durability of FRPs are critically dependent on it. Moisture and temperature are the environmental factors that most adversely affect the structural integrity and mechanical performance of FRPs. Moisture ingression and changing humidity along with change of temperature eventually change the T_g of a polymer matrix [104]. The changes in temperature have to be measured with utmost accuracy and reproducibility by means of an

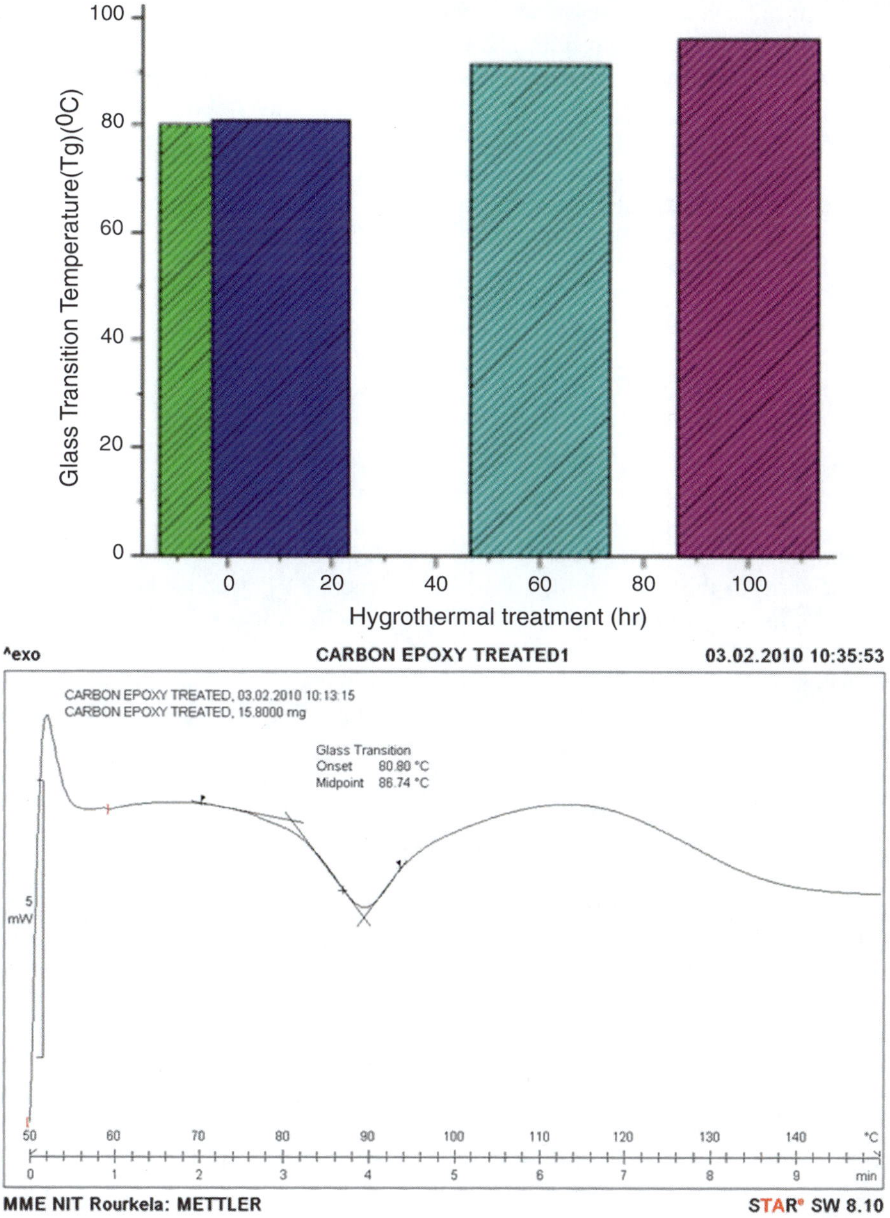

Fig. 4.13 (a) Variation in T_g with different time interval of hygrothermal treatment. (b) T_g of carbon/epoxy polymer composites. T_g visibly increases with short treatment times. This may be because of formation of double H-bond formation. The absorption and diffusion of water in interface is related to the free volume and polymer–water affinity. The amount of free volume depends on the molecular packing and is also affected by crosslink densities. The increase in T_g value is indicative of the increase in crosslink density

advanced thermal analysis system like alternating DSC. The changes in chemistry at the interphase due to moisture ingression need to be further assessed by FTIR microscopy by selected area imaging. The stress transfer efficiency and consistency are decisively controlled by the physical and chemical nature of the interphase area [105]. An approach may be suggested here to evaluate the status and strength of the phase by AFM. With more precise and critical analysis of the environmentally induced health of the interphase, the better can the potential and promise of the FRPs be realized and explored.

4.5 Possible Failure Modes of FRP Composites at Low Temperatures

Fibrous composites are now have huge and widespread application in aircraft, space shuttles, motor vehicles, sporting goods, and a wide range of military equipment. They are extremely versatile materials. While we migrate from a metal world to the FRP world, structural designers still face challenges when it comes to the mechanical properties of FRP composites under various environmental conditions. Advanced polymer composites consist of polymer matrix reinforced with high strength and high stiffness fibers. The effects of environment can dramatically change the instantaneous response of the composite. In general, to achieve optimum performance, composite designers strive for a volume fraction (V_f) of 60%, but it is generally about 55%. With 60% volume fiber content, unidirectional CFRP shows transverse failure strain at only 0.7–0.9%, which is lower than the 5–6% strain of the pure anhydride cured DGEBA epoxy resin [106]. The classic definition of the interface in the fiber composites is a boundary formed by reinforcing fiber and matrix, which is crucial to the overall composite portrayal. Too weak an interface gives low composite strength and stiffness, whereas too strong an interface results in brittleness [107]. In addition, fracture surfaces and fracture paths vary with the nature of the fiber–matrix interface and depend on the environment and temperature. Understanding composite failure underpins the development of failure criteria. Factors such as temperature, moisture, and loading rate can have considerable effects on the fractography behavior. Fracture analysis is complicated for polymers because, in addition to temperature and time dependence, there are effects from plastification, chain orientation, and adiabatic temperature rise of the matrix polymer [8].

A judicial control of the mechanical behavior of the interface region leads to high toughness of the composite material. Brittle fiber/brittle matrix requires strong interfacial adhesion, which leads to low fracture energy absorption [7]. Due to thermal contraction, a tighter packing and thus higher bond strength at decreasing temperature results. At very low temperatures, chain atoms have no thermal energy for counteracting the binding potentials. Induced fracture processes are crazing, necking, or plastification. Due to the homogeneous nature of composites, stress distributions leads to stress contraction, which reduces fracture strength.

At high temperatures, the matrix-dominated properties of CFRP will be influenced by the softening of the matrix when the temperature reaches the value of matrix T_g. However, at low temperatures, polymer matrices are brittle. Fiber-dominated properties and mechanical properties are generally unaffected by temperature.

4.5.1 *Matrix Deformation and Cracking*

Thermosetting polymer matrices are amorphous in nature. They follow the state of non-equilibrium, which is responsible for the sensitivity to various environmental factors such as temperature, humidity, and moisture as well as to experimental parameters and loading rate. Application of different loading rates may interact with the above parameters and accelerate the failure and fracture of PMC materials. Various failure modes seem to occur during degradation of composite materials. The role of matrix in PMC has three major important aspects: it is used to protect fibers against environmental attack; it serves as a load-transferring medium, and it acts as an energy-dissipation outlet for arresting crack propagation. The properties of matrix materials in FRP are extensive and varied. These thermosetting resins are converted from low molecular weight polyfunctional monomers or oligomers with low viscosity to 3D crosslinked networks that are rigid in nature and quite brittle. The matrix materials usually have relatively low modulus and strength values and comparable or higher ductility values [7]. Epoxy resins are generally used for high performance fiber composites. In liquid state, they can be used as impregnation material for the wet-layup technique. The mechanical properties of composites at low temperatures are influenced by the matrix. At low temperatures, thermal pre-stress is crucial. This reduces effective strain to failure and is the source of microcracks in the matrix [8]. The various fracture strains of matrix at 4.2 K and fibers are: epoxy resin 2%, fiber glass 3%, carbon fiber 1.2–2.2%, Kevlar fiber 2%, and thermoplastic matrices 4% [8]. The author describes the fracture morphologies associated with matrix-dominated fractures associated with epoxy-based PMCs with glass, carbon, and Kevlar fiber reinforcement.

Figure 4.14 shows matrix microcracking and macrocracking of epoxy matrix resin. Matrix microcracking is the source of damage in laminates and comprises intralaminar or ply cracks that transverse the thickness of the ply and run parallel to the fibers in that ply. Microcracks in carbon fiber/epoxy laminates always formed instantaneously across the entire cross-sectional area of the 90° plies.

The first form of microcracks causes very mild effects in the thermomechanical behavior of composites. With increasing the applied load and loading speed, additional microcracks are generated and they are the incentive for failure of

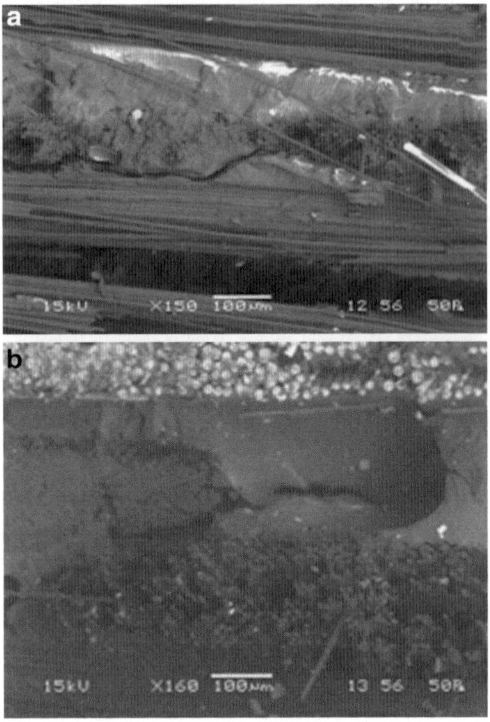

Fig. 4.14 SEM micrograph shows (**a**) epoxy matrix cracking at high temperature and (**b**) multiple matrix cracking (macrocracking) with fiber pull-out at high temperature

PMCs. We can assume that microcrack density is a function of loading conditions and applied load. After the initial phase, microcrack density typically increases very rapidly with applied load. At high crack density, the saturation damage state is approached. Most microcracking experiments have been conducted on laminates under uniform stress such that the stress state in the 90° plies is uniform prior to microcracking. As discussed for low temperatures, the matrix behaves like a brittle material. Morphology such as riverline marking and texture microflow can be observed in the resin. These failure marks shows the sign of local crack growth directions. These marks are less apparent when the toughness of the matrix increases.

Figure 4.15 shows a visible riverline marking with scarps, potholes, and textured microflow of the matrix resin. We consider the riverline markings as the most valuable feature for diagnosing crack growth direction. Other important matrix features are scarps and ribbons. When multiple microcracks are formed and begin to propagate in several planes, they subsequently converge into one plane. When two crack planes converged, the boundary between these planes shows a sharp step form. This is an example of the formation of scarps. The global growth direction in Fig. 4.15b goes from top to bottom. It should be noted that the local crack growth inferred from the textured microflow is from the fibers into the surrounding resin. These are indications of the absorption of energy by local

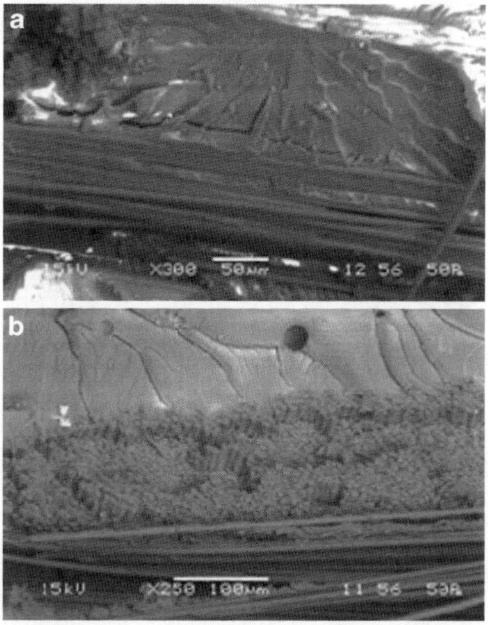

Fig. 4.15 Riverline marking with (**a**) fiber imprint in epoxy matrix resin and (**b**) scarps, potholes, and textured microflow in epoxy matrix resin

deformation. Figure 4.15b shows four converging crack planes in a resin-rich region within a glass/epoxy polymer composite at low temperature. Ribbon-like structures also form with low temperature treatment. These developed if the crack planes overlap before they coalesce to form scarp [107]. Considering the influence of low temperature on the fracture micromechanism in composites; the matrix is brittle and residual tensile stresses in the matrix are promoted. At very low temperatures, resin embrittlement dominates the phenomena and thus composite toughness decreases. As temperature increases very close to T_g of the matrix, it becomes ductile and behaves as a softer material. As shown in Fig. 4.16, at high temperature fiber imprints are formed with stresses generated at the edge point. The matrix behaves as softer, meaning that the process zone is spread over a large depth, and the degree of bridging increases [107]. The matrix failure modes at high as well as low temperatures are responsible for the integrity and durability of PMCs. It is accepted that voids and porosity have prejudicial effects on the composite performance. There are many factors responsible for formation of these voids,. primarily trapping of volatile materials including air and inclusions. These volatiles materials are either entrapped during fabrication or are the water or residual solvents from the pre-pregging process (shown in Fig. 4.17). When the matrix behaves as brittle, the energy of fracture is fairly low and there is little step marking formed during failure, which is referred to as cleavage marking. The characteristic feature of cleavage fracture is flat facets that generally are about 9.22–20.6 μm. Identification of the cause of fracture through fractography has become a standard investigation procedure.

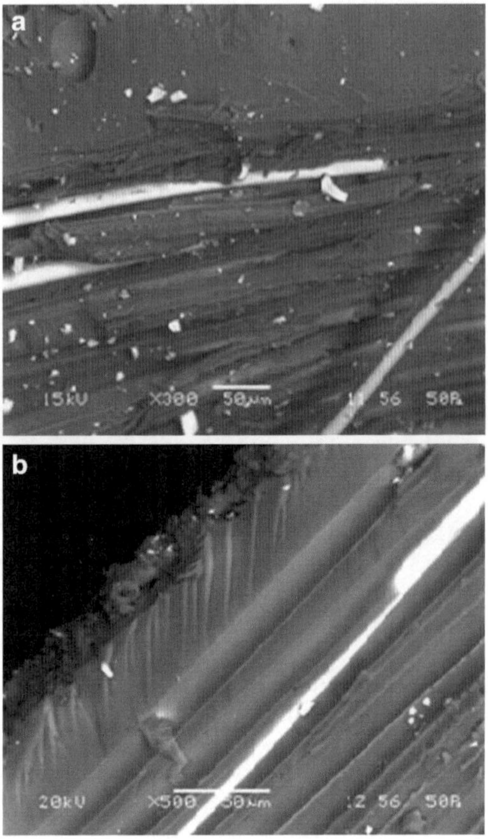

Fig. 4.16 Thermally conditioned GFRP composite showing (**a**) fiber imprint region in epoxy matrix and (**b**) stresses generated at the edge of the fiber in microcomposites

4.5.2 Interfiber Fracture and Fiber Pull-Out

Fibers are of different forms, i.e., natural, regenerate, and synthetic. A polymer composite consists of fibers embedded into a matrix with prominent interface (2D) or interphase (3D) region. The fibers act as load-bearing members with high strength and modulus. Due to this, the composites have an enormous number of novel applications. Research in this area is very active at present. The most commonly used fibers are glass fibers, carbon fibers, and Kevlar fibers. A key feature of this interface bonding is due to formation of strong ionic and covalent bonds. As the fibers are brittle in nature, their mechanical properties are controlled by their fracture behavior. Thus, there is a need for a deeper grasp of this failure mechanism of fiber fracture. The reinforcement fibers may be continuous, filaments, chopped fibers, or whiskers. Advanced PMC are being used in many sectors such as the automobile and aerospace industries. A superior combination of high strength, stiffness, and density makes the carbon fiber eligible for these advanced applications. Whereas research on glass fiber has seen an explosive

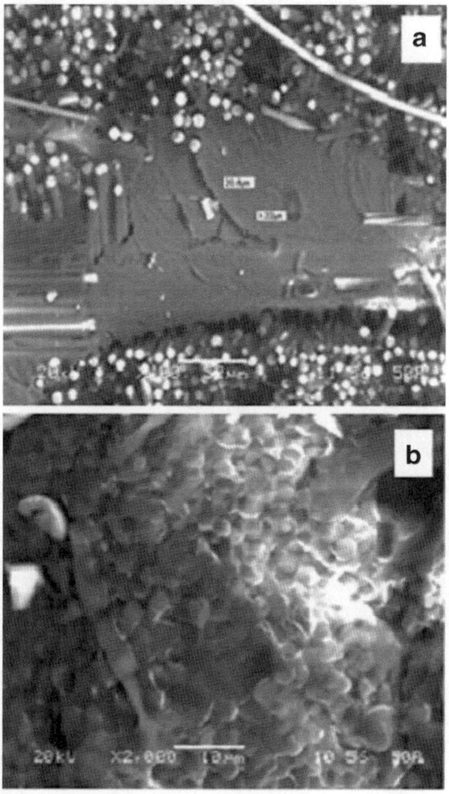

Fig. 4.17 Scanning micrograph shows (**a**) cleavage marking of glass epoxy prepreg composite after cryogenic treatment and (**b**) tiny bubbles in epoxy matrix of glass epoxy prepreg composite after cryogenic treatment

growth due to its application in areas requiring high tensile strength, in general, highly oriented fibers such as aramid fail in a fibrillar fashion. The term fibrillar fracture here signifies that the fracture surface is not transverse to the axis but runs along a number of planes parallel to the fiber axis. Some of the failure modes of discussed fiber-dominated morphologies are as follows: The bond strength of glass/epoxy reinforced composites is reported to be around 33 MPa. E-glass/epoxy with silane treatment has a bond strength of 49 MPa. The fiber–matrix interface is usually quite poor and shows brittle failure at the end. The bond strength of carbon/epoxy polymer composite is 57 MPa, and that of Kevlar/epoxy polymer composite is about 38 MPa. In Fig. 4.18a, we can assume that the fibers tend to fail homogenously with proper interfacial bonding. However, in Fig. 4.18b the fibers tend to fracture independently at different points along their lengths. Fracture surfaces seem to be flat. Carbon fiber and glass fibers are inevitably brittle materials, so radial patterns appear at the fiber breaking end. Aramid family fibers (Kevlar) exhibit different failure modes due to their non-brittleness.

Brittle fibers have low fracture strain and a low energy absorbing capability [7]. The flexure response of PMC is a complex phenomenon. The material behavior and the loading condition may influence the behavior of energy-absorbing mechanisms in composites.

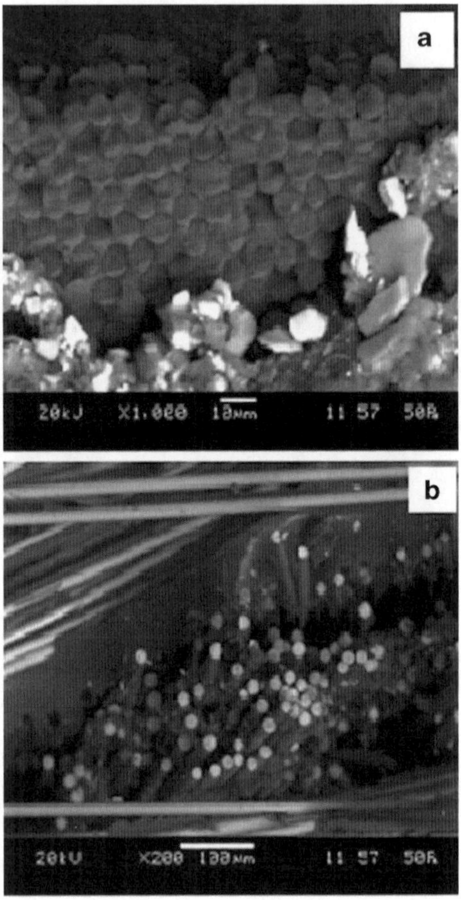

Fig. 4.18 SEM micrograph showing (**a**) uniform failure of fiber of GFRP composites and (**b**) fiber pull-out and brittle failure of fiber in GFRP composite

Fiber pull-out appears when the fibers are embedded in a tough matrix. Fibers can fracture early at the weak cross-section point, which is not necessarily the direction of crack propagation. As the applied loading rate increases, the fracture proceeds and the broken fibers will likely pull out from the matrix. This is the reason for the mechanism of fiber pull-out in PMC composites. Figure 4.19a shows fiber pull-out and the fiber fracture ends in hygrothermally treated GFRP composite after 10 h. In Fig. 4.19b, a strong adhesive failure is expected to occur. This failure is confirmed by the cohesive failure produced by the GFRP composite. The weak boundary layer (WBL) is a region of low cohesive strength between fiber and matrix regions. The WBL provides a weak interface bonding, promoting interface debonding and fiber pull-out as shown in Fig. 4.19 [108].

Such a crack running parallel to the fibers through the entire thickness of a layer is called an inter fiber fracture (IFF). IFF comprises both cohesive matrix fracture and adhesive fracture of the fiber–matrix-interface. The crack itself does not grow slowly but is generated spontaneously and stopped only by the fibers of the neighboring layers

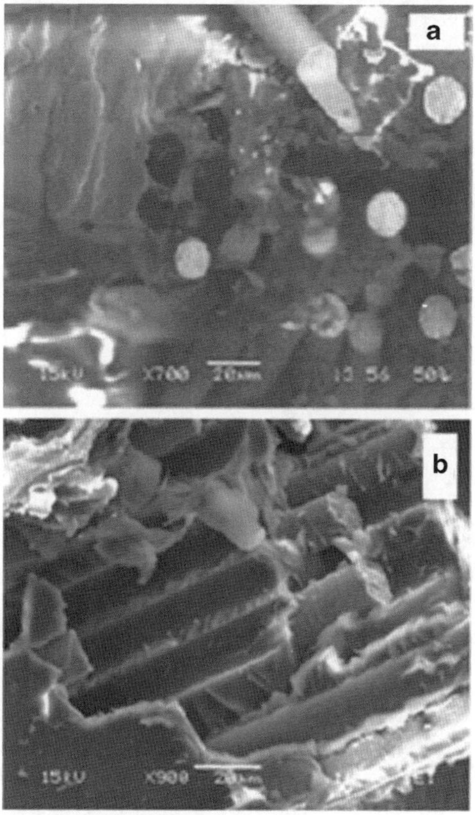

Fig. 4.19 SEM micrograph of hygrothermally treated GFRP composite showing (**a**) fiber pull-out and fiber fracture and (**b**) strong adhesion between the interphase regions

of different fiber direction. Typically, at the tip of the IFF crack, small delaminated zones can be observed. The cohesion between the broken and the neighboring layers is affected locally but the integrity of the laminate is still preserved. It does not fall into parts as would naturally be the case if the laminate consisted of layers with only one fiber direction. Fibers are among the stiffest and strongest materials present in nature or manufactured by man. They are used as structural components embedded in a matrix that maintains the fibers oriented in the optimum direction, distributes the concentrated loads, protects the fibers against wear or chemical attack from the environment, and provides the transverse stiffness to avoid buckling in compression. Figure 4.20 presents fracture of fibers during processing or in service, which is generally an undesirable feature. In polymeric fibers, the fundamental processes leading to failure are chain scission and/or chain sliding or a combination of both. Service environment can be a major determining factor in the failure process of fibers. In general, because of the high surface-to-volume ratio of fibers, the incidence of a surface flaw leading to fracture is greater in fibers than in bulk materials. Fractography is the study of the fracture surface of fibers and can be a useful technique for obtaining

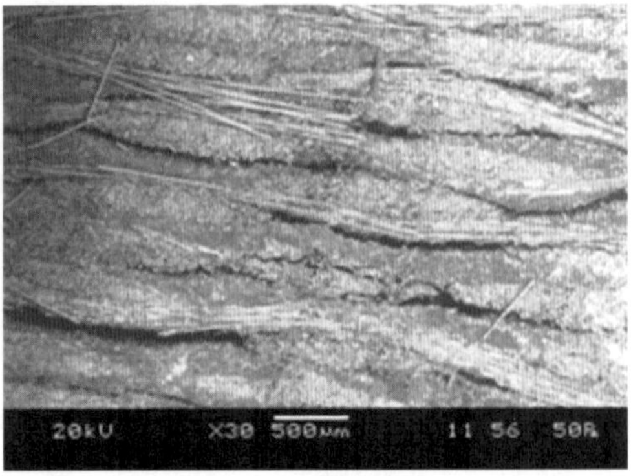

Fig. 4.20 Interfiber fracture

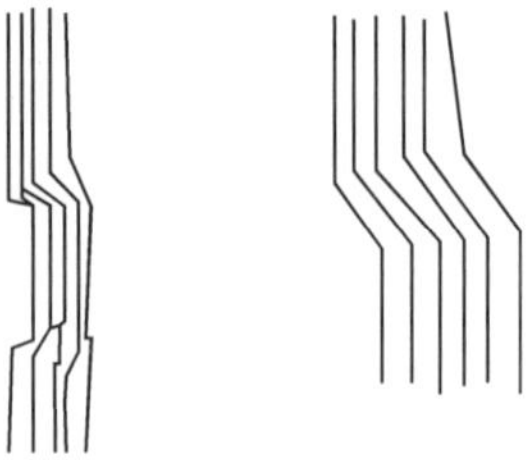

Fig. 4.21 Two compressive failure modes: fiber microbuckling of polymeric chain (*left*) and misorientation failure (*right*)

fracture parameters and for identifying the sources of failure. In general, the strength of a fiber decreases as its diameter increases.

Rigid-rod polymeric fibers such as aramid fibers show very high strength under axial tension. The failure in tension brings into play covalent bonding along the axis, which ultimately leads to chain scission and/or chain sliding or a combination of both. However, these fibers have poor properties under axial compression, torsion, and in the transverse direction.

As the orientation of chains in a fiber become more parallel to its axis, its axial tensile modules (E) increases but the shear modulus (G) decreases, i.e., the ratio E:G increases rapidly. During failure involving compressive stresses, fibrillation occurs, which results in a large degree of new surface area. This fibrillation process results in high-energy absorption during the process of failure, which makes these fibers useful for resistance against ballistic penetration [7].

Figure 4.21 shows two types of compressive failure modes that are important for failure and fracture of polymer composites: first, microbuckling or fiber orientation of polymeric chains and, second, kink band formation. Microbuckling includes closely spaced chains in a small region of fiber and this may be the reason for the formation of kink band failure under compressive loading.

Environmental factors such as humidity, temperature, ultraviolet radiation, and microorganisms can affect the strength and the fracture process in polymeric fibers. Figure 4.22 shows various failure modes such as fiber–matrix debonding, fiber fracture, resin-rich region, and brittle failure of fibers that play dominant roles in affecting the mechanical behavior and durability of the polymer composite material.

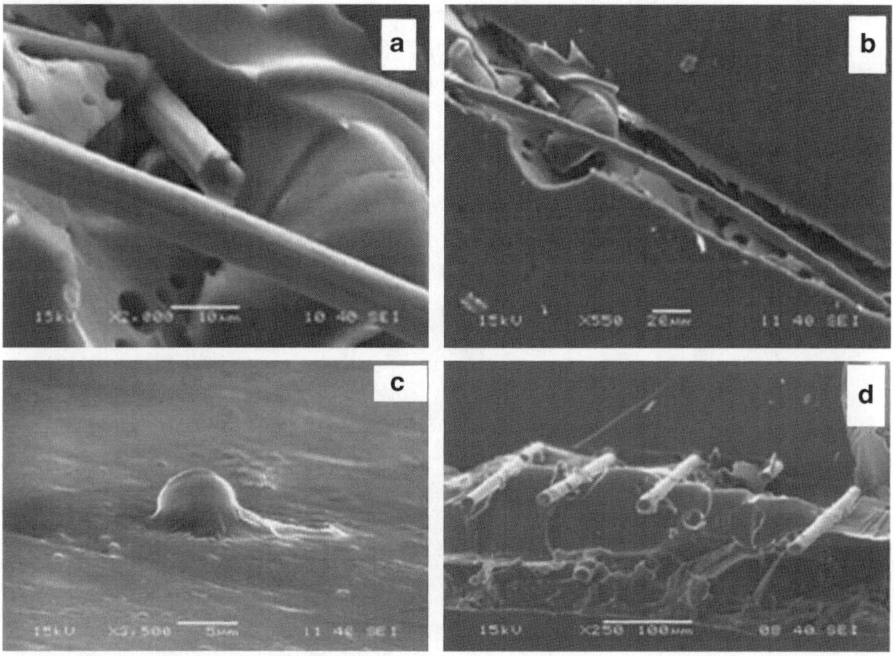

Fig. 4.22 Various failure modes: (**a**) fiber matrix debonding, (**b**) fiber fracture, (**c**) resin-rich region, and (**d**) brittle failure of fiber

4.5.3 Delamination

Delamination is a critical failure mode in PMCs. Interfacial debonding (delamination) may cause premature buckling of the laminate. It occurs due to excessive moisture absorption, loss of fatigue life, or decrease in stiffness of the composite. In many practical situations, this problem can be overcome by finite element methods (FEM). Research on this failure mode is very extensive and growing rapidly. In the short-term, delamination causes damage growth and premature failure. However, in the long-term, it leads to load-bearing layers for different environmental factors [107]. Delamination of a laminate occurs when the transverse shear force resultant exceeds a threshold value [7]. Low interfacial shear strength may be the reason for initiation and propagation of delamination. It was reported that fatigue delamination growth rate at 77 K and at 4 K is lower than that of room temperature. The delamination rate at 4 K is higher than that at 77 K [109].

Delamination is the life-limiting failure process in a composite material. It induces great loss of stiffness, local stress concentration, and buckling failure of polymer composites. Figure 4.23 reveals delamination of fiber-reinforced composites. This failure mode arises due to interlaminar stresses from impact or shear loading, foreign particles, or some discontinuities in the structure.

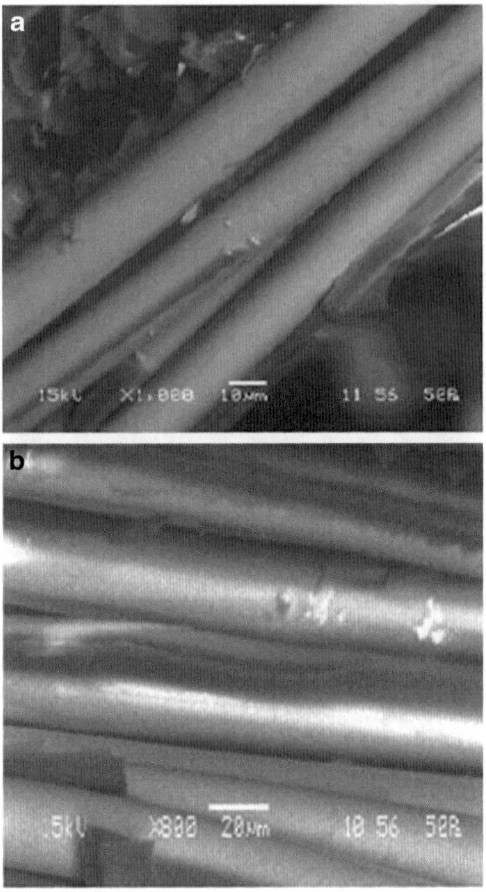

Fig. 4.23 Scanning micrographs showing (**a**) delamination and fiber pull-out of fiber reinforced composite and (**b**) delamination of fiber-reinforced composite

The difference in CTE between the fiber and the matrix may be the cause of residual thermal stress that develops in the material when worked at various temperatures. However, moisture absorption and desorption are the reason behind the formation of residual stress. This would modify the local stress threshold required for interfacial debonding, a potential precursor to delamination nucleation.

Causes of delamination include the high interlaminar stresses at the free edge, impacts, and fabrication defects. Low velocity impact of foreign objects is considered one of the most important causes of delamination. Delamination appears at several interfaces through the thickness away from the point of impact and significantly reduces the compressive strength. Blistering is an example of delamination under load (local stress concentration); under compressive loading the initial void area can spread rapidly, which tends to produce buckling. The strain energy required to separate the bonded layers is nucleation of delamination. The boundary between two plies of different orientation which the region of delamination where

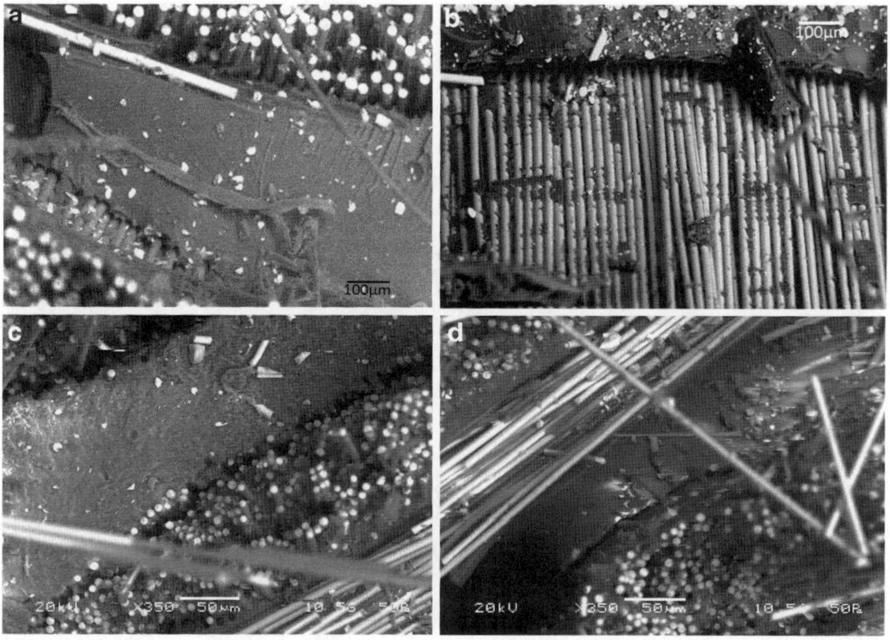

Fig. 4.24 Failure modes during thermal treatment of polymer composites: (**a**) matrix roller in toughened epoxy matrix, (**b**) fiber matrix failure in FRP, (**c**) weak interface region in FRP, and (**d**) toughened epoxy matrix region

shear deformation forms. This occurs preferentially along the axis of the fiber, since it is the source of least resistance [110]. However, at low temperatures under mode II loading a hackle pattern on the fracture surface is highlighted and delamination occurs due to fiber–matrix debonding [111].

4.5.4 *Interfacial Damage and Debonding*

The polymer composite is assumed to have excellent bond strength between fiber and matrix. When it is treated at low temperatures, the original bond strength of the matrix is of very little importance since it is increased by shrinkage of the matrix onto the fiber surface. Carbon and Kevlar fibers have high transverse thermal expansion due to which they have very less shrinkage effect [8].

Figure 4.24 shows matrix roller (Fig. 4.24a) and fiber–matrix failure (Fig. 4.24b) during thermal treatment of polymer composite. We can assume that various failure modes are formed due to inclusion of small defects. During the crack propagation, large transverse tensile stress and interfacial shear stress

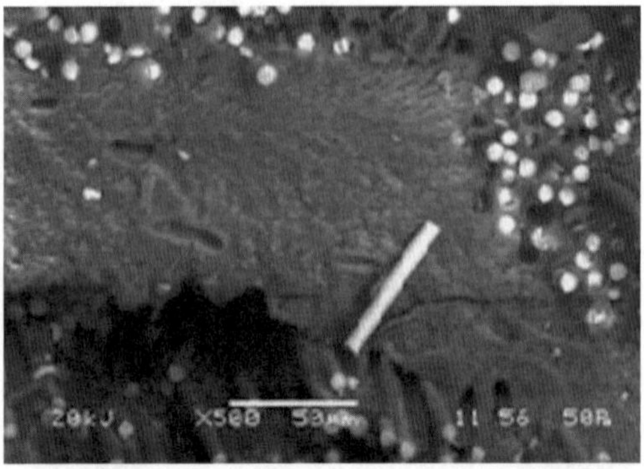

Fig. 4.25 Scanning micrograph showing fiber pull-out and matrix cracking

can cause fiber–matrix interfacial debonding, delamination, and interfacial cracking slightly ahead of the crack tip. The energy imposed on a composite can be absorbed by two mechanisms: elastic and plastic deformation of the material. Plastic deformation mechanisms such as shear yielding can be promoted by various techniques to toughen the epoxy matrix. The energy absorbing capability or toughness of a material can be enhanced by increasing either the total area of new surfaces created or the material deformation capability. Near the crack tip region, a matrix plastic deformation zone occurs. Immediately behind the crack tip, the broken fibers can pull out of the matrix. Weak interface strength can promote extensive debonding, leading to an increase in impact energy. Figure 4.24 shows a weak interface region (Fig. 4.24c) and toughened matrix (Fig. 4.24d) of the fiber–matrix interface region. From a macroscopic point of view, poor fiber–matrix bond strength is associated with brush-like failure. This means that there is a lack of matrix on the fiber surface. Factors such as excessive exposure to moisture, temperature, or humidity can contribute to the weak fiber–matrix interface strength. In this state, load transfer between the fiber and matrix is reduced [112]. At high temperatures, fiber–matrix debonding and matrix ductility plays the dominant role and the residual stress effects are negligible. If the fiber–matrix interface strength is high, the degree of fiber debonding will be limited before the fiber failure occurs and cracks propagate across the fibers. This will lead to a flat and relatively smooth surface and is often referred to as brittle failure. If the interface strength is weak, the degree of fiber debonding will be extensive before fiber failure occurs. This leads to broom-like failure.

Figure 4.25 shows matrix cracking as well as fiber pull-out of the polymer matrix material. Delamination and microcracking are some of the most frequently observed damage phenomena that may develop in polymer composites exposed to cryogenic temperatures. The mechanical performances of FRPs are strongly dependent upon the quality of fiber–matrix adhesion. Cryogenic hardening may modify the local threshold required for breaking of adhesion bondage at the fiber–matrix interface. Most polymers lose their ductile properties below their T_g. Cooperative chain motions involving main chain bond rotation become extremely restricted.

4.6 Issues Concerning the Interface, Matrix, and Fibers of FRP Composites at Low Temperatures

4.6.1 Experimental Explorations

Composite materials are tailored from fiber reinforcement and a resin matrix, which are significantly different in their physical and chemical properties [113]. They have distinct macroscopic or microscopic scales of structure. The fibers of high strength and stiffness are dispersed in a matrix material, which acts as a trimmer and transfers the load across the interphase region. The interphase in composites is the region that forms in the vicinity of the fiber surface and possesses distinct properties to those of the bulk resin [7]. One mechanism of interphase formation, based on thermodynamic considerations, predicts that small monomers preferentially segregate to the fiber surface [114]. Previous investigations have shown that the interphase has a significant effect on composite properties such as strength, toughness, ballistic resistance, and durability. Modification of the interphase could lead to improvements and a balance of properties required for a given application; therefore, significant efforts are directed at developing a fundamental understanding of the role of the interphase.

These composites are modern engineering materials that have wide applications in a range of areas from aerospace, automobiles, and boats to cryogenic equipment such as cryogenic fuel tanks, cryogenic fuel delivery lines, cryogenic wind tunnels, and parts of the cryogenic side of turbo-pumps because of their ease of handling, low fabrication cost, and excellent mechanical properties [114]. When the temperature is decreased down to cryogenic temperatures, internal stresses are generated in the epoxy matrix due to thermal contraction. The fracture toughness of the matrix at cryogenic temperatures can be improved by controlling the chemical structure, network structure, and morphology. Delamination and matrix microcracking are the two common failure modes observed at cryogenic temperatures. The amount or density of microcracking depends mainly on the tensile modulus of the fibers and the properties of the matrix used in the composite. Addition of toughening agents in the matrix decreases the microcracking propensity of these laminates [78]. Glass fibers and epoxy resins are known to be highly sensitive to loading rate [115]. The mechanical properties of glass-reinforced epoxy composites are found to be rate sensitive even at low strain rates. CFRP structural components used in aircrafts can suffer high and low velocity impact at cryogenic temperatures. Fiber reinforcement architecture and stacking sequence play an essential role in the behavior of composites under such thermal and loading conditions [116]. At lower speeds, there will be more deterioration in the matrix because more time is available for failure to take place, resulting in lower ILSS values, which increase with the crosshead speed. At higher loading rates, the reduction in ILSS values may be due to the restriction and minimization of relaxation processes at the crack tip, leading to the growth of stress-induced cracks without blunting. The matrix is unable to transfer the load properly due to less availability of time at higher speeds, i.e., the load on the matrix is like an impact [117]. Several damage mechanisms may appear, such as fiber fracture, matrix cracking, fiber pull-out, and delamination. Hence, it can be concluded that the laminate requires an optimum time to transfer load effectively through the interface.

These interphases posses unique properties that are different to those of the bulk matrix [118]. The matrix can include impurities, unreacted polymer components, non-polymerized additives etc. The thickness and properties of the interphase have crucial impact on the composite properties. The interphase not only allows load transfer between fibers through matrix but also provides a matching of chemical and thermal compatibility between the constituents. A thin rigid interphase leads to a low fracture resistance, whereas a thick soft interface results in a better fracture resistance but a lower composite stiffness [30]. Epoxy resin and E-glass fiber are reported to be sensitive to loading rate. The ductility of a matrix resin may become a limiting factor for composite strength at high strain rate. Epoxy resin is more ductile than its composite at low strain rate [119]. The cryogenic conditioning causes differential contraction and increases the resistance to debonding by mechanical keying factors [3–6]. When cooling from the curing temperature, differences in thermal contraction between the matrix and fiber will generate shear stresses in the resin. When a stress is applied, shear stresses greater than the shear strength of the resin are readily generated and failure of the resin phase will result. When the composite is stressed further by cooling and loading in the cold state, it is likely that there will be resin–fiber debonding. A plastic deformation zone ahead of the crack tip region may possibly be formed by matrix deformation and microcracking. A weak interfacial bond may result in a low flexural strength of the laminate [120]. The deteriorated integrity can cause low strength at high loading.

Although the mechanical strength of most polymers increases or remains the same as temperature is decreased, the elongation to failure decreases to extremely low values at cryogenic temperatures. This behavior restricts the use of most polymeric materials at low temperatures. It has been proposed that the local intermolecular rearrangement results in relaxation at the low temperature region. The relaxation phenomena indicate a considerable dependence on the morphology of the polymer. There appears to be no definite relationship between the chemical structure of a polymer and its cryogenic behavior. Polymers that are able to change their main chain bond angles seem to excel in cryogenic properties. These polymers are flexible and may undergo deformation even when their segmental motions are frozen at cryogenic temperatures. The mechanical properties of E-glass/epoxy composites are ratesensitive at a low range of strain rate. It seems that the greater the strain rate and the loading velocity, the greater the stiffness and ultimate strength of the composite material [121]. The failure strength of glass/epoxy composites increases manifold, and failure strain reduces sharply at a high range of strain rate. Loading rate sensitivity seems to be controlled by the area of interface and the percentage of polymer matrix phase present in the composite [122]. The interphase is defined as a region that is manifested as a result of bonding and reactions between the fiber and the matrix. This region is the site of synergy in composite materials and it has an influence on the overall mechanical properties.

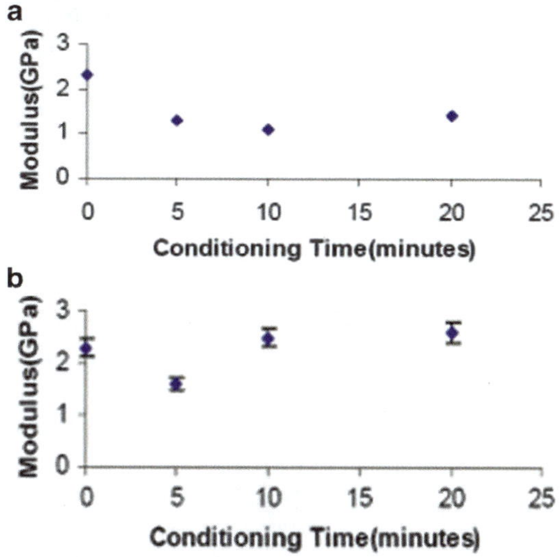

Fig. 4.26 Effect of thermal shock on modulus of (**a**) thermally conditioned Kevlar polyester samples and (**b**) cryogenically conditioned Kevlar polyester composites. Samples were kept at cryogenic temperature for the indicated times

Environmental attack can degrade the fiber–matrix interface mostly by mechano-chemical processes.

Figure 4.26 shows the effect of thermal shock on the modulus of Kevlar and polyester samples when treated with thermal and cryogenic temperatures. The toughness value increases with increased period of time at the cryogenic temperature due to the greater amount of shrinkage compressive stress.

The higher ILSS values for cryogenically conditioned samples is attributed to an enhanced mechanical keying factor by the generation of cryogenic compressive stresses, which enhance the friction at the interface due to contraction of epoxy matrix at low temperature. The composite laminate needs an optimum time to transfer load effectively through the interface [123, 124]. There is an increase in T_g after cryogenic conditioning of the laminates, which may be due to incomplete reversibility of molecular chain mobility resulting in reduced chain flexibility.

4.6.2 *Deformation of the Resin Matrix and the Mechanical Behavior of FRP Composites*

New technologies demand new materials. The spectrum of composite end-users is very broad and encompasses nearly all industrial fields. Polymer composites with their wide range of possible fillers and polymers open the way to an enormous range of materials with differing chemical, physical, and mechanical properties. While good energy absorption, light weight, and design flexibility (function integration) are positive points in favor of polymer composites, problem with mass manufacturing and surface finish still hinder the widespread use of composite materials as structural elements. These matrix materials are characterized by a constant molecular motion in the amorphous state, even under static loads. The re-arrangement rate is a function of many parameters including stress level, length of macromolecules, and temperature. As the material is intrinsically heterogeneous, damage and degradation in the material will therefore often be strongly influenced by local processes. The anisotropy contributes to more complexity in the assessment of the damage mechanisms and in their impact on the composite responses [119]. The use of composites will continue to grow in coming years with emerging applications in large bridge structures, offshore platforms, engine machinery, computer hardware, and biomedical devices. This wide variety of applications is due to the many outstanding physical, thermal, chemical, and mechanical properties of the composites. However, application is plagued by problems such as low through-thickness mechanical properties, poor impact damage tolerance, and anisotropic properties. Polymer composites soften, creep, and distort when heated to high temperatures (above 100–200 °C), which can result in buckling and failure of load-bearing composite structures.

All structural materials exhibit some degree of viscoelasticity, and the extent of such behavior often depends on environmental conditions such as temperature. Polymeric materials are viscoelastic at room temperature, and this effect become stronger as the temperature approaches T_g. Viscoelastic effects in PMCs are most pronounced in the matrix-dominated response to off-axis or shear loading. This deformation and plastic deformation are similar in that both are driven by shear stress [114].

This deformation process is slow and depends on many factors, but it is also known to alter the physico-mechanical, dielectric, and diffusion characteristics of polymers in the glassy state. At the molecular level, this drive towards equilibrium takes place via molecular relaxations and, in some instances, molecular rearrangements. The rearrangements are envisioned as a form of morphological changes (e.g., ordering).

During thermal fatigue, additional chemical reactions may occur (further crosslinking) between unreacted groups, leading to variations in crosslink density that in turn lead to an increase in stiffness and viscosity of the composites.

Temperature effects on performance and fracture morphology are usually apparent only when the temperature approaches the T_g of the material. Below T_g, physical aging occurs in a polymer and this aging process changes the viscoelastic response of the material during a long-term test. In a short-term test, where the test duration is much less than the aging time, no significant aging occurs during the test. Physical aging in polymers is associated with a slow loss of free volume that has been trapped in the polymer microstructure after cooling below T_g. As the temperature of a polymer increases, its total volume consists of V_0 (the volume occupied by the polymer molecules) and V_F (the free volume between the molecules). When it is cooled to room temperature, a significant amount of free volume is "locked in." This is a thermodynamically unstable condition because the polymer will slowly give up free volume with time to approach a more stable thermodynamic condition. When polymer free volume is given up, the chain mobility decreases and the relaxation times increases, thus reducing the speed of the relaxation or creep.

Fiber–matrix interfacial adhesion plays an important role in determining the mechanical properties of polymer composites. The time dependency of the matrix material is much more significant than that of the fiber, so the fiber modulus can be assumed to be elastic, and the time dependency of $E_1(t)$ is governed by $E_m(t)$ alone.

A strong interface displays an exemplary strength and stiffness, but it is very brittle in nature with easy crack propagation through the interface. A weak interface reduces the stress transmissibility and consequently decreases the strength and stiffness. Here, a crack is more likely to deviate and grow at the weak interface resulting debonding and/or fiber pull-out and contributes to improved fracture toughness.

Residual stresses play an important role in the microcracking process of laminates. In addition to residual stresses due to thermal shrinkage, the presence of moisture can induce residual stresses that influence microcracking. Delamination or interlaminar fracture is a very important failure mode in composite laminates. This provides one of the few examples of self-similar crack growth in laminates.

Assessment of resistance to delamination is generally considered to be synonymous with characterization of interlaminar fracture toughness. The short beam shear test is commonly used to measure the ILSS of composites. Here, the specimen could fail by fiber rupture, microbuckling or interlaminar shear cracking, or a combination of fracture modes. In the case of the four-point bending test, the failure occurs by matrix shear yielding, interfacial debonding, or some combination of both.

4.7 Conclusion

Failure is a detour, not a dead-end street. The structural integrity and durability of fibrous composites more than compensate for their disadvantages. Composites, the preamble to a lighter world, have an enormous number of applications from ground-based materials to the space shuttle. In contrast, a significant mismatch in the environmentally induced degradation of matrix and fiber leads to the evolution of localized stress and strain fields in the FRP composites. The hygroscopic nature of polymeric composites necessitates a complete understanding of the interaction between structural integrity and hygrothermal environments. Composites absorb moisture in humid environments and undergo dilational expansion. Hygrothermal exposure reduces the T_g of polymers and also causes plasticization, resulting in a reduction in mechanical properties. The presence of moisture and stresses associated with moisture-induced expansion may cause lowered damage tolerance and structural durability. Mechanical properties of material depend not only on the substrate strength but also on the interface strength. However, moisture absorption and desorption are also the reason behind the formation of residual stress. This modifies the local stress threshold required for interfacial debonding, a potential precursor to delamination nucleation. Interfacial debonding (delamination) may cause premature buckling of the laminate. Composites are available that incorporate increasingly diverse material forms and are manufactured by an extensive array of molding and forming processes They are poised to take the spotlight in manufacturing arenas throughout the world.

In the future, composite technology will play an important role for the development of energy-saving vehicles and technologies for environment and climate protection. Carbon fiber-reinforced composite is one of the potential materials for replacing the pre-stressed steel tendons at low temperatures (−27 °C) in Canada. Both in the aeronautic and the automotive industry, composites are more and more used for crash structures. Advantages of FRP structures for crash applications are their ability to absorb energy by means of numerous available absorbing sites. The mechanical properties of composites at low temperatures are influenced by the matrix. At these temperatures, thermal pre-stress is crucial. This reduces the effective strain to failure and is the source of microcracks in the matrix. The first formation of microcracks causes very mild effects in the thermo-mechanical behavior of composites. Considering the influence of low temperatures on the fracture micromechanisms in composites, the matrix is brittle and residual tensile stresses in the matrix are promoted. At very low temperatures, resin embrittlement dominates the phenomena and thus composite toughness decreases. As the temperature increases very close to the T_g of the matrix, it becomes ductile and behaves as a softer material. Composites at low temperatures, despite their decades of development, are young by comparison and are therefore not as well known by design and manufacturing engineers. The implication thus far is that composites at low temperature must adapt to meet the full potential of their inherent properties.

Acknowledgements The authors would like extend sincere thanks and an appreciation to the National Institute of Technology, Rourkela, India for supporting and funding instrumental facilities to carry out investigations on composite materials. We are indebted furthermore to the same for an extensive literature support from the library. It is also our privilege to acknowledge the devotion and dedication received from many graduate, undergraduate and doctoral students and scholars over the years. The investigation has also been funded from different sponsoring agencies.

References

1. Lee SM (1993) Handbook of composite reinforcements. VCH, New York
2. Dorgham MA (1986) The economic case for plastics. In: Dorgham MA, Rosta DV (eds) Designing with plastics and advanced plastic composite. Interscience, Geneva, pp 10–13
3. Ray BC (2005) Effect of hydrothermal shock cycles on shear strength of glass fiber/polyester composite. J Reinf Plast Compos 24:1335–1340
4. Ray BC (2005) Freeze-thaw response of glass-polyester composites at different loading rate. J Reinf Plast Compos 24:1771–1776
5. Ray BC (2005) Effects of thermal and cryogenic conditionings on mechanical behavior of thermally shocked glass fiber/epoxy composites. J Reinf Plast Compos 24:713–715
6. Ray BC (2005) Thermal shock and thermal fatigue on delamination of glass fiber reinforced polymeric composites. J Reinf Plast Compos 24:111–116
7. Jang BZ (1994) Advanced polymer composites: principle and applications. ASM International, Materials Park
8. Hartwig G (2004) Polymer properties at room and cryogenic temperatures. Plenum, New York

9. Bunsell AR, Renard J (2005) Fundamentals of fibre reinforced composite materials. Institute of Physics, London
10. Hull D (1996) An introduction to composite materials. Cambridge University Press, Cambridge
11. Kaw AK (2006) Mechanics of composite materials. Taylor & Francis, London
12. Kelly A, Zweben C (2000) Comprehensive composite materials. Elsevier Science, Oxford
13. Moncrieff RW (1979) Man-made fibers, 6th edn. Butterworth, London
14. Thomson JL (1995) The interface region in glass fiber reinforced epoxy resin composites: 3. Characterization of fiber surface coatings and the interphase. Composites 26:487–498
15. Plonka R, Mader E, Gao SL, Bellmann C, Duschk V, Zhandarov S (2004) Adhesion of epoxy/glass fiber composites influenced by aging effects on sizing. Adhesion 35:1207–1216
16. Watts AA (ed) (1980) Commercial opportunities for advanced composites, STP 704. ASTM, Philadelphia
17. Kim JK, Mai YW (1992) Interfaces in composites. In: Chou TW (ed) Structure and properties of composites, vol 13, Materials science and technology. VCH, Weinheim, pp 239–289
18. Ishida H, Koening JL (1980) Effect of hydrolysis and drying on the coupling agent/matrix interface of fiber glass reinforced plastic by Fourier transform infrared spectroscopy. Polym Sci B Polym Phys 17:615–626
19. Moussawi E, Drown EK, Drzal LT (1993) The silane sizing composites interphase. Polym Compos 14:195–200
20. Kim K, Mai YW (1998) Engineered interfaces in fiber reinforced composites. Elsevier, Kidlington
21. Plueddemann EP (ed) (1974) Interfaces in polymer matrix composites, vol 6, Composite materials. Academic, New York
22. Plueddemann EP, Stark GL (1980) In: Proceedings 35th Annual Technical Conference Reinforced Plastics/Composites Institute, New Orleans. SPI, Washington
23. Tsai HC, Arocho AM, Gause LW (1990) Prediction of fiber-matrix interphase properties and their influence on interface stress, displacement and fracture toughness of composite materials. Mater Sci Eng A126:295–304
24. Williams JG, Donnellan ME, James MR, Morris WL (1990) Properties of the interphase in organic matrix composites. Mater Sci Eng A 126:305–312
25. Frank O, Tsoukleri G, Riaaz I, Papagelis K, Parthenios J, Andre C, Geim AK, Novoselov S, Galootis C (2011) Development of universal stress tensor for graphene and carbon fiber. Nature Commun 2:255. doi:10-1038/ncomms1247
26. Ochola RO, Marcus K, Nurick GN, Franz T (2004) Mechanical behaviour of glass and carbon fibre reinforced composites at varying strain rates. Compos Struct 63:455–467
27. Marshall P, Price J (1991) Topography of carbon fiber. Composite 22:388–393
28. Horie K, Murai H, Mita I (1977) Bonding of epoxy resin to graphite fibers. Fiber Sci Technol 9:253–264
29. Donnet JB, Bansal RC (1984) Carbon fibers. Marcel Dekker, New York
30. Jones C (1991) The chemistry of carbon fiber surfaces and its effect on interfacial phenomena in fiber/epoxy composites. Compos Sci Technol 42:275–298
31. Thorne DJ (1970) Strength-modulus relation for carbonized acrylic fibres. Nature 225: 1039–1040
32. Kaelble DH, Dynes PJ, Maus L (1976) Hydrothermal aging of composite materials. Part 1: interfacial aspects. Adhesion 8:121–144
33. Gilat A, Goldberg RK, Roberts GD (2002) Experimental study of strain-rate-dependent behavior of carbon/epoxy composite. Compos Sci Technol 62:1469–1476
34. Goan JC, Martin TW, Prescot R (1973) The influence of interfacial bonding on the properties of carbon fiber composites. In: Proceedings 28th Annual Technical Conference on Reinforced Plastics/Composites Institute, Washington, DC. Society of the Plastics Industry (SPI), Washington
35. Chiao CC, Chiao TT (1977) Aramid fibers and composites. Lawrence Livermore National Laboratory Report UCRL-80400

36. Ganczakowski HL, Beaumont PWR (1989) The behaviour of Kevlar fibre-epoxy laminates under static and fatigue loadings. Part I-experimental. Compos Sci Technol 36:299–319
37. Allred RE (1984) In: Proceedings of the 7th annual meeting of the Adhesion Society, New York.
38. Allred RE, Merrill EW, Roylance DK, Ishida H, Kumar G (eds) (1985) Molecular characterization of composite interfaces. Plenum, New York, pp 333–375
39. Northolt MG (1974) X-ray diffraction study of poly(p-phenylene terephthalamide) fiber. Eur Polym J 10:799–804
40. Morgan RJ, Kong FM, Lepper JK (1988) Laser induced damage mechanismof Kevlar-49/ epoxy composites. J Compos Mater 22:1027
41. Happe JA, Morgan RJ, Walkup CM (1985) ^{1}H, ^{19}F and ^{11}B nuclear magnetic resonance characterization of BF_3:amine catalysts used in the cure of C fiber-epoxy prepregs. Polymer 26:827-836
42. Morgan RJ, Pruneda CO, Butler N, Kong FM, Caley LE, Moore RL (1984) ^{1}H, ^{19}F and ^{11}B Nuclear magnetic resonance characterization of BF_3: amine catalysts used in the cure of C fiber-epoxy prepregs. Proceedings of the 29th National SAMPE Symposium, London, pp. 891–902
43. Penn LS, Ishida H, Kumar G (eds) (1985) Molecular characterization of composite interfaces. Plenum, New York
44. Moss M (1978) Effect of water-bearing fiber in Kevlar 49/epoxy composites. Sandia National Laboratories report SAND-77-0444
45. Hahn HT, Pagano NJ (1975) Curing stresses in composite laminates. J Compos Mater 9: 91–106
46. Updegraff I (1982) Unsaturated polyester resins. In: Lubin G (ed) Handbook of composites. Van Norstrand Reinhold, New York
47. Camino G, Luda MP, Polishchuk AY, Revellino M, Blancon R, Merle G, Martinez-Vega JJ (1997) Kinetic aspects of water sorption in polyester. Compos Sci Technol 57:1469–1482
48. Launikitis MB (1982) Handbook of composites. Van Norstrand Reinhold, New York
49. Bascom WD, Drzal LT (1987) The surface properties of carbon fibers and their adhesion to organic polymers. NASA contract report 4084
50. Kaiser T (1989) Highly cross-linked polymers. Prog Polym Sci 14:373–450
51. Mijovic J, Lin KF (1985) The effect of hygrothermal fatigue on physical/mechanical properties and morphology of neat epoxy resin and graphite/epoxy composite. J Appl Polym Sci 30:2527–2549
52. Hartwig G, Knaak S (1984) Fibre-epoxy composites at low temperatures. Cryogenics 24: 639–647
53. Hiltner A, Baer E (1974) Mechanical properties of polymers at cryogenic temperature; relationships between relaxation, yield and fracture process. Polymer 15:805–813
54. Dodiuk H, Kenig S, Liran I (1991) Low temperature curing epoxies for elevated temperature composites. Composites 22:319–327
55. Boller A, Schick C, Wunderlich B (1995) Modulated scanning calorimetry in the glass transition region. Thermochim Acta 266:97–111
56. Nielsen LE (1975) Mechanical properties of polymers and composites. Marcel Dekker, New York
57. Lu X, Jiang B (1990) Glass transition temperature and molecular parameters of polymers. Polymer 32:471–478
58. Min BG, Stachurshi ZH, Hodgkin JH (2003) Cure kinetics of elementary reactions of a DGEBA/DDS epoxy resin: 1. Glass transition temperature. Polymer 34:4908–4912
59. Barjasteh E, Bosze EJ, Tsai YI, Nutt SR (2009) Thermal aging of fibreglass/carbon-fiber hybrid composites. Compos Part A: Appl Sci 40:2038–2045
60. Ahlborn K, Knaak S (1988) Cryogenic mechanical behavior of a thick-walled carbon fiber reinforced plastic structure. Cryogenic 28:273–277

61. Kellogg KG, Patil R, Kallmeyer AR, Dutta PK (2005) Effect of load rate on notch toughness of glass FRP subjected to moisture and low temperature. Int J Offshore Polar Eng 15:54–61
62. Ray BC (2006) Temperature effect during humid ageing on interfaces of glass and carbon fibers reinforced epoxy composites. J Colloid Interf Sci 298:111–117
63. Ray BC, Surendra Kumar M, Sharma N (2009) Microstructural and mechanical aspects of carbon/epoxy composites at liquid nitrogen temperature. J Reinf Plast Compos 28:2013–2023
64. Ray BC (2006) Adhesion of glass/epoxy composites influenced by thermal and cryogenic environments. J Appl Polym Sci 102:1943–1949
65. Ray BC (2006) Loading rate effects on mechanical properties of polymer composites at ultralow temperatures. J Appl Polym Sci 100:2289–2292
66. Yano O, Yamaoka H (1995) Cryogenic properties of polymers. Prog Polym Sci 20:585–613
67. Naruse T, Hattori T, Miura H, Takahashi K (2001) Evaluation of thermal degradation of unidirectional CFRP rings. Compos Struct 52:533–538
68. Colin X, Verdu J (2005) Strategy for studying thermal oxidation of organic matrix composites. Compos Sci Technol 65:411–419
69. Ray BC (2004) Thermal shock on interfacial adhesion of thermally conditioned glass fiber/epoxy composites rates. Mater Lett 58:2175–2177
70. Ray BC (2004) Effect of crosshead velocity and sub-zero temperature on mechanical behavior of hygrothermally conditioned glass fiber reinforced epoxy composites. Mater Sci Eng: A 397:39–44
71. Mader E, Gao SL, Plonka R (2004) Enhancing the properties of composites by controlling their interphase parameters. Adv Eng Mater 6:147–150
72. Wang Y, Hahn TH (2007) AFM characterization of the interfacial properties of carbon fiber reinforced polymer composites subjected to hygrothermal treatments. Compos Sci Technol 67:92–101
73. Aktas M, Karakuzu R (2009) Determination of mechanical properties of glass-epoxy composites in high temperatures. Polym Compos 30:1437–1441
74. Morioka K, Tomita Y, Takigawa K (2001) High-temperature fracture properties of CFRP composite for aerospace applications. Mater Sci Eng A 319:675–678
75. Reed RP, Golde M (1997) Cryogenic composite supports: a review of strap and strut properties. Cryogenic 37:233–250
76. Salin I, Seferis JC (1996) Anisotropic degradation of polymeric composites: from neat resin to composite. Polym Compos 13:430–433
77. Borje A, Aders S, Lars B (2000) Micro- and meso-level residual stresses in glass/vinyl-ester composite. Compos Sci Technol 60:2011–2028
78. Timmerman JF, Tillman MS, Matthew S, Hayes BS, Seferis JC (2002) Matrix and fiber influences on the cryogenic micro cracking of carbon fiber/epoxy composites. Compos Part A: Appl Sci 33:323–329
79. Ueki T, Nishijma S, Izumi Y (2005) Designing of epoxy resin systems for cryogenic use. Cryogenics 45:141–148
80. Hartwig G (1994) Polymer properties at room and cryogenic temperatures. Plenum, New York
81. Sawa F, Nishijima S, Onada T (1995) Molecular design of an epoxy for cryogenic temperature. Cryogenic 35:767–769
82. Kanchanomai C, Rattananon S, Soni M (2005) Effects of loading rate on fracture behavior and mechanism of thermoset epoxy resin. Polym Test 24:886–892
83. Zhou J, Lucas JP (1999) Hygrothermal effects of epoxy resin. Part II: variations of glass transition temperature. Polymer 40:5513–5522
84. Ngono Y, Marechal Y, Mermilliod N (1999) Epoxy-amine reticulates observed by infrared spectroscopy. 1: Hydration process and interaction configurations of embedded water molecules. J Phys Chem B 103:4979–4985
85. Liang L, Zhang SY, Chen YH, Liu MJ, Ding YF, Luo XW, Pu Z, Zhou WF, Li S (2005) Water transportation in epoxy resin. Chem Mat 17:839–845

86. Maxwell ID, Pethrick RA (2003) Dielectric studies of water in epoxy resin. J Appl Polym Sci 28:2363–2379
87. Eidelman N, Raghavan D, Forster AM, Amis EJ, Karim A (2004) Combinatorial approach to characterizing epoxy curing. Macromol Rapid Comm 25:259–263
88. Xiao GZ, Shanahan MER (1997) Water absorption and desorption in an epoxy resin with degradation. J Polym Sci Part B 35:2659–2670
89. Zheng Q, Morgan RJ (1993) Synergistic thermal-moisture damage mechanisms of epoxies and their carbon fiber composites. J Compos Mater 27:1465–1478
90. Tsenoglou CJ, Pavlidou S, Papaspyrides CD (2006) Evaluation of interfacial relaxation due to water absorption in fiber-polymer composites. Compos Sci Technol 66:2855–2864
91. Browning CE, Husman GE, Whitney JM (1977) Moisture effects in epoxy matrix composites. In: Davis JG (ed) Composite materials: testing and design ASTM STP 617. American Society for Testing and Materials, Philadelphia, pp 481–496
92. Wu P, Siesler HW (2003) Water diffusion into epoxy resin: a 2D correlation ATR-FTIR investigation. Chem Phys Lett 374:74–78
93. Lin YC, Chen X (2005) Moisture sorption–desorption–resorption characteristics and its effect on the mechanical behaviour of the epoxy system. Polymer 46:11994–12003
94. Morozova EM, Sokolova NP, Bulgakova RA, Karpova IV (2004) The investigation of the interaction at the glass-fiber-polymer matrix interface. Colloids Surf A Physicochem Eng Asp 239:81–83
95. Abdelkader AF, White JR (2005) Water absorption in epoxy resins: the effects of the crosslinking agent and curing temperature. J Appl Polym Sci 98:2544–2549
96. Xiaoping H, Shenliang H, Liang Y (2003) A study on dynamic fracture toughness of composite laminates at different temperatures. Compos Sci Technol 63:155–159
97. Garton A, Daly JH (1985) Characterization of the aramid: epoxy and carbon: epoxy interphases. Polym Compos 6:195–200
98. Okoli OI, Smith GF (1998) Failure modes of fiber reinforced composites: the effect of strain rate and fiber content. J Mater Sci 33:5415–5422
99. Okoli OI, Smith GF (1999) Aspects of the tensile response of random continuous glass/epoxy composites. J Reinf Plast Compos 18:606–613
100. Okoli OI (2001) The effects of strain rate and failure modes on the failure energy of fibre reinforced composites. Compos Struct 54:299–303
101. Barre S, Chotard T, Benzeggagh ML (1996) Comparative study of strain rate effects on mechanical properties of glass fibre reinforced thermoset matrix composites. Compos Part A-Appl S 27A:1169–1181
102. Gonzalez-Benito J (2003) The nature of structure gradient in epoxy curing at a glass fiber/epoxy matrix interface using FTIR imaging. J Colloid Interf Sci 267:326–332
103. Jurgen E, Schawe K (2002) About the changes of heat capacity, glass transition temperature and relaxation time during polymerization reaction of thermosetting systems. Thermochim Acta 391:279–295
104. Marom G, Broutman LJ (1981) Moisture penetration into composites under external stress. Polym Compos 2:132–136
105. Detassis M, Pegoretti A, Migliaresi C (1995) Effect of temperature and strain rate on interfacial shear stress transfer in carbon/epoxy model composites. Compos Sci Technol 53:39–46
106. Hughes D, Buck B, Cornwell.A. (1982) In: Lubin G (ed) Handbook of composite materials. Van Norstrand Reinholds, New York
107. Greenhalgh ES (2009) Failure analysis and fractography of polymer composites. CRC, Cambridge
108. Brewis DM (1993) Adhesion to polymers: how important are weak boundary layers? Int J Adhes Adhes 13:251–256
109. Miura M, Shindo Y, Narita F, Watanable S, Suzuku M (2009) Mode III fatigue growth of glass fiber reinforced polymer woven laminates at cryogenic temperature. Cryogenic 49: 407–412

110. Purslow D (1986) Matrix fractography of fibre reinforced epoxy composites. Composites 17: 289–303
111. Shindo Y, Takano S, Horiguchi K, Sato T (2006) Cryogenic fatigue behavior of plain weave glass/epoxy composite laminates under tension-tension cycling. Cryogenics 46:794–798
112. Abdel-Magid B, Ziaee S, Gass K, Schneider M (2005) The combined effects of load, moisture and temperature on the properties of E-glass/epoxy composites. Compos Struct 71:320–326
113. Hamouda AMS, Hashmi MSJ (1998) Testing of composite materials at high rates of strain: advances and challenges. J Mater Process Technol 77:327–336
114. Anashkin OP, Keilin VE, Patrikeev VM (1999) Cryogenic vacuum tight adhesive. Cryogenics 39:795–798
115. Shokrieh MM, Omidi J (2009) Tension behavior of unidirectional glass/epoxy composites under different strain rates. Compos Struct 88:595–601
116. Rio TG, Zaera R, Barbero E, Navarro C (2005) Damage in CFRPs due to low velocity impact at low temperature. Compos Part B: Eng 36:41–50
117. Saniee FF, Majzoobi GH, Bahrami M (2005) An experimental study on the behaviour of glass–epoxy composite at low strain rates. J Mater Process Technol 162:39–45
118. Ishida H (1990) Controlled interphases in composite materials. Elsevier Science, New York, pp 431–440
119. Padmanabhan K (1996) Time-temperature failure analysis of epoxies and unidirectional glass/epoxy composites in compression. Compos Part A-Appl S 27A:585–596
120. Tanoglu M, McKnight SH, Palmese GR, Gillespie JW (2000) A new technique to characterize the fiber/matrix interphase properties under high strain rates. Compos Part A: Appl Sci 31: 1127–1138
121. Jacob GC, Starbuck JM, Fellers JF, Simunovic S, Boeman RG (2004) Strain rate effects on the mechanical properties of polymer composite materials. J Appl Polym Sci 94:296–301
122. Naik NK, Yernamma P, Thoram NM, Gadipatri R, Kavala VR (2010) High strain rate tensile behavior of woven fabric E-glass/epoxy composite. Polym Test 29:14–22
123. Harding J, Li YL (1992) Determination of interlaminar shear strength for glass/epoxy and carbon/epoxy laminates at impact rates of strain. Compos Sci Technol 45:161–171
124. Harding J, Welsh LM (1983) A tensile testing technique for fibre-reinforced composite at impact rates of strain. J Mater Sci 18:1810–1826

Chapter 5
Interlaminar Delamination Fracture and Fatigue of Woven Glass Fiber Reinforced Polymer Composite Laminates at Cryogenic Temperatures

Yasuhide Shindo, Tomo Takeda, and Fumio Narita

Contents

Abstract This chapter describes the results of our studies on the interlaminar delamination fracture and fatigue of woven glass fiber reinforced polymer composite laminates under Mode I, Mode II, and Mode III loadings at cryogenic temperatures. Delamination fracture tests were carried out at cryogenic temperatures, and the critical energy release rate at the onset of delamination propagation, i.e., fracture toughness, was evaluated based on a finite element analysis coupled with damage. In addition, cryogenic fatigue delamination tests were performed, in order to obtain the delamination growth rate as a function of the range of the energy release rate. After the tests, fractographic observations were made to assess the delamination mechanisms at cryogenic temperatures.

Y. Shindo (✉) • T. Takeda • F. Narita
Department of Materials Processing, Graduate School of Engineering, Tohoku University, Aoba-yama 6-6-02, Sendai 980-8579, Japan
e-mail: shindo@material.tohoku.ac.jp; takeda-t@material.tohoku.ac.jp; narita@material.tohoku.ac.jp

S. Kalia and S.-Y. Fu (eds.), *Polymers at Cryogenic Temperatures*,
DOI 10.1007/978-3-642-35335-2_5,

Keywords Cryomechanics • Material testing • Fractography • Finite element method • Polymer-matrix composites • Glass fibers • Delamination • Fracture • Fatigue • Cryogenic devices

5.1 Introduction

Woven composite materials have been used in many kinds of engineering applications because of their unique architectural features, ease of handling, low fabrication cost, and excellent mechanical properties [1]. In particular, superconducting magnets of cryogenic power systems, such as the International Thermonuclear Experimental Reactor (ITER), may use large quantities of woven glass fiber reinforced polymer (GFRP) composite laminates as electrical and thermal insulation, and structural support [2]. Hybrid materials consisting of woven GFRP composites and electrical barrier materials, such as polyimide films, have been considered as the ITER candidate insulation systems [3]. The demand for lightweight structures in reusable launch vehicles (RLVs) may also result in the use of carbon fiber reinforced polymer (CFRP) composites in the design of cryogenic storage tanks [4], and woven CFRP materials have been identified as possible candidates for such systems [5]. For successful application of woven composites in cryogenic systems, a thorough understanding of their performance in a variety of conditions is of great importance.

Several review articles were published which summarize the research work on the physical and mechanical properties of fiber reinforced composites at cryogenic temperatures [6–8]. Also, our research group has extensively investigated the cryogenic behavior of woven composite laminates. For woven GFRP composite laminates, the interlaminar shear [9, 10], tensile [11, 12] and compressive [13] responses at cryogenic temperatures were investigated. Studies on the interlaminar shear failure of hybrid composite laminates at cryogenic temperatures were also conducted [14, 15]. In addition, the cryogenic translaminar crack behavior in woven GFRP laminates was characterized [16]. With regard to the composite response to cyclic loading at cryogenic temperatures, the tensile fatigue behavior of woven GFRP laminates was studied [17]. The cryogenic translaminar fatigue crack growth in woven GFRP laminates was also examined [18].

In fiber reinforced composite materials, there are several failure modes: fiber breakage, matrix cracking, fiber/matrix debonding, fiber pull-out, interlaminar delamination, etc. Among these failure modes, interlaminar delamination is one of the major failure modes in laminated composites because such composites lack through-thickness reinforcement. Furthermore, delamination in composites often results in the loss of their stiffness and strength, which may lead to safety and reliability problems. Therefore, understanding of the delamination behaviors in woven composite laminates under static and cyclic loadings at cryogenic temperatures is very useful for the design of cryogenic composite structures and future material development. However, the majority of studies on delamination in composite laminates has been concerned with the room temperature behaviors, and as a result, the knowledge of the cryogenic delamination behaviors is very limited.

Using fracture mechanics to characterize the onset and growth of delamination has become a generally accepted practice [19–21]. In general, there are three modes of fracture: opening Mode I, sliding shear Mode II, and scissoring shear Mode III. All three modes may be present at the delamination front in practical composite structures. Thus, the complete understanding requires the investigation of delamination behaviors under Mode I, Mode II, and Mode III conditions and combinations of these modes. The delamination fracture of woven GFRP laminates under combined Mode I and Mode II loading at cryogenic temperatures was examined experimentally and numerically [22, 23]. The mixed-mode I/II fatigue delamination growth behavior in woven GFRP laminates at cryogenic temperatures was also characterized [24].

In this chapter, we present our research findings on the delamination fracture and fatigue of woven GFRP composite laminates under Mode I, Mode II, and Mode III loadings at cryogenic temperatures. Cryogenic delamination fracture tests were performed on the composite specimens, and a finite element analysis coupled with damage was used to determine the critical energy release rate at the onset of delamination propagation, i.e., fracture toughness. Cryogenic fatigue tests were also conducted to obtain the delamination growth rate data for woven GFRP laminates under cyclic loadings. In addition, the delamination mechanisms at cryogenic temperatures were discussed based on fractographic observations.

5.2 Interlaminar Delamination Fracture

5.2.1 Mode I Fracture

In this section, we report the results on the delamination fracture behavior of woven GFRP composite laminates subjected to Mode I loading at cryogenic temperatures. The temperature dependence of the delamination fracture toughness of woven GFRP laminates, designated SL-E (Nitto Shinko Corporation, Japan), for Mode I was investigated by performing the double cantilever beam (DCB) test and the three-dimensional finite element analysis [25]. The DCB test has emerged as the preferred configuration for measuring the Mode I delamination fracture toughness of fiber reinforced composite materials [26–28]. The DCB specimen geometry is shown in Fig. 5.1. The woven GFRP panel for the specimens was of 3.5 mm thickness. The specimen length and width were 70 mm and 22 mm, respectively. A nonadhesive insert, i.e., a polymer film, was placed at the specimen midplane which serves as a delamination initiator. To produce an initial delamination, each

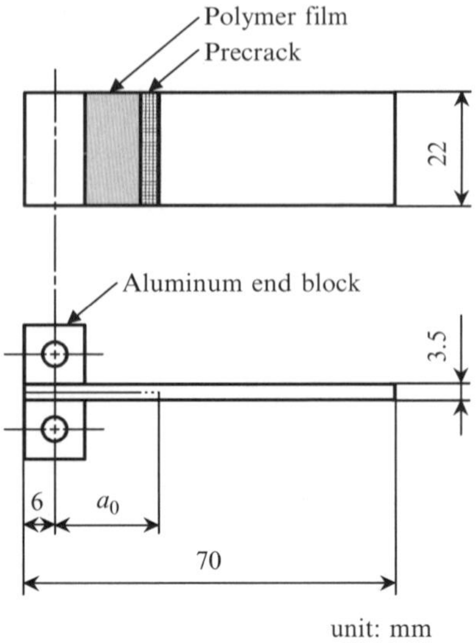

Fig. 5.1 DCB specimen configuration

specimen was precracked. The initial delamination length a_0 was about 20 mm. Aluminum end blocks were adhesively bonded to the specimens to enable load application. The DCB tests were performed under displacement control at room temperature, liquid nitrogen temperature (77 K), and liquid helium temperature (4 K). The delamination opening displacement was measured from the testing machine's displacement. A three-dimensional finite element analysis was also carried out to evaluate the Mode I delamination fracture toughness of the woven GFRP laminates. The delamination fracture toughness increased between room temperature and 77 K, further cooling to 4 K produced a toughness decrease due to the suppression of molecular motion of polymer resins at 4 K [29].

It is well known that composites undergo damage when subjected to mechanical loads, especially at cryogenic temperatures. The effect of damage on the Mode I delamination fracture properties of woven GFRP laminates at cryogenic temperatures was examined numerically [30]. A damage analysis based on a finite element method was conducted to predict the initiation and growth of damage in the DCB specimens. Damage in the woven GFRP laminates was addressed using failure criteria and stiffness degradation technique. The energy release rate was slightly affected by the damage near the delamination front.

5.2.2 Mode II Fracture

The three-point bend end-notched flexure (3ENF) test has been used for the determination of the Mode II delamination fracture toughness of laminated

Fig. 5.2 4ENF test setup

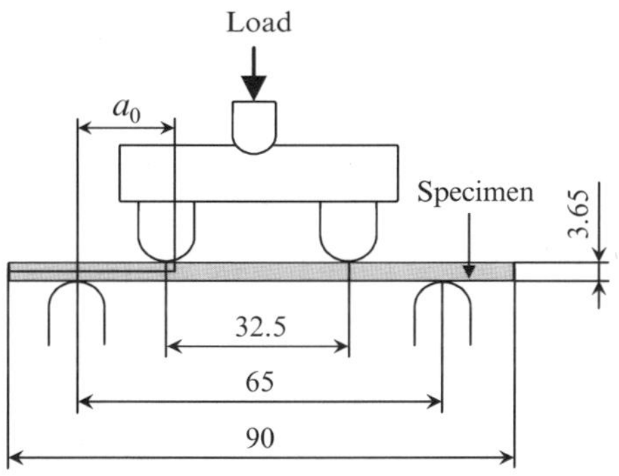

composites [28]. However, the drawback of the 3ENF test is that delamination growth is unstable. The delamination fracture of woven GFRP composite laminates under Mode II loading at cryogenic temperatures was examined by both 3ENF test and three-dimensional finite element method [31]. The Mode II delamination fracture and damage behavior of woven GFRP laminates at cryogenic temperatures was also investigated using a finite element analysis of the 3ENF specimen [32].

As a modified version of the 3ENF test, the four-point bend end-notched flexure (4ENF) test has been introduced [33]. The 4ENF test has certain advantages over the 3ENF test and appears to be gaining in popularity [34, 35]. Unlike the 3ENF test, delamination growth is stable in the 4ENF test. Here, we describe 4ENF test and analysis on woven GFRP laminates, designated G-11 (Arisawa Mfg. Co., Ltd., Japan), under cryogenic conditions [36]. A G-11 type product is fabricated to meet performance criteria established by the National Electrical Manufacturers Association (NEMA).

The 4ENF test setup is shown in Fig. 5.2. The composite specimen contained a polymer film delamination starter at the specimen midplane and was precracked to an initial delamination length a_0 of about 18 mm. The specimen length, width, and thickness were 90 mm, 20 mm, and 3.65 mm, respectively. The 4ENF uses a four-point bend configuration where the supports comprise the outer span and the inner span consists of the loading noses. The outer span length and the inner span length were, respectively, 65 mm and 32.5 mm. The specimens were tested under displacement control at room temperature, 77, and 4 K. The testing machine transducer was used to record the displacement of the center of the inner span. In addition, using a finite element analysis coupled with damage, the initiation and growth of damage in the specimens were predicted, and the Mode II delamination fracture toughness of the woven GFRP laminates was evaluated. Matrix microcracking developed on the upper delamination surface and underneath the load point on the bottom surface of the specimen, and the fiber-dominated failure occurred near the delamination front. The predicted damage pattern was consistent

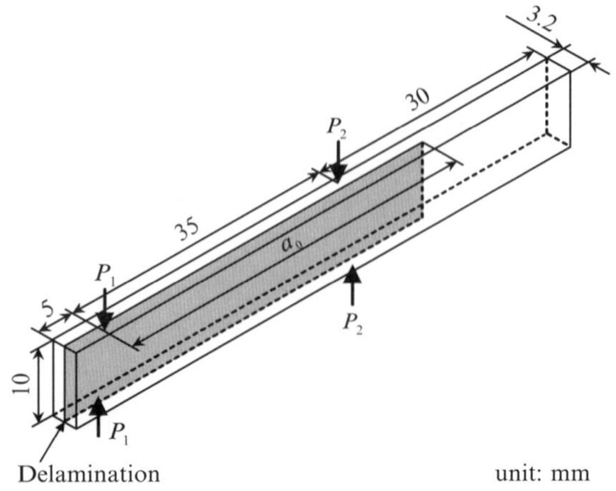

Fig. 5.3 Schematic representation of modified SCB test

with the observed damage zone in the tested specimens. It was also found that the Mode II delamination fracture toughness increased with decreasing temperature from room temperature down to cryogenic temperatures.

5.2.3 Mode III Fracture

The determination of the delamination fracture toughness of laminated composites under Mode III conditions remains a difficult task. Several test techniques have been proposed for characterizing Mode III delamination fracture [37], and the two most mature methods are likely the edge crack torsion (ECT) [38] and modified split cantilever beam (SCB) [39] tests. Of the two methods, the modified SCB test would seem to hold promise because the test specimens are similar to those used for existing Mode I and Mode II tests and are fairly simple to manufacture. Our effort in Mode III delamination fracture of G-11 woven GFRP composite laminates at cryogenic temperatures is presented here, using the modified SCB test technique [40].

Figure 5.3 presents a schematic representation of the modified SCB test geometry. The specimen length was 70 mm; the width was 10 mm; and the thickness was 3.2 mm. A polymer film was inserted at the specimen midplane to provide a starter delamination. A precrack was used to extend the delamination length and the initial delamination length a_0 was about 45 mm. Loads of P_1 and P_2 were applied to each arm of the specimen. The fracture tests on the SCB specimens were carried out under displacement control at room temperature, 77 K, liquid hydrogen temperature (20 K), and 4 K. The displacement was measured by the testing machine's displacement. A three-dimensional finite element analysis was also performed to determine the critical Mode III energy release rate at the onset of delamination

Fig. 5.4 Mode I fatigue fracture surface at 4 K

growth, i.e., Mode III fracture toughness. The delamination fracture toughness increased between room temperature and 20 K. Further cooling to 4 K was characterized by a toughness decrease.

5.3 Interlaminar Delamination Fatigue

5.3.1 Mode I Fatigue

In this section, we present our investigation on the delamination growth behavior in woven GFRP composite laminates, designated SL-EC (Nitto Shinko Corporation, Japan), subjected to Mode I fatigue loading at cryogenic temperatures [41]. The configuration of a DCB specimen, as shown in Fig. 5.1, was used for the experiments. Mode I fatigue delamination tests were performed under load control with constant amplitude at room temperature, 77, and 4 K. The load cycle was sinusoidal and the ratio of the minimum applied load P_{min} to the maximum applied load P_{max} (load ratio) was equal to 0.1. The applied load was measured by the testing machine load cell. Also, the cyclic frequency was 2 Hz. The range of the Mode I energy release rate during cyclic loading was calculated from a three-dimensional finite element analysis, and the delamination growth rate data were expressed in terms of the energy release rate range.

The fatigue results showed that higher resistance to fatigue delamination growth was obtained for cryogenic conditions and the delamination growth rate at 77 K was lower than that at 4 K. In addition, the dominant fatigue delamination growth mechanisms were different at room temperature and cryogenic temperatures. At room temperature, fiber/matrix debonding was the main mechanisms. On the other hand, at cryogenic temperatures, both fiber/matrix debonding and brittle fracture of polymer matrix were the dominant fatigue delamination growth mechanisms (see Fig. 5.4).

Fig. 5.5 Mode II fatigue fracture surface at 4 K

5.3.2 Mode II Fatigue

This section presents the characterization of the cryogenic fatigue delamination growth behavior in G-11 woven GFRP composite laminates under Mode II loading [42]. Mode II fatigue delamination tests were conducted at room temperature, 77, and 4 K using the 4ENF test setup (see Fig. 5.2). The composite specimen was loaded cyclically under displacement control at a frequency of 2 Hz and a displacement ratio of 0.1. The displacement ratio is defined as $\delta_{min}/\delta_{max}$, where δ_{max} and δ_{min} are the maximum and minimum applied displacements, respectively. The applied displacement was measured by the testing machine's displacement. The Mode II energy release rate range was determined by a three-dimensional finite element method.

The delamination growth rate appeared to decrease as the temperature decreased from room temperature to 4 K. The fatigue fracture surfaces were characterized by the presence of hackles and fiber/matrix debonding. Hackles are formed by microcracking in the resin material just ahead of the delamination front [43–45]. The dependence of the delamination growth rate on the temperature can be explained by the decrease in the hackle spacing (defined in Fig. 5.5) with decreasing temperature.

5.3.3 Mode III Fatigue

This section describes a study on the Mode III fatigue delamination growth in G-11 woven GFRP composite laminates at cryogenic temperatures [46]. The modified SCB test geometry, as shown in Fig. 5.3, was employed and the composite specimens were tested at room temperature, 77, and 4 K. All fatigue tests were performed in displacement control at a test frequency of 5 Hz and a constant displacement ratio, i.e., ratio of the minimum applied displacement δ_{min} to the maximum applied displacement δ_{max}, of 0.1. The applied displacement was measured by the testing machine's displacement. A three-dimensional finite element analysis was also conducted to determine the Mode III energy release rate range.

The delamination growth rates at 77 and 4 K were lower than that at room temperature. It was also noticed that the delamination growth rate at 4 K was higher

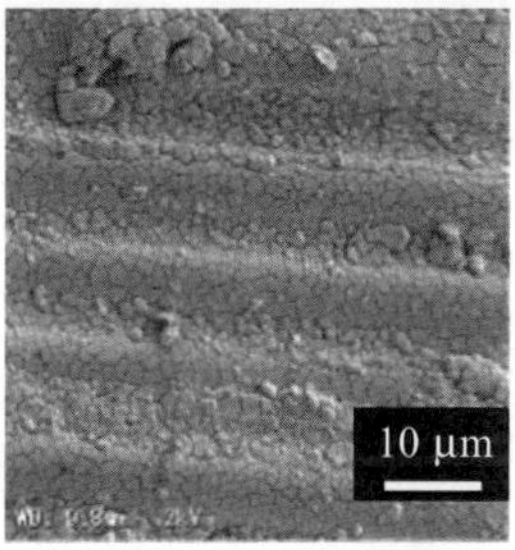

Fig. 5.6 Mode III fatigue fracture surface at 4 K

than that at 77 K. In addition, matrix resin debris was the significant feature of the fatigue fracture surface, as shown in Fig. 5.6, and the size of the resin debris decreased as the temperature decreased.

5.4 Conclusions

Contributions of our research group are summarized in this chapter, focusing on the interlaminar delamination fracture and fatigue of woven GFRP composite laminates under Mode I, Mode II, and Mode III conditions at cryogenic temperatures. It was found that delamination behaviors in woven GFRP laminates under static and cyclic loadings were strongly dependent on temperature. The observed trend was an increase in the resistance to delamination fracture and fatigue as temperature decreased from room temperature to cryogenic temperatures. Furthermore, fractographic observations showed the profound difference in the cryogenic delamination mechanisms between the three modes. Our studies provide some basic insights into delamination characteristics in woven GFRP laminates at cryogenic temperatures. Such knowledge for composite materials is essential to establish design allowable and damage tolerance guidelines for cryogenic composite structures.

References

1. Chou TW, Ishikawa T (1989) Analysis and modeling of two-dimensional fabric composites. In: Chou TW, Ko FK (eds) Textile structural composites. Elsevier, New York
2. Mitchell N, Bauer P, Bessette D, Devred A, Gallix R, Jong C, Knaster J, Libeyre P, Lim B, Sahu A, Simon F (2009) Status of the ITER magnets. Fusion Eng Des 84:113–121
3. Bittner-Rohrhofer K, Humer K, Weber HW, Hamada K, Sugimoto M, Okuno K (2002) Mechanical strength of an ITER coil insulation system under static and dynamic load after reactor irradiation. J Nucl Mater 307–311:1310–1313
4. Freeman DC Jr, Talay TA (1997) Reusable launch vehicle technology program. Acta Astronaut 41:777–790

5. Choi S, Sankar BV (2008) Gas permeability of various graphite/epoxy composite laminates for cryogenic storage systems. Compos Part B 39:782–791
6. Hartwig G (1988) Overview of advanced fibre composites. Cryogenics 28:216–219
7. Kasen MB (1981) Cryogenic properties of filamentary-reinforced composites: an update. Cryogenics 21:323–340
8. Reed RP, Golda M (1994) Cryogenic properties of unidirectional composites. Cryogenics 34:909–928
9. Shindo Y, Wang R, Horiguchi K, Ueda S (1999) Theoretical and experimental evaluation of double-notch shear strength of G-10CR glass-cloth/epoxy laminates at cryogenic temperatures. ASME J Eng Mater Technol 121:367–373
10. Shindo Y, Wang R, Horiguchi K (2001) Analytical and experimental studies of short-beam interlaminar shear strength of G-10CR glass-cloth/epoxy laminates at cryogenic temperatures. ASME J Eng Mater Technol 123:112–118
11. Kumagai S, Shindo Y, Horiguchi K, Narita F (2004) Experimental and finite-element analysis of woven glass-cloth/epoxy laminate tensile specimen at room and low temperatures. Mech Adv Mater Struct 11:51–66
12. Takeda T, Takano S, Shindo Y, Narita F (2005) Deformation and progressive failure behavior of woven-fabric-reinforced glass/epoxy composite laminates under tensile loading at cryogenic temperatures. Compos Sci Technol 65:1691–1702
13. Shindo Y, Tokairin H, Sanada K, Horiguchi K, Kudo H (1999) Compression behavior of glass-cloth/epoxy laminates at cryogenic temperatures. Cryogenics 39:821–827
14. Miura M, Shindo Y, Takeda T, Narita F (2010) Effect of damage on the interlaminar shear properties of hybrid composite laminates at cryogenic temperatures. Compos Struct 93:124–131
15. Shindo Y, Takeda T, Narita F, Miura M, Watanabe S, Koizumi N, Idesaki A, Okuno K (2010) Interlaminar shear properties of composite insulation systems for fusion magnets at cryogenic temperatures. Cryogenics 50:36–42
16. Shindo Y, Sumikawa M, Narita F, Sanada K (2005) Acoustic emission and fracture behavior of GFRP woven laminates at cryogenic temperatures. Cryogenics 45:439–449
17. Kumagai S, Shindo Y, Inamoto A (2005) Tension–tension fatigue behavior of GFRP woven laminates at low temperatures. Cryogenics 45:123–128
18. Shindo Y, Inamoto A, Narita F (2005) Characterization of mode I fatigue crack growth in GFRP woven laminates at low temperatures. Acta Mater 53:1389–1396
19. Brunner AJ, Blackman BRK, Davies P (2008) A status report on delamination resistance testing of polymer-matrix composites. Eng Fract Mech 75:2779–2794
20. O'Brien TK (1998) Interlaminar fracture toughness: the long and winding road to standardization. Compos Part B 29:57–62
21. Tay TE (2003) Characterization and analysis of delamination fracture in composites: an overview of developments from 1990 to 2001. ASME Appl Mech Rev 56:1–31
22. Shindo Y, Shinohe D, Kumagai S, Horiguchi K (2005) Analysis and testing of mixed-mode interlaminar fracture behavior of glass-cloth/epoxy laminates at cryogenic temperatures. ASME J Eng Mater Technol 127:468–475
23. Shindo Y, Takahashi S, Takeda T, Narita F, Watanabe S (2008) Mixed-mode interlaminar fracture and damage characterization in woven fabric-reinforced glass/epoxy composite laminates at cryogenic temperatures using the finite element and improved test methods. Eng Fract Mech 75:5101–5112
24. Shindo Y, Miura M, Takeda T, Saito N, Narita F (2011) Cryogenic delamination growth in woven glass/epoxy composite laminates under mixed-mode I/II fatigue loading. Compos Sci Technol 71:647–652
25. Shindo Y, Horiguchi K, Wang R, Kudo H (2001) Double cantilever beam measurement and finite element analysis of cryogenic mode I interlaminar fracture toughness of glass-cloth/epoxy laminates. ASME J Eng Mater Technol 123:191–197
26. ASTM D 5528-01 (2001) Standard test method for mode I interlaminar fracture toughness of unidirectional fiber-reinforced polymer matrix composites. In: Annual book of ASTM standards. ASTM International, West Conshohocken, PA

27. ISO 15024:2001(E) (2001) Fibre-reinforced plastic composites—determination of mode I interlaminar fracture toughness, G_{IC}, for unidirectionally reinforced materials. The International Organization for Standardization
28. JIS K 7086 (1993) Testing methods for interlaminar fracture toughness of carbon fibre reinforced plastics. Japanese Standards Association
29. Nishijima S, Okada T, Honda Y (1994) Evaluation of epoxy resin by positron annihilation for cryogenic use. Adv Cryog Eng 40:1137–1144
30. Sumikawa M, Shindo Y, Takeda T, Narita F, Takano S, Sanada K (2005) Analysis of mode I interlaminar fracture and damage behavior of GFRP woven laminates at cryogenic temperatures. J Compos Mater 39:2053–2066
31. Horiguchi K, Shindo Y, Kudo H, Kumagai S (2002) End-notched flexure testing and analysis of mode II interlaminar fracture behavior of glass-cloth/epoxy laminates at cryogenic temperatures. J Compos Technol Res 24:239–245
32. Shindo Y, Narita F, Sato T (2006) Analysis of mode II interlaminar fracture and damage behavior in end notched flexure testing of GFRP woven laminates at cryogenic temperatures. Acta Mech 187:231–240
33. Martin RH, Davidson BD (1999) Mode II fracture toughness evaluation using four point bend, end notched flexure test. Plast Rubber Compos 28:401–406
34. Davies P, Sims GD, Blackman BRK, Brunner AJ, Kageyama K, Hojo M, Tanaka K, Murri G, Rousseau C, Gieseke B, Martin RH (1999) Comparison of test configurations for determination of mode II interlaminar fracture toughness results from international collaborative test programme. Plast Rubber Compos 28:432–437
35. Sun X, Davidson BD (2006) Numerical evaluation of the effects of friction and geometric nonlinearities on the energy release rate in three- and four-point bend end-notched flexure tests. Eng Fract Mech 73:1343–1361
36. Shindo Y, Sato T, Narita F, Sanada K (2008) Mode II interlaminar fracture and damage evaluation of GFRP woven laminates at cryogenic temperatures using the 4ENF specimen. J Compos Mater 42:1089–1101
37. Farshad M, Flüeler P (1998) Investigation of mode III fracture toughness using an anti-clastic plate bending method. Eng Fract Mech 60:597–603
38. Lee SM (1993) An edge crack torsion method for mode III delamination fracture testing. J Compos Technol Res 15:193–201
39. Sharif F, Kortschot MT, Martin RH (1995) Mode III delamination using a split cantilever beam. In: Martin RH (ed) Composite materials: fatigue and fracture—fifth volume, ASTM STP 1230. American Society for Testing and Materials, Philadelphia, pp 85–99
40. Rizov V, Shindo Y, Horiguchi K, Narita F (2006) Mode III interlaminar fracture behavior of glass fiber reinforced polymer woven laminates at 293 to 4 K. Appl Compos Mater 13:287–304
41. Shindo Y, Inamoto A, Narita F, Horiguchi K (2006) Mode I fatigue delamination growth in GFRP woven laminates at low temperatures. Eng Fract Mech 73:2080–2090
42. Shindo Y, Takeda T, Narita F, Saito N, Watanabe S, Sanada K (2009) Delamination growth mechanisms in woven glass fiber reinforced polymer composites under mode II fatigue loading at cryogenic temperatures. Compos Sci Technol 69:1904–1911
43. Asp LE, Sjögren A, Greenhalgh ES (2001) Delamination growth and thresholds in a carbon/epoxy composite under fatigue loading. J Compos Technol Res 23:55–68
44. Gilchrist MD, Svensson N (1995) A fractographic analysis of delamination within multidirectional carbon/epoxy laminates. Compos Sci Technol 55:195–207
45. Lee SM (1997) Mode II delamination failure mechanisms of polymer matrix composites. J Mater Sci 32:1287–1295
46. Miura M, Shindo Y, Narita F, Watanabe S, Suzuki M (2009) Mode III fatigue delamination growth of glass fiber reinforced polymer woven laminates at cryogenic temperatures. Cryogenics 49:407–412

Chapter 6
The Behavior of Polymer-Based Dielectrics Under Cryogenic Conditions

H. Rodrigo

Contents

Abstract Dielectrics is ubiquitous in all electrical systems. In this chapter we introduce many aspects in a systematic manner in order for the reader to be able to follow it in a sequence. Firstly, the various media and the salient features of each are discussed. The media range from vacuum to highly compressed gases, liquids, and solids. The basic mechanism of dielectric behavior is discussed in each different case when subject to

H. Rodrigo (✉)
Center for Advanced Power Systems, Florida State University, 2000 Levy Avenue, Tallahassee, FL 32310, USA
e-mail: hrodrigo@ieee.org

S. Kalia and S.-Y. Fu (eds.), *Polymers at Cryogenic Temperatures*,
DOI 10.1007/978-3-642-35335-2_6, © Springer-Verlag Berlin Heidelberg 2013

electric field. The imposed electric field can be of different forms such as steady-state AC and DC voltages. It can also be due to transients in the system brought about by switching operations or they could be due to naturally occurring phenomena namely lightning. The configuration of electrode geometries and the polarity of the electrodes bring about phenomena within the media that need to be understood in order to design electrical systems for a variety of applications. We also focus our attention particularly when the media are at cryogenic temperatures. The electronic and ionic reaction mechanics change drastically at low temperatures. In the early part of this chapter the discussion is centered around the basics such as partial discharge, electrical breakdown, dielectric losses, and permittivity. This is followed by applications of power cable design and operation at or near the boiling point of liquid helium. The reader is directed to some real life experiences of such cable systems. Once again the emphasis here has been on the dielectric aspects and the role that materials play in enabling such technology. With the advent of high temperature superconductors (HTS) the outlook was more promising as superconductivity could be achieved at around the boiling point of liquid nitrogen (77 K).

Most dielectric systems were designed with the cryogen playing a dual role of being the cooling medium and also being an integral part of the dielectric. As HTS became more widespread cold compressed gaseous helium is now considered viable for some special applications. Advances in the development of polymeric materials for cryogenic applications have largely kept up with HTS technology. However, there are problems that have to be overcome, in particular mechanical strength at low temperatures. Another problem inherent to devices such as cables and coils is winding gaps and inclusions. This presents opportunities for partial discharges to start and if not avoided leads to aging of the device and finally failure. Important dielectric properties such as permittivity and loss tangent have been discussed at some length. The measurement of these parameters not only gives their numerical values but also provides insight into the behavior of the material properties in a general sense. Composites is another area that is being actively pursued and this topic is discussed in particular the differences between micro- and nano-fillers in polymer resins. The large increase in surface area with the reduction in particle size to the nano-scale has opened up great opportunities for advancement. The interfacial region is one that holds the key to future advancement in this technology. Cryogenic nanocomposites are a fascinating technology that has opened up new horizons, and many laboratories are making great progress in understanding the behavior of these materials both theoretically and with experiments.

Keywords Dielectric • Cryogenic • Polymer • Nanoparticles • Electrical breakdown • Partial discharge

6.1 Introduction

Dielectrics or electrical insulation spans the whole of electrical engineering. In this chapter the author refers to the subject material as dielectric and electrical insulation interchangeably to mean the same thing. The area of dielectrics takes many forms, namely vacuum, gases, liquids, and solids. Where any one of the above or a combination is to be used depends on the application. In this chapter the main focus will be on the application of dielectrics in electrical systems. The parameters of importance for the study of dielectrics are (a) the availability of free electrons, (b) the electric field configuration and its magnitude and the medium in which the activity takes place. In the vast majority of cases that one encounters in electrical engineering, the field configuration is nonuniform. The degree of nonuniformity will depend to a large extent on the application. So gaining an understanding of the electric field configuration is of the utmost importance. Understanding the underlying basic physics of how the electric field could occur is critical. The mechanics of the various processes that occur within a given medium will be dependent on the medium itself, temperature, pressure electrode material, current, and voltage. The discussion here will be confined to solid polymeric materials operating at low temperature. However, one has to have a thorough understanding of the various means by which these low temperatures can be achieved, and what part they play in the overall dielectric system. Some areas where cryogenic materials find usage are listed below:

(a) Cables
(b) Capacitors
(c) Transformers
(d) Motors
(e) Magnets
(f) Coils
(g) Fault current limiters

The most commonly used cryogenic fluid is liquid nitrogen. Liquid helium has high dielectric strength; however, it is not widely used in the electric power industry due to its high cost in comparison to liquid nitrogen, and the difficulties of handling. More importantly with the advent of high temperature superconductivity (HTS) the need to operate at or near the boiling point of liquid helium, 4.2 K, has largely diminished.

6.2 Cryogenic Applications

The cryogenic media most commonly used in many electrical engineering applications are liquid and compressed gas. Often the cryogen in addition to being the cooling medium forms an integral part of the dielectric system, examples being:

(a) Liquid nitrogen and liquid helium
(b) Gaseous helium cold and compressed

Generally the dielectric medium is rendered useless after a fault. However, media such as vacuum, gases, and some liquids are self-healing. Self-healing

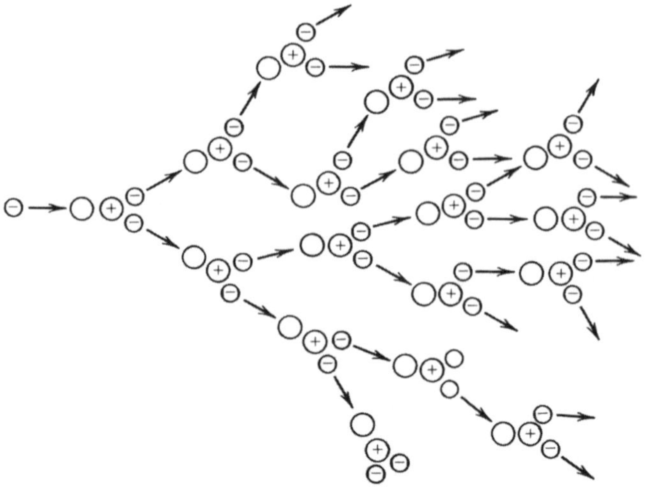

Fig. 6.1 Illustration of the formation of an electron avalanche which leads to ionization of the medium and finally dielectric breakdown [23]

media can fail in the future due to contamination of the medium and electrode damage.

On the other hand solid dielectrics fall into the first category, where the failure is irreversible.

Let us turn our attention to the basic mechanics of dielectric failure in general. In order for dielectric failure to occur in any medium two fundamental parameters have to be present:

(a) Availability of free electrons
(b) An electric field sufficiently large to accelerate these free electrons

Figure 6.1 shows the sequence of events that leads to dielectric breakdown.

In the gas phase, the nascent stage of discharge is due to corona which leads to streamers and finally leaders.

In the liquid phase, the initiation is similar but it is dominated by streamers.

In the solid phase, the onset is triggered by partial discharge due to voids in the solid. The electric field is higher in the void (gas) than in the solid as often the relative permittivity of the gas is lower than that of the solid. Partial discharge, which is essentially corona, eventually leads to failure of the insulation—damage is permanent.

For a solid material to be useful as a dielectric the operating voltage must be less than the voltage at which partial discharge begins.

In vacuum depending on how high the vacuum is, the medium for any activity that can trigger an electrical discharge is absent. Hence although there will be a sufficiently large electric field, the mean free path for collision with other atoms and molecules can be very long. As the voltage of the high voltage electrode is raised,

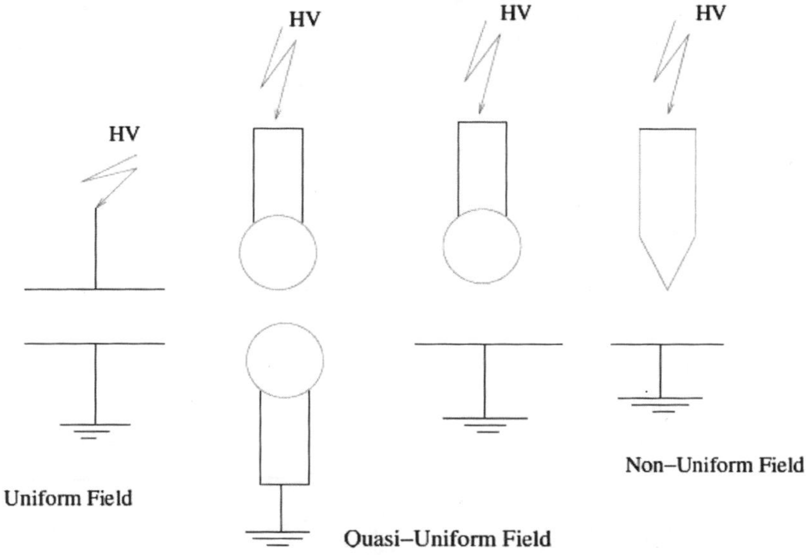

Fig. 6.2 Field configurations encountered in high voltage engineering

the work function of the electrode material will be exceeded which gives rise to emissions from the electrode; therefore a medium is created. The processes that ensue are now similar to those with other media which gives rise to collisions with free electrons leading to electrical breakdown of the vacuum gap.

Since the scope of this chapter does not extend to a full-scale discussion of vacuum breakdown, the reader is directed to a few references which will be useful [24, 35].

6.3 Field Configurations

It is not usual to encounter purely uniform field conditions in the practice of high voltage engineering. Figure 6.2 gives some commonly used electrode geometries.

6.4 Polarity Effects

6.4.1 Positive Polarity

In a nonuniform field electrode geometry when the high voltage electrode has positive polarity, as shown in Fig. 6.3, electrons are attracted towards the point electrode. Since the mobility of electrons is much greater than that of the positive and negative ions, the ion cloud behind the electrons forms a space charge field [1, 14, 38].

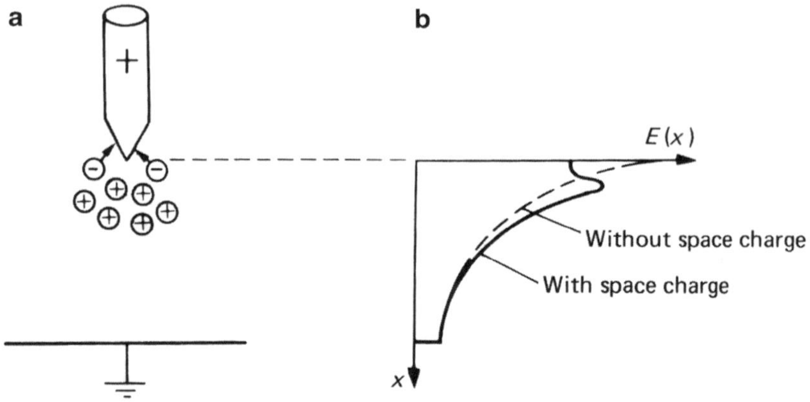

Fig. 6.3 (**a**) Space charge build-up in positive point–plane gap. (**b**) Field distortion by space charge [34]

This space charge field combines with the applied field leading to enhancement further out into the gap. This facilitates increased ionization which eventually leads to breakdown of the gap.

6.4.2 Negative Polarity

For negative polarity as shown in Fig. 6.4, the electrons are repelled away from the high voltage electrode, which forms a space charge cloud around it. The only way for breakdown of the gap to occur is by increasing the electric field which is achieved by increasing the voltage [8, 43–46, 49, 53, 60].

For nonuniform field geometries negative polarity breakdown is higher than positive polarity.

6.5 Dielectric Losses

When an insulator is subjected to an electric field, as is the case with a capacitor, losses occur due to the passage of current. The current consists of:

(a) Conduction current
(b) Displacement current

In a good insulator under DC electric fields the conduction current is very small and hence the losses are also small.

However, under AC electric field conditions there is an appreciable component of the field in phase with the applied electric field. This gives rise to dielectric losses

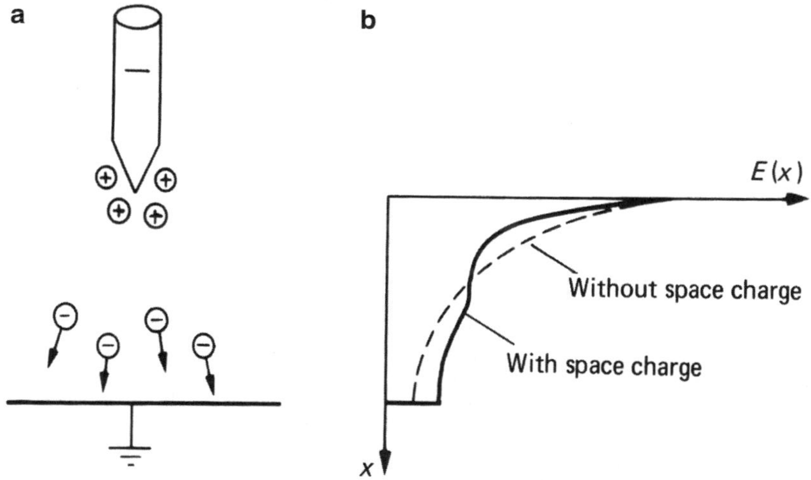

Fig. 6.4 (**a**) Space charge build-up in negative point–plane gap. (**b**) Field distortion by space charge [34]

in the form of heat. The losses are dependent on the frequency of the applied field. Thus the permittivity of the medium (dielectric) which is a measure of the polarization of the medium becomes a complex quantity.

The following analysis is well known from Maxwell's equations and is given here for the sake of completion.

$$\nabla \times H = J + \frac{\partial D}{\partial t} = \sigma E + j\omega\varepsilon E$$

where, ε is the permittivity.

Hence

$$\nabla \times H = \sigma E + j\omega(\varepsilon' - j\varepsilon'')E$$

Here

$$\varepsilon = \varepsilon' - j\varepsilon''$$

ε' = real part or lossless part of ε
ε'' = imaginary or the lossy part of ε.

$$\begin{aligned}\nabla \times H &= [(\sigma + \omega\varepsilon'') + j\omega\varepsilon']E \\ &= (\sigma' + j\omega\varepsilon')E\end{aligned}$$

where

$$\sigma' = \sigma + \omega\varepsilon'' = \text{equivalent conductance}$$

$$\nabla \times H = J_{\text{total}}E$$

The total current density J_{total} is the vector sum of conduction current density $(\sigma' + j\omega\varepsilon')E$ which is in quadrature with the displacement current density $j\omega\varepsilon' E$.

Thus the dielectric loss is given by

$$\tan\,\delta = \frac{\sigma'}{\omega\varepsilon'}$$

where tan δ is called the loss tangent of the dielectric.

The conduction current gives the energy loss in the dielectric and the displacement current represents the energy storage capacity.

6.5.1 Method Used for Dielectric Loss Measurement

The Schering bridge circuit [29, 34] is one that is used widely in high voltage laboratories. In the modern version of this circuit the detector and the low voltage arms are incorporated into a single instrument so that the high voltage transformer, the standard discharge free capacitor, and the test object are placed in an enclosure. Instrument cables from the enclosure are terminated at the detector. The instrument is self-adjusting and reads the dielectric loss (tan δ) and the capacitance of the test object simultaneously. It can be used at power frequencies typically 50–60 Hz. The bridge is sensitive to the fundamental frequency of operation.

6.6 Partial Discharge in Solids

6.6.1 Electrical Breakdown in Solids

The electrical breakdown in solids is almost always preceded by partial discharge activity. Partial discharge is a complex process involving many physical aspects. The scope of this chapter does not extend to a full-scale discussion of partial discharge. Here the discussion is confined to the most fundamental details.

In the fabrication of any solid insulation in high voltage applications it is inevitable that manufacturing imperfections would find their way into the final product. The most common of these are voids in the solid material [29, 30]. Many devices such as transformers, cables, and capacitors use solid insulation in addition to liquids and gases. In general the solid is the strongest dielectric and as the voltage

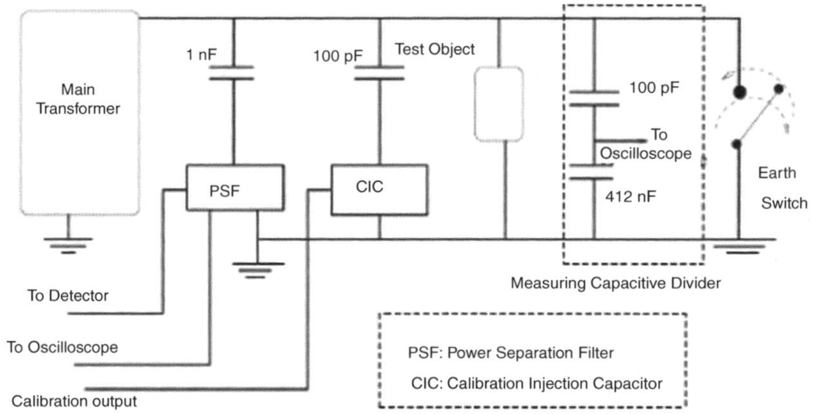

Fig. 6.5 Partial discharge measurement system

is increased partial discharge begins in either a void within the solid or at a triple point within the device, which eventually leads to dielectric breakdown.

6.6.2 *Measurement of Partial Discharge*

The measurement of partial discharge is conducted within a well-constructed Faraday enclosure. Figure 6.5 shows a schematic of the partial discharge measurement system. It is important that this enclosure has an entirely separate grounding system to all other activities in the laboratory situation. In addition all power inputs into the equipment within the Faraday enclosure must be filtered for extraneous signals and harmonics of higher order. The reasons for such elaborate precautions are that the charge levels that have to be detected are in the pico-Coulomb (pC) range and of course one has to ensure that the signals detected by the equipment within the enclosure can only come from the activities that take place in the object under test. The voltage levels in many of these activities are typically in the kV or higher ranges.

It is generally accepted that the partial discharge inception voltage of test objects subject to high voltage is reached when the level is $\approx$10 pC. It is critically important that the equipment to be used in a high voltage application must be operated at a voltage below this threshold of the inception voltage for partial discharge.

6.7 Cryogenic Dielectrics

The most commonly used cryogenic liquid that is used today is liquid nitrogen (LN_2); this is specially true with the advent of HTS. However, there are limitations of using liquid nitrogen at atmospheric pressure. The advantage of LN_2 is that it forms part of the dielectric system and by itself is a strong dielectric. In addition it is abundant and relatively inexpensive. Another cryogenic fluid that is used widely and not without limitations is liquid helium (LHe). Beginning around 1970 and for a decade or so after there were serious efforts at using low temperature materials cooled by LHe for transmission of electric power. Liquid helium is the coldest of all the cryogenic fluids known, boiling at 4.2 K under atmospheric conditions. It is also quite a strong dielectric both in the liquid and high density gaseous state [2, 9, 12, 19, 20, 40, 54]. The dielectric properties of liquid helium are as good as transformer oil [18] as the density of gaseous helium goes up the dielectric properties improve significantly [18]. However, it is relatively expensive. A third cryogen that has been considered is liquid hydrogen. However, it is not used in the electric power industry as it requires very special handling. Liquid neon which has a boiling point of 27 K has been considered from time to time. It is not a serious candidate as it is rare and very expensive; additionally it is not a strong dielectric. In the gaseous state its dielectric strength weakens considerably with even trace quantities of other gases such as Argon [38]. Liquid helium is used widely in low temperature superconducting systems such as magnetic resonance imaging (MRI).

The use of gaseous helium at high pressure and temperatures where HTS becomes viable is being pursued actively in some specialized applications of power transmission, such as on board ships [21, 48].

6.8 Polymeric Dielectrics

Some of the early attempts of harnessing the advantages of superconductivity for transmission of electric power were carried out in the early 1970s and the following decade or so. There was a concerted effort at Brook Haven National Laboratory in the United States towards implementing AC high voltage transmission lines [11]. In this work the temperature of operation of the cables was near the boiling point of LHe (4.2 K). The insulating materials used were plastics whose dielectric strengths were of the order of 10 kV/mm. They had tested several short length model cables for dielectric integrity. Figure 6.6 shows an isometric view of the cable construction.

The dielectric loss factor of the material that was used for this cable was 2×10^{-5}. Minimizing the dielectric loss is an important aspect of cable design as this loss manifests itself in the form of heat which has to be removed. Thus the refrigeration costs increase, which if not tightly controlled results in making the operation unviable.

The work at Brook Haven resulted in a cable being fabricated. The cable was designed with a view of operating under AC voltage at 138 kV. One of the most important factors in the implementation of this cable was to ensure that it was free of partial discharge at the operating voltage. Helium gas is by far the weakest link in this system; the gas gets trapped in the winding gaps of the insulation tape. The polymeric dielectric material typically had a relative permittivity (ε_r) in the range of 2.2, whereas gaseous helium has an ε_r of about 1.0 so due to this the electric field is enhanced in the gas which results in the gas partially breaking down electrically.

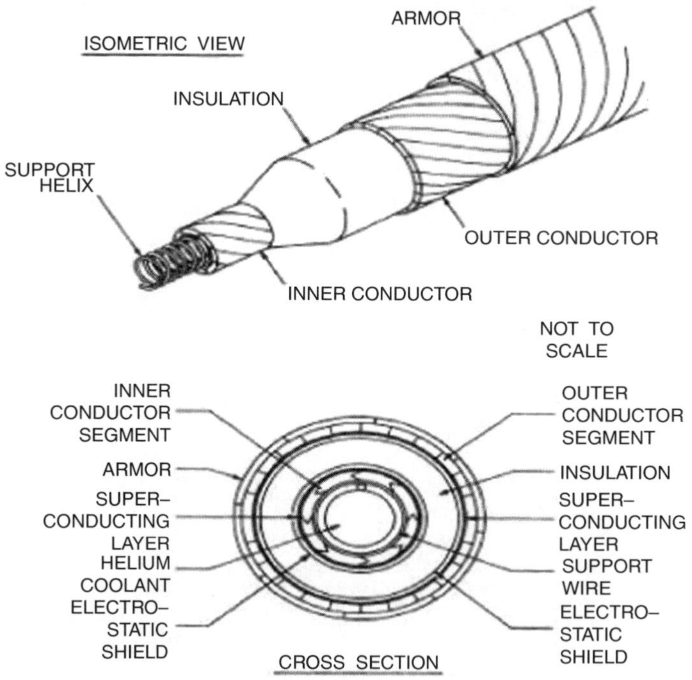

Fig. 6.6 Isometric view of flexible superconducting coaxial cable [10]

In the early 1980s there were attempts at building DC cables that were to operate at 100 kV DC at Los Alamos National Laboratory [6, 7]. The cable was designed for operating conditions of 1.38 MPa in supercritical helium at 12 K. DC was chosen as the preferred voltage mode as at superconducting temperatures the materials used have zero resistance and there are no AC losses due to frequency effects. Also under steady-state conditions the cable is resistive and under transient conditions such as during faults it is capacitive. These fault conditions can be due to lightning strikes, and also switching the cable on and off can create conditions akin to faults. These two conditions are the ones that are generally accepted for a superconducting cable. However, due to the inevitable presence of harmonics on the lines which is a result of converters at the ends of the cable, the voltage is not a pure DC. The harmonics that travel on the cable form standing wave patterns with successive maxima and minima [7]. These harmonic currents give rise to two types of losses: hysteresis losses in the superconductor and resistive losses in the former which is made of Copper. In the studies at Los Alamos they had studied over a dozen materials ranging from Cellulose paper impregnated with oil to plastics such as Kapton and Valeron.

Rather an important concern of these studies had been the formation of ice crystals. The strength of ice had been studied prior to these studies [32, 39]. The work at Los Alamos [6] had studied the breakdown characteristics of many materials under uniform field conditions in a cryostat under the same conditions as given above before the final selection of suitable materials was made. It is noteworthy that plastics such as Kapton and Mylar both exhibited breakdown characteristics which were similar. This fact is important because under the right conditions of operation failure of the dielectric could well be due to surface effects than the dielectric being punctured. In recent work Graber et al. [21] have also shown that dielectric breakdown of cryogenic cables can occur due to surface burns on the dielectric material as against clear punctures. Also interesting is the fact that at 1.38 MPa the variation of electric field as a function of the reciprocal of temperature is linear [6].

6.8.1 Mechanical Properties

For superconducting cables designed to operate between 4.4 and 8 K under pressures ranging from 0.4 to 1.5 MPa [11, 62] the dielectric materials were subject to arduous mechanical stresses. These stresses usually result in the development of cracks which gave rise to partial discharges. This only adds and compounds the problems of partial discharge taking place in winding gaps which occur due to tape insulation being wound on the conductors and the formers. The partial discharge inception voltage is dependent on the polymeric material used [62]. This at first seems counter intuitive; however it stands to reason when one considers the fact that different materials have different coefficients of contraction. There is also an effect which is dependent on the helium gas pressure in these cases together with the material of the former on which the insulation is wound [27]. The aging of insulation is attributed mainly to partial discharge development which occurs due to gaps and bubbles being formed, this latter in particular where a liquid cryogen is involved [4, 61, 63].

The coefficient of contraction for plastics used in these applications is greater than for the metal of the former [62]. Thus the mechanical stresses give rise to cracks once again. The glass transition temperatures for these plastics occur at much higher temperatures than those at which the cables have to operate; hence the mechanical properties of the materials are not robust enough leading to crack propagation.

In addition to the above there are other factors contributing to degradation in subcritical helium environment. Helium gas trapped in the winding gaps experiences the same electric field as the insulating material lapped on the cable. The gas being dielectrically weaker gives rise to partial discharge which over time erodes the insulation which can eventually lead to electrical breakdown. Most of the damage is brought about by the electrons and not the helium ions. Electron impact on the insulation is the most likely source [17]. In an AC cable where the winding gaps can be thick in comparison to the insulation tape thickness so the chances of high energy electrons being trapped is very real. If a sufficient number of these electrons agglomerate in one location, then their density is increased. If this were to be the case in the positive cycle of the AC wave, voltage reversal causes the intrinsic strength of the polymer to be exceeded over a small thickness leading to dielectric failure [62]. It has been shown that due to thermal effects on polymers the covalent bonds get excited in the vibration mode and in the Van der Waals potential [50]. Similar discussion of the thermal stresses within fiber–epoxy composites at low temperatures has been given by Harrwig and Knaak [26] The asymmetries give rise to negative and positive coefficients of expansion again leading to weakening of the material mechanically. Other parameters such as surface finish of the formers have bearing on solid dielectrics impregnated with both gas and liquid helium.

A material widely used in the electric power industry for insulation of power cables is extruded polyethylene (PE). There have been attempts at using PE at cryogenic temperatures [51]. It has been shown that the dielectric strength of PE improves by a factor of 20 % from the value at room temperature [28]. Similar results have been reported for other polymer materials [47]. While it is possible, and indeed both the electrical and mechanical strengths are significantly enhanced at cryogenic temperatures, extremely careful handling is very important. The model cables that were used for tests have shown that the cooling rate for short sample lengths of 20 cm had to be of the order of 0.3 K/min [51]. The contraction of the PE was similar for comparable short lengths of model cables. Once again it has to be borne in mind that the metal of the former does not contact as much as the dielectric which is wound on it. This situation almost always leads to cracking of the dielectric. The situation is of course exacerbated when the length of the cable is increased.

In spite of these drawbacks there are some real gains to be achieved by using PE at cryogenic temperatures such as the voltage at which dendrite formation begins. The voltage at which this phenomenon occurs is increased by a factor of 6 [33]. In addition it has been reported that the loss tangent of PE at cryogenic temperatures gets reduced by half when compared to room temperature [51].

Some of the most important factors to consider in longer lengths of cable systems are the bending radii and notches that are invariably present in the semiconducting layers together with drill holes. All these factors have real potential to cause the insulation to develop cracks.

6.9 Composite Dielectrics

6.9.1 Introduction

A composite material is defined as one that has two or more components. For dielectric materials this generally is the addition of some nonconducting ceramic or clay compound to a host resin. The additive material is in the form of spheres, tubes, rods, or planes. The type of application largely determines the shape and composition of the additive. In the present discussion we will confine ourselves to spherical particles whose dimensions are in the μm and nm scale. The term nanodielectrics was coined by Lewis [36] in a seminal paper where he delved into the beneficial aspects of adding filler materials whose dimensions were in the nanometric scale to a host polymer. As can be gathered this area of nanodielectrics is a young and emerging technology. The usage and study of nanodielectrics under cryogenic conditions is even more recent. However, there seems great promise for this area of expertise to grow, in particular with advances in HTS. The electrical power industry seems to indicate an interest in this area for the promise it holds for future transmission of power [55]. It has been shown that even if the host material such as Kraft paper is impregnated with polypropylene, although in this case the result is not a composite as no particles are embedded in the original paper, the resulting material performs very well at 77 K for DC power transmission [3]. In addition superconductors find usage in electrical motors, generators, cables, transformers, and magnets. In all these devices a common aspect is insulation. So it stands to reason to concentrate on making materials which can be used in these many areas of activity.

6.9.2 The role of Nanomaterial in the Composite

The enhancement of a variety of properties has been achieved by the addition of foreign materials into a host. For the records the earliest such examples can be found in artifacts dating back to several centuries [22]. In the present discussion we confine ourselves to the beneficial aspects of the resulting materials with respect to electrical power. The enhanced properties that are sought after in a new dielectric material for cryogenic applications are enhanced dielectric breakdown strength [58, 59], pliability, mechanical strength [31], and decrease of relative permittivity [42]. It has been shown that as the particle size is reduced there is substantial improvements to be gained in breakdown electric field strength and the onset of electroluminescence [41]. Their studies have shown that there is an optimum limit of 10 wt.% of both nano- and micro-size with regard to breakdown field enhancement of the composites. However, the nanocomposite has a significantly higher breakdown field strength. Electroluminescence studies conducted using highly nonuniform field of point plane geometry have shown once again that the nanocomposite has an initiation field strength of twice that for the microcomposite and the host resin [41]. A study on electroluminescence on an epoxy resin under both uniform and nonuniform field conditions has shown that the polarity of the excitation voltage has very little bearing on the level of light intensity [25]. An explanation of possible mechanisms of this electroluminescence advanced by the authors is that the emissions are due to the existence of hot carriers that can be accelerated to high kinetic energies or by bipolar injection from anode and cathode, charge migration, and the formation of recombination domains (Griseri et al.). Alternatively, Nelson et al. [41] point to the mechanism for electroluminescence to be from the downward transition of excited species formed by electron injection in the high field tip. In this discussion it is quite evident that the large increase in surface area as the particle size is decreased plays a vital role. The nanoparticles have created a mechanism for electron scattering which will skew the energy distribution with beneficial results, which means a higher voltage is required for initiation of electroluminescence [41].

Nanocomposites have been shown to enhance the surface properties of the new materials in addition to the improvements in the bulk [13]. In this work they had examined the triple junction effect of a nanocomposite in an atmosphere of compressed air at 5 bar. The geometry used was plane–point–plane. It has been shown that the surface damage on the nanocomposite was negligible in comparison to the damage suffered by the base resin under the same AC voltage conditions. This work if extended will reveal many more interesting aspects of surface corona on nanocomposites.

Although the nanomaterial forms a crucial part of the nanocomposite, it should not be overlooked that the host resin is key in determining the dielectric properties of the composite. The role of the host resin as to whether it is of polar or nonpolar has very complex behavior with regard to the loss characteristics as a function of frequency [64]. They studied the dielectric performance of two different sets of composites made up with Alumina. One host resin was polar and the other nonpolar. The particle size was of the order of 10 nm. Dielectric spectroscopy was so complex that it was difficult to decipher the mechanisms underlying the observations.

6.9.3 Dielectric Constant of a Composite

The very early attempts at calculating the effective dielectric constant of a two component of a composite were done by Maxwell Garnett [16]. Since then others have made useful contributions [15]. In this latter case the authors had taken the approach that the additive material had an ellipsoidal shape distribution. It is argued [37] that the model advanced by Gao and Gu does not take into account the finite interfaces between the particles and the host matrix. The interface that is formed within the composite is a highly complex region where many forces come into play such as covalent bonding, charge transfer reactions and to account for polarization due to dipoles Van der Waals forces [37].

6.9.4 Nanodielectrics Under Cryogenic Conditions

Nanodielectrics will be an important area of study in the superconductivity industry in the not-too-distant future. There have already been considerable efforts at trying to understand the basic processes and mechanics of nanocomposites at cryogenic temperatures [56, 57]. There is evidence that it is not only the particle size and the base resin that determine the performance of the composite as a dielectric but also the method of fabrication [58]. Again there are more problems that one encounters such as particles fusing to give rise to agglomeration which contribute to the lowering of desirable properties like enhancement of dielectric strength [58]. Work performed by using polyvinyl alcohol, which is water soluble, incorporating additives of barium titanate and titanium dioxide has been shown to have increases in both real and imaginary parts of the permittivity and correspondingly increase in dielectric losses [57]. This however is not unexpected as the increase in permittivity is associated with the bound water and interfacial polarization [57]. The role of water in a two-component composite playing a significant role in increasing both the value of permittivity and dielectric loss as a function of frequency at a fixed temperature was reported by Sillars [52]. It is noteworthy that an insulator containing a small quantity of a semiconducting material exhibits a higher conductivity under AC voltage than under DC voltage [52]. The energy loss being attributed to dipole rotation at radio frequencies, but under DC voltage in the same material is said to be due to some manifestation of ionic conduction [52]. The reduction of particle size from micrometer scale to nanometer scale once again has shown that the polarization that occurs for the former due to moisture absorption is not present in the latter case, and the interfacial region is critical in the mechanism of charge trapping [5]. Controlling the particle size in the composite is critical for enhanced electrical and mechanical performance. It has been shown that in situ precipitation of titanium dioxide particles in clear epoxy yielded good uniform dispersion of nanoparticles. The method used was a drop casting technique [44]. This method while giving desired results for particular applications may not always be practical. The advantage of this method is that the nanocomposite can be molded into any desired shape and is suitable for cryogenic applications.

6.10 Conclusions

Dielectrics in its many forms have been studied over a long period of time using a variety of experimental and theoretical methods. However, we are yet to device predictable models that could be used across even a narrow set of variables that present themselves in an electrical system. In this chapter we have drawn attention to polymeric materials under cryogenic conditions, which adds yet another dimension to an already complex situation. The rapid development of high temperature superconductors and their potential application in the electric power industry makes it even more important to embark into the area of dielectrics under cryogenic conditions.

References

1. Allen NL, Mikropoulos PN (1999) Streamer propagation along insulating surfaces. IEEE Trans Dielectr Electr Insul 6(3):357–362
2. Blank C, Edwards MH (1960) Dielectric breakdown of liquid helium. Phys Rev 119(1):50–54
3. Bochenek E, Franke H, Wimmershoff R (1975) Manufacture and initial tests of a high power dc cable with superconductors. IEEE Trans Magn 11:366–369
4. Buliniski A, Densley J, Sudersham TS (1980) The ageing of electrical insulation at cryogenic temperatures. IEEE Trans Electr Insul EI-15:83–88
5. Cao Y, Irwin PC (2003) The electrical conduction in polyimide nanocomposites. In: Annual report conference on electrical insulation and dielectric phenomena. Alburquerque, New Mexico, pp 116–119
6. Chowdhuri P (1981) Some characteristics of dielectric materials at cryogenic temperatures for HVDC systems. IEEE Trans Electr Insul EI-16(1):40–51
7. Chowdhuri P (1982) Electrical characteristics of a DC superconducting cable. Cryogenics 22: 171–180
8. Cook CM (1982) Charging of insulator surfaces by ionization and transport in gases. IEEE Trans Electr Insul EI-17:172–178
9. Fallou B, Galand J, Bouvier B (1970) Dielectric breakdown of gaseous helium at very low temperatures. Cryogenics 10:142–146
10. Forsyth EB, Blewett RB, Garber M et al (1973) Flexible superconducting power cables. IEEE Trans Power App Syst 92(2):494–505
11. Forsyth EB, McNerney AJ, Muller AC, Rigby SJ (1978) Progress in the development of gas-impregnated lapped plastic film insulation. IEEE Trans Power App Syst 97(3):734–743
12. Fowler WB, Dexter DL (1968) Electronic bubble states in liquid helium. Phys Rev 176: 337–343
13. Frechette MF, Larocque RY, Trudeau MI et al (2005) Nanodielectric surface performance when submitted to partial discharges in compressed air. In: Annual report conference on electrical insulation and dielectric phenomena. Nashville, Tennessee, pp 727–731
14. Gallimberti I, Marchesi G, Niemeyer L (1991) Streamer corona at an insulating surface. In: Seventh international conference on high voltage engineering, Dresden, paper 41.10
15. Gao L, Gu JZ (2002) Effective dielectric constant of a two-component material with shape distribution. J Phys D Appl Phys 35:267–271
16. Garnett JCM (1904) Colours in metal glasses and in metallic films. Philos Trans R Soc Lond A 203:385–420
17. Garton CG (1962) The energy of discharges and their interaction with solid dielectrics. Gas discharges and the electrical supply industry. Butterworth, London, pp 412–419
18. Gerhold J (1972) Dielectric breakdown of helium at low temperatures. Cryogenics 12(5): 370–376
19. Gerhold J (1988) Helium breakdown near the critical state. IEEE Trans Electr Insul 23(4): 765–768
20. Goldschvartz JM, Ouwerk AC, Blaisse BS (1970) On the electrical breakdown of liquid helium. IEE Conf Publ 67:218–222
21. Graber L, Rodrigo H, Salmhofer F et al (2011) Experimental determination of dielectric properties of Cryoflex in high pressure cold helium gas. In: International symposium on high voltage engineering, Hannover, Germany, D 007
22. Green C, Vaughan A (2008) Nanodilectrics—how much do we really understand. IEEE Insul Mag 24(4):6–16
23. Greenwood A (1991) Electrical transients in power systems. Wiley Interscience, New York, p 492
24. Greenwood A (1994) Vacuum switchgear. Institution of Electrical Engineers Publishing, UK, London

25. Griseri V, Dissado LA, Fothergill JC et al (2002) Electroluminescence mechanisms in an epoxy resin under divergent and uniform field. IEEE Trans Dielectr Electr Insul 9(1):150–160
26. Harrwig G, Knaak S (1984) Fibre–epoxy composites at low temperatures. Cryogenics 24: 639–646
27. Hiley J, Dhariwal RS (1985) Dielectric breakdown in high density helium and in helium impregnated solid dielectrics. Cryogenics 25:334–337
28. Ieda M, Sawa G, Miyairi K (1972) Dielectric breakdown of polyethelene film at cryogenic temperatures. In: International symposium on high voltage engineering, Munich, pp 414–421
29. Kind D (1978) An introduction to high voltage experimental technique. Vieweg, Braunschweig, Germany
30. Kind D, Kaerner H (1985) High voltage insulation technology. Vieweg, Braunschweig, Germany
31. Koo JH (2006) Polymer nanocomposites; processing, chacterization and applications. McGraw Hill, New York
32. Kohno T et al (1980) Electrical breakdown of ice at cryogenic temperatures. IEEE Trans Electr Insul EI 15(1):27–32
33. Kosaki M, Shimizu N, Horii K (1977) Treeing of polyethelene at 77 K. IEEE Trans Electr Insul EI-12(1):40–45
34. Kuffel E, Zaengl WS (1984) High voltage engineering fundamentals. Pergamon Press, Oxford
35. Latham R (ed) (1995) High voltage vacuum insulation. Academic, New York
36. Lewis TJ (1994) Nanometric dielectrics. IEEE Trans Dielectr Electr Insul 1(5):812–825
37. Lewis TJ (2005) Interfaces: nanometric dielectrics. J Phys D Appl Phys 38:202–212
38. Loeb LB (1965) Electrical coronas their basic physical mechanisms. University of California Press, Berkeley, pp 30–46
39. Mathes KN (1967) Dielectric properties of cryogenic liquids. IEEE Trans Electr Insul EI 2(1): 24–32
40. Meats RJ (1972) Pressurised—helium breakdown at very low temperatures. Proc IEE 119(6): 760–766
41. Nelson JK, Hu Y, Thiticharoenpong J (2003) Electrical properties of TiO_2 nanocomposites. In: IEEE annual report conference on electrical insulation and dielectric phenomena. Boulder, Colarado, pp 719–722
42. Nelson JK, Fothergill JC (2004) Internal charge behaviour in nanocomposites. Nanotechnology 15:586–595
43. Poli E (1982) A comparison between positive and negative corona. In: Seventh international conference on gas discharges and their applications, London, UK, pp 132–135
44. Polizos G et al (2010) Properties of a nanodielectric cryogenic resin. Appl Phys Lett 96(15): 152903
45. Rodrigo H, Allen NL, Tan BH (2002) Negative corona development along insulator surfaces under impulse voltages. In: Proceedings of international conference on gas discharges and their applications, vol 1, Liverpool, UK, pp 263–266
46. Rodrigo H, Tan BH, Allen NL (2005) Negative and positive impulse corona development along cylindrical insulator surfaces. IEE Proc Sci Meas Technol 152(5):201–206
47. Rodrigo H, Baumgartinger W, Ingrole A, et al (2008) Breakdown of transformer insulation materials under cryogenic and room temperature. Mater Res Soc Symp Proc, vol 1134, BB03.10.
48. Rodrigo H, Salmhofer F, Kwag DS, Pamidi S, Graber L, Crook DG, Ranner SL, Dale S, Knoll D (2012) Dielectric studies of a novel high pressure helium gas cooled DC power cable. Cryogenics 52:310–314
49. Sato S, Zaengl WS, Knecht A (1987) Analysis of accumulated surface charge on DC epoxy resin spaces. IEEE Trans Electr Insul EI-22:3333–3340
50. Schwarz G (1988) Thermal expansion of polymers from 4.2 to room temperature. Cryogenics 28:248–254

51. Shimizu N, Nagata T, Horii K et al (1986) Thermal contraction and cracking of extruded polyethlene electrical insulation at cryogenic temperatures. Cryogenics 26:459–466
52. Sillars RW (1937) The properties of a dielectric containing semiconducting particles of various shapes. J Inst Electr Eng 80:378–394
53. Stoop Th, Tom J, Verhaart HFA et al (1985) The role of electrostatic in insulator flashover in SF_6. In: Eighth international conference on gas discharges and their applications, Oxford, UK, pp 286–289
54. Thoris J, Leon B, Dubois A, Bobo J (1970) Dielectric breakdown of cold gaseous helium in large gaps. Cryogenics 10:147–149
55. Tuncer E, Polizos SI (2011) Electrical insulation paper and its physical properties at cryogenic temperatures. IEEE Trans Appl Superconduct 21(3 Part 2):1438–1440
56. Tuncer E, Sauers I, James DR et al (2009) Nanodielectrics for cryogenic applications. IEEE Trans Appl Supercond 19(3):2354–2358
57. Tuncer E, Sauers I, James DR et al (2008) Nanodielectric system for cryogenic applications: barium titanate filled polyvinyl alcohol. IEEE Trans Dielectr Electr Insul 15(1):236–241
58. Tuncer E, Sauers I, James DR et al (2007) Enhancement of dielectric strength in nanocomposites. Nanotechnology 18:325704
59. Tuncer E, Sauers I, James DR et al (2007) Electrical properties of epoxy resin based nanocomposites. Nanotechnology 18:025703
60. Verhaart HFA, Tom J, Verhage A et al (1987) Avalanches near solid insulators. In: Fifth international conference on gas discharge and their applications, Braunschweig, paper 13.01
61. Weedy BM, Swingler SG (1979) Life expectancy of liquid nitrogen taped cable insulation. IEEE Trans Electr Insul EI-14:222–228
62. Weedy BM, Swingler SG, Shaikh S (1982) Life expectancy and failure mechanisms of lapped synthetic tape impregnated with supercritical helium. IEE Proc 129 Pt A(5):328–331
63. Weedy BM, Swingler SG (1987) Review of tape materials for cable insulation at liquid nitrogen temperatures. Cryogenics 27:668–672
64. Zhang C, Mason R Stevens GC (2005) Dielectric properties of alumina polymer nanocomposites. In: Annual report conference on electrical insulation and dielectric phenomena. Nashville, Tennessee, pp 721–724

Chapter 7
Medical Applications of Poly(vinyl alcohol) Cryogels

S. Reiter, R. Mongrain, M. Abdelali, and J.-C. Tardif

Contents

Abstract Cryo-treatment of polymeric hydrogels, such as poly(vinyl alcohol) cryogels (PVA-C), have long been developed for a variety of medical applications. In the case of PVA-C the simplicity of its fabrication in its most basic form, obtained through a series of freezing and thawing cycles from low to room temperatures, along with its highly tailored microstructure, soft-tissue-like mechanical properties, and excellent biocompatibility makes it one of the most promising medical hydrogels researched today. In this chapter, we discuss the properties of soft tissues, with an emphasis on human vasculature, as a preamble for our

S. Reiter • R. Mongrain
Department of Mechanical Engineering, McGill University, Montreal, QC, Canada

Department of Atherosclerosis, Montreal Heart Institute, Montreal, QC, Canada
e-mail: steven.reiter@mail.mcgill.ca; rosaire.mongrain@mcgill.ca

M. Abdelali • J.-C. Tardif (✉)
Department of Atherosclerosis, Montreal Heart Institute, Montreal, QC, Canada

Départment de Médicine, Université de Montréal, Montreal, QC, Canada
e-mail: maria.abdelali@icm-mhi.org; jean-claude.tardif@icm-mhi.org

S. Kalia and S.-Y. Fu (eds.), *Polymers at Cryogenic Temperatures*,
DOI 10.1007/978-3-642-35335-2_7,

investigation of current techniques in PVA-C manufacturing. The mechanical properties of PVA-C are then presented, outlining its behavior with varying cryo-treatments and PVA concentrations. We then highlight an example of PVA-C used for a coronary imaging phantom, and lastly provide brief examples of PVA-C and its derivatives in a variety of biomedical applications.

Keywords Poly(vinyl alcohol) cryogel • Vasculature elasticity • Freeze–thaw cycles • Coronary phantoms • Tissue scaffolds

7.1 Introduction

The improvement of clinical diagnostic technology, medical training, and surgical solutions is critical to providing a more reliable and sustainable healthcare. Advances in the former two especially contribute to the efficiency of primary health care services and preventative medicine. Imaging phantoms—synthetic models that mimic the geometry, imaging physical properties and/or mechanical behaviors of organs—are frequently used to optimize current and leading-edge diagnostic imaging modalities in ultrasound [1], magnetic resonance imaging (MRI) [2], and high-resolution optical coherence tomography (OCT) [3] technologies. As for surgical tools, the scarcity of organ donations [4], as well as difficulties in harvesting certain allograft (human) and xenograft (animal) tissues such as coronary arteries [5], fuel the necessity of scientific research into materials that harbor similar mechanical and structural characteristics as their physiological counterparts. These tissues must be suitable for surgical implantation and subsequent nascent cell adhesion, differentiation, and protein proliferation in a non-cytotoxic and preferentially bio-absorbable scaffold [6].

Biocompatible hydrogels are one of the most common scaffold materials used today for tissue grafts and are prevalent in compliant phantoms [7]. Hydrogels are rubber-like, viscoelastic polymers that are capable of absorbing much of their weight in water. They are constructed from long-chain hydrophilic molecules that form semicrystalline matrices. As living tissue largely consists of organized water-imbibed cells within fibrous extracellular matrices, hydrogels are nearly optimal solutions for organ replacement scaffolds [6]. Some current clinical examples include hydrogels for skin wound dressing [8], vocal fold restoration [9], knee and joint lubrication [10], and ocular implants. Modifying the imaging properties of the hydrogels with inclusions can make them very suitable for optimizing emerging medical image modalities [1].

One of the most promising hydrogels for several other clinical applications is poly(vinyl alcohol) cryogel, or PVA-C. Unlike most compliant hydrogels which are manufactured using potentially toxic chemical cross-linkers, biocompatible PVA-C's long-chain matrix network is formed by slowly rearranged hydrogen bonds during thermal cycles of freezing and thawing. Not only is PVA-C non-cytotoxic, but its matrix configuration, and thus its mechanical properties, may also be easily tailored by adjusting the number of freeze–thaw cycles (FTCs) to which it is subjected [11]. The mechanical properties in question are PVA-C's hyperelastic (rubber-like) behavior, its viscoelasticity—hysteresis and time-dependent stress–strain—and its fracture toughness.

In this chapter, we will discuss the methodology used to manufacture PVA cryogels, with emphasis on their utility in the aforementioned medical applications. An introduction to typical manufacturing processes of creating PVA-C will first be presented. Afterwards, we provide a detailed analysis of a PVA-C mock coronary atherosclerotic phantom with tailored mechanical properties used to improve current imaging technologies and to validate numerical studies of diseased arterial failure mechanisms. To demonstrate the relevance of PVA-C in this phantom, we begin with a study of arterial composition and mechanical behavior. The chapter ends with a discussion of relevant PVA-C literature and the cryogenic techniques used to make them, along with future trends for PVA-C bio-implants.

7.2 Soft-Tissue Biomechanical Behavior

Soft tissues, as compared to hard tissues, are those that are capable of withstanding significant strains prior to failure. The vast majority of organs in the human body, with exception to bone and teeth, are classified as soft tissue and consist of two principle components: structural proteins and cell content [12]. The organization, interaction, and type of these basic components provide the framework for all soft biological tissues and their mechanical behaviors. As our research applications of PVA-C are primarily concerned with healthy and atherosclerotic (diseased) coronary arteries, the following study of soft-tissue mechanics will be done with regard to vasculature. However, general principles revealed therein may be readily applied to the biomechanics of all soft tissues.

7.2.1 Composition of a Normal Artery

Mammalian arteries are highly heterogeneous structures with three distinct layers of organized cells and proteins: the innermost intimal layer, the media, and the outermost adventitial layer. Figure 7.1 illustrates the various layers of the arterial wall [13]. The intima is a mechanically weak layer essentially tasked with providing a non-thrombogenic barrier between the intraluminal blood-flow and the cell-and-protein structure of the artery. It consists of an innermost endothelial cell layer organized upon a basal elastic lamina. Around this is found the medial layer consisting largely of vascular smooth-muscle cells (SMCs) within a helicoidally-oriented network of elastic fibrils that provides a stabilizing matrix. Finally, the adventitia is composed of the axially oriented tough protein, Type I collagen, admixed with elastin, fibroblasts, nerves, as well as small off-shooting branches

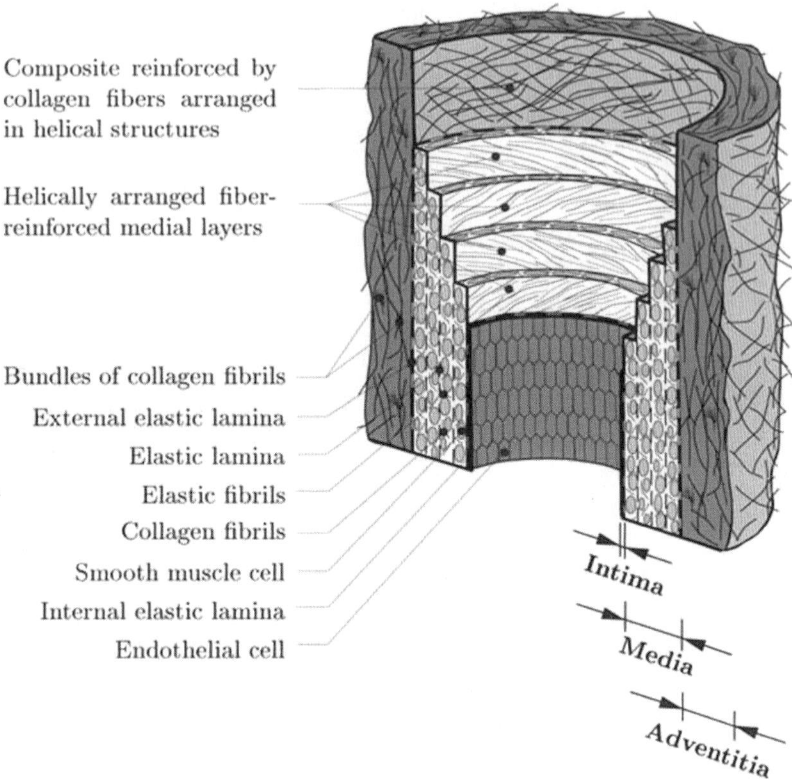

Fig. 7.1 Layered structure of the arterial wall, along with its constituents and their organization [13]

referred to as the vasa vasorum [5]. The adventitia prevents the artery from overdistension under more severe physiological exertions and acts as a protective layer.

Globally, due to the water content of the vascular SMCs, arteries are composed of nearly 70 % (wet weight when fully hydrated) and 45 % collagen and 20 % elastin (dry weight when dehydrated) [5].

7.2.2 Mechanical Behavior of Arterial Tissue

Figure 7.2 depicts a typical loading response of arterial tissue under uniaxial stress in the circumferential direction of a coronary artery. The most characteristic mechanical behavior of arterial tissue is its J-shaped stress–strain curve. At relatively low strains (typically below 5 %), the mechanical behavior of arterial tissue is nearly linear, with a Young's modulus on the order of 300 kPa. At higher strains, the crimped fibers described above straighten in the direction of the tensile stress,

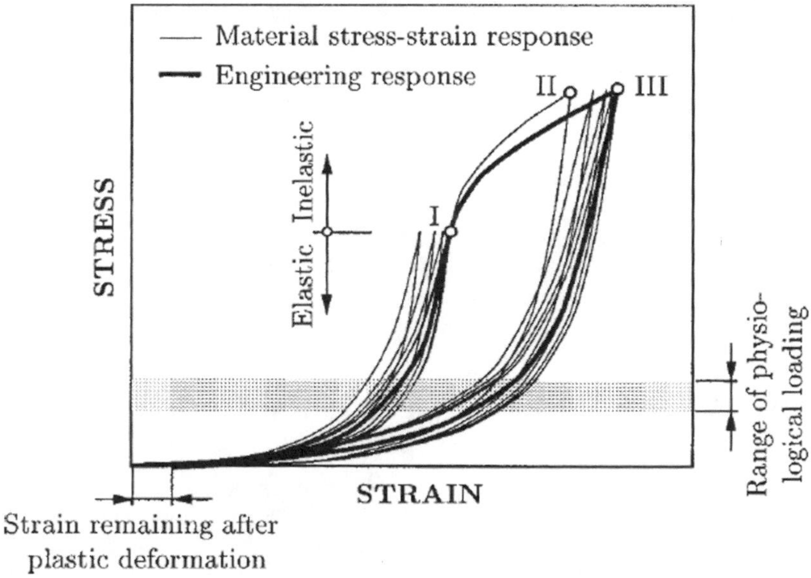

Fig. 7.2 Stress–strain characteristics of bovine arterial (media) sample, exhibiting hysteresis, preconditioning, and plasticity [13]

and the strength of the fibers—collagen has a Young's modulus on the order of 1 GPa—is exhibited and strain stiffening occurs. This stress–strain response may be expressed in the form of a strain-energy density function W, such as that proposed by Mooney and Rivlin:

$$W = C_1(I_1 - 3) + C_2(I_2 - 3) \tag{7.1}$$

where C_1 and C_2 are material constants and I_1 and I_2 are the first and second strain invariants of the right Cauchy-Green deformation tensor, C, with principal strains $\lambda_1 = 1, 2, 3$

$$I_1 = \text{Trace}(C) = (\lambda_1)^2 + (\lambda_2)^2 + (\lambda_3)^2 \tag{7.2}$$

$$I_2 = \frac{1}{2}(\text{Trace}(C^2) - \text{Trace}(C)^2) = (\lambda_1)^2(\lambda_1)^2 + (\lambda_1)^2(\lambda_1)^2 + (\lambda_1)^2(\lambda_1)^2 \tag{7.3}$$

For instance, for the left porcine coronary artery, the material constants are $C_{10} = 16.7$ kPa and $C_{11} = 12.7$ kPa [14].

Another salient feature of passive arterial tissue, from in vitro uniaxial or biaxial testing, is the effect of softening (known as the Mullins effect) that occurs during cyclic testing. Preconditioning is established when the elastic response no longer changes with equivalent loading–unloading. The viscoelastic response of arterial tissue is seen in the hysteresis unloading curves in each cycle. This is largely attributed to the SMC content of the artery. Note that arterial tissue is also nearly incompressible (Poisson's ratio $0.42 < \nu < 0.48$) because of its large water content [5]. Beyond strain point *I* in Fig. 7.2, at non-physiological strains such as those experienced during angioplasty intervention, inelastic (plastic) deformations occur [13].

Fracture toughness is another parameter of arterial tissue that has been seldom investigated. Mechanical tissue rupture may be related to a variety of pathologies, such as atherosclerotic plaque rupture, as well as aortic dissection and aneurysm rupture. Tests by [15] determined the Mode *I* fracture toughness of porcine aorta tissue and obtained an average toughness of 300 J/m^2.

7.3 Preparation of PVA-C and Its Salient Physical Properties

In this section, we present how various physical properties of arteries can be replicated in PVA cryogels by tailoring their microstructure with cryo-treatment.

7.3.1 PVA-C Microstructure

PVA-C is a semi-crystalline polymer. Images of the PVA-C microstructure have been obtained via scanning electron microscope (SEM) [16], such as that seen in Fig. 7.3b.

The microstructure is formed by the cryogenic treatment of PVA in solution, during which phase separation occurs and hydrogen bonding between hydroxyl groups results in crystallization. Repeated freeze–thaw cycling results in further inter and intramolecular bonding of the hydroxyl groups and a more complex (dense) microstructure [17]. Indeed, we can observe these high-resolution topological features, measured through the use of atomic force microscopy (AFM) in Fig. 7.4, which are of importance to cellular adhesion in various tissue engineering applications [17].

Generally, the crystallinity—related to the degree of cross-linking—is correlated with the material stiffness of the PVA-C. Gupta et al. [11] performed swelling tests on PVA-C samples of varying PVA concentration and demonstrated that higher concentrations resulted in higher cross-linking, and thus higher moduli. More cross-linking also results in lower swelling ratios, as per Fig. 7.5.

7.3.1.1 PVA-C Manufacturing

The solution of PVA used to make the cryogel is prepared in nearly the same fashion across the literature: powdered PVA is incorporated into distilled or tap water, which is subsequently heated to 80 °C with continuous stirring until the

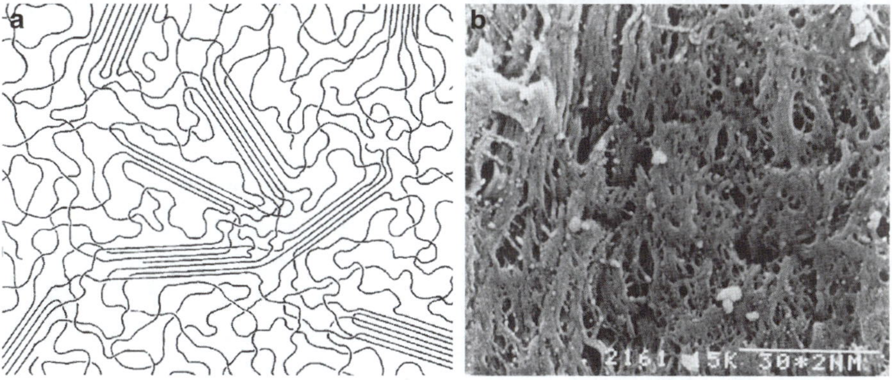

Fig. 7.3 Portrayal of semi-crystalline porous structure of PVA-C in (a) theoretical drawing form [16], (b) from scanning electron microscope (SEM) [21]

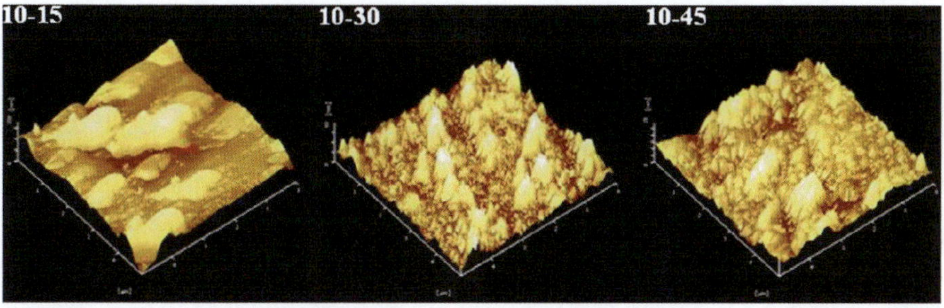

Fig. 7.4 Atomic force microscopy (AFM) topological plots of 10 % PVA/weight cryogels after (a) 15 cycles, (b) 30 cycles, and (c) 45 freeze–thaw cycles [17]

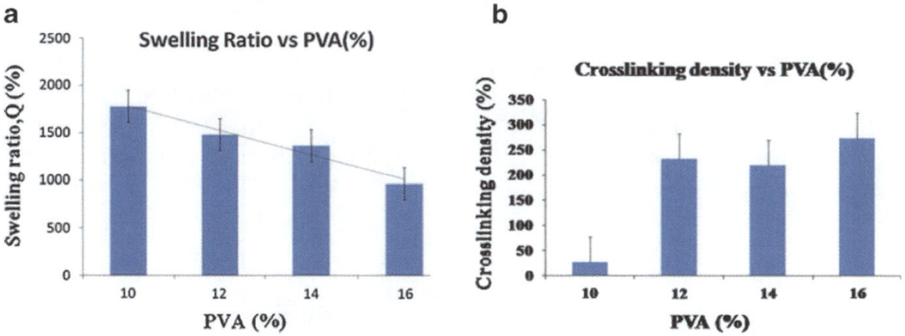

Fig. 7.5 Swelling ratio (a) and cross-linking density (b) as a function of PVA concentration [11]

initially cloudy solution becomes colorless and transparent. Once the solution has cooled and all trapped air bubbles have escaped, the solution is poured into a mould and placed into a temperature-controlled cryo-chamber to prepare the desired hydrogel shape. Typically, the PVA solution is gradually brought down to cold

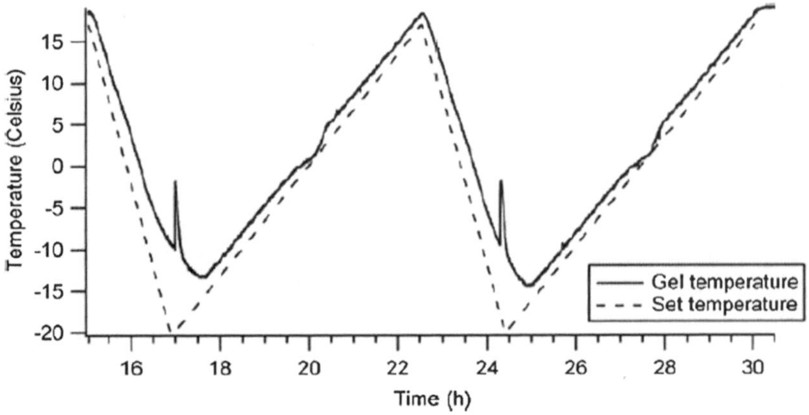

Fig. 7.6 Temperature profile that PVA is exposed to in order to prepare the PVA-C [14]

temperatures on the order of −20 °C, and then brought back to near-ambient temperature (around 20 °C). Each freeze–thaw cryogenic cycle takes several hours to complete, depending on the rate of temperature change. Figure 7.6 depicts a typical freeze–thaw temperature profile used during cryogel formation.

According to Pazos et al. [1], the freezing ramp may be adjusted with little effect to the overall mechanical behavior of the PVA-C: a reasonable freeze ramp is 0.333 °C/min. However, the thawing ramp may be optimized to obtain stiffer gels in less time. Pazos states that a thaw ramp of 0.111 °C/min over 4 cycles results in a similar stiffness as a gel exposed to a ramp of 0.333 °C/min over 5 cycles: the latter providing the shorter cryo-treatment time [14]. Repeated freeze-thaw cycles may then be performed to alter the properties of the cryogel. Two pertinent temperature deviations occur at −8 °C, when crystallization occurs and the temperature drops sharply to −1 °C, and at 0 °C during the thawing cycle as the ice in the PVA-C begins to melt.

Disregarding the additives that some research groups have introduced into PVA-C for various clinical applications [11, 18] there are three manners to change the overall mechanical characteristics of the basic PVA-C hydrogel: change the concentration of dissolved PVA powder, modify the temperature profile, and increase the number of freeze–thaw cryo-treatments.

7.4 Examples of Tailoring PVA-C Through Cryo-Treatment Variations

7.4.1 Tailoring Mechanical Behavior of PVA-C

As concerns the cryogenic processing of PVA-C, two articles provide adequate fundamentals of tailoring PVA-C mechanical properties for soft tissue applications.

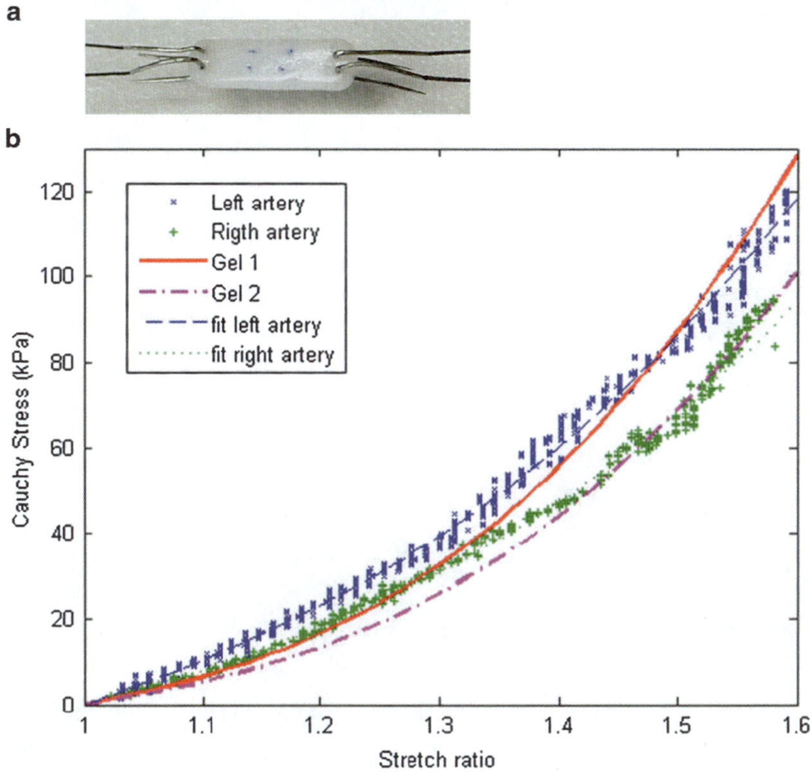

Fig. 7.7 (**a**) Suture assembly for uniaxial testing of PVA-C tissue; (**b**) good correlation between stress–strain curves of left and right porcine coronaries and the PVA cryogels prepared to mimic them [14]

The first is a study of the uniaxial (tensile) properties of PVA-C [14] under various cryo-treatment cycles (Fig. 7.6), mimicking porcine coronary tissue. Uniaxial tests were performed using a Bose Enduratec Electroforce 3200 mechanical tester on rectangular strips of both arterial and PVA tissues. Figure 7.7a depicts the sutures used to pull the sample and obtain the stress–strain curves in Fig. 7.7b.

From Fig. 7.7b, there is obvious evidence of correlation between the mechanical properties of passive porcine coronary arteries and the prepared PVA cryogels. The coefficients of (7.1) for the PVA-C mimicking the left coronary artery, for instance, are $C_1 = 9.8$ kPa and $C_2 = 21.3$ kPa. Within the strain range of Fig. 7.7b, we can see that the low-strain Young's modulus for the cryogel undershoots the modulus of the artery, and vice versa at higher strain rates, as exemplified in the difference between the coefficient values. These optimized cryogels were produced in accordance with data on the cryo-treatment of PVA-C under different freeze–thaw cycling (Fig. 7.8). Once again, monotonic increase of the J-shaped elasticity is seen, where the tensile properties quickly stiffen with increasing cycles from 2 to 8, and slowly increase afterwards.

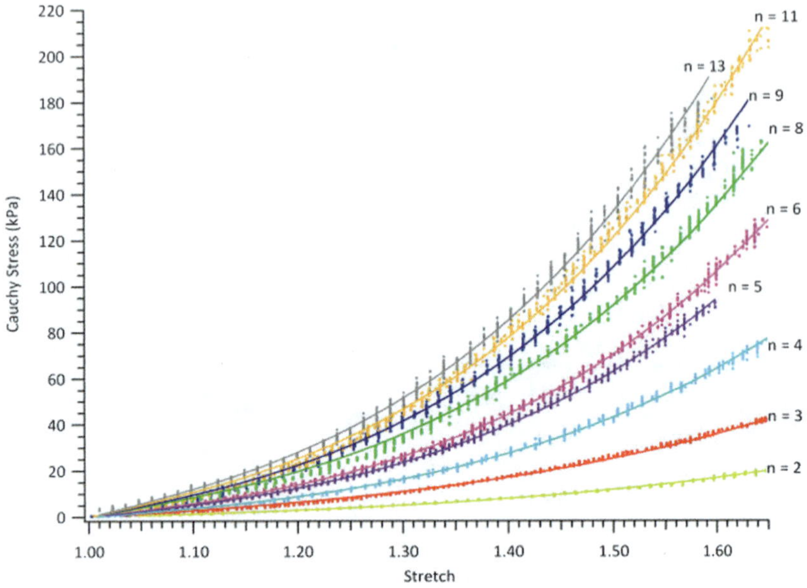

Fig. 7.8 Plots depicting an increase in elasticity with an increase in cryogel cycling [14]

Recent fracture testing performed by our group, using an in-house Mode I fracture tester for the Enduratec 3200, has demonstrated that the fracture toughness of PVA-C, after 4–6 cycles, correlates well with data previously obtained for normal porcine aorta tissue gathered by [15].

The second study is that of Wang and Campbell, which investigates the compressive properties of a PVA cryogel for supplementing lumbar intervertebral discs [19]. Using a similar methodology for making the PVA solution as previously described, this team prepared several samples of PVA-C with FTCs between one and six and additionally by varying the PVA concentrations, ranging from 3 to 40 % (w/w). The temperature range used for their samples was between −20 and 20 °C, with a ramp of 0.1 °C/min in both the freezing and thawing cycles. Elastic modulus values were obtained from unconfined compressive testing, and stress relaxation and creep strain measurements were also acquired during mechanical testing to determine the viscoelastic nature of the hydrogels. Monotonically increasing elasticity was observed for increasing PVA concentration and FTCs, which is expected as higher degrees of cross-linking occur as a result of both parameters [11]. However, as concerns the viscoelastic response, negligible differences in relaxation rate were observed throughout the range of concentrations and cycles, with an average stress relaxation percentage of 10 %. In all cases, hysteresis in loading and unloading also indicates energy dissipation through heat losses, which resembles that of native soft tissue.

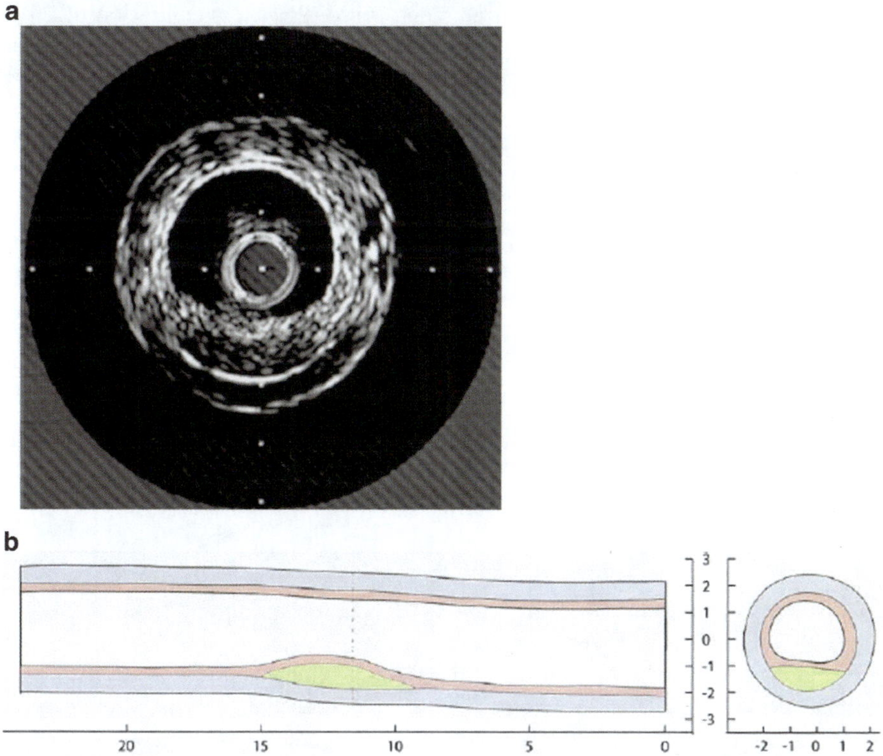

Fig. 7.9 (**a**) Intravascular ultrasound (IVUS) image of a two-layered PVA-C atherosclerotic coronary phantom with trapped lipid pool visualized with cellulose scatterer [1], and (**b**) full longitudinal (*left*) and cross-sectional (*right*) reconstruction of the phantom geometry with custom Matlab code [22]

7.4.2 Modification of Imaging Properties of Coronary Phantom

For the purpose of using the mock PVA-C coronary as a deformable medical imaging phantom, 1 % (w/w) cellulose is incorporated into the PVA solution prior to cryo-treatment to act as an acoustic scatterer [1]. In Fig. 7.9 below, a two-layered PVA-C coronary phantom with atherosclerotic lipid-pool inclusion is imaged with intravascular ultrasound (IVUS) (Fig. 7.9a) and reconstructed along its length using several IVUS cross-sectional photos (Fig. 7.9b).

7.5 Further PVA-C Applications

There are many other examples of PVA-C and its derivatives in biomedical material literature. Kehoe et al. [20] review a variety of peripheral nerve conduits to repair damaged nerve bundles one of which, Salubridge™ by Salumedica™

L.C.C. uses a similar PVA-C cryo-treatment, with a PVA concentration of 20 % (w/w) mixed with 0.9 % (w/w) sodium chloride. Dawson et al. [18] developed an anisotropic PVA structure, more closely resembling soft fibrous tissue with micropores on the order of 10–30 μm, by manufacturing a mould holding PVA slurry of 5–10 % (w/w) PVA solution and slowly lowering it into a vat of liquid nitrogen. Ice crystals formed during the extreme cooling process provide the scaffold for the heterogeneous structure to form, as concentrated PVA networks are trapped between the ice pillars.

Finally, a technique for manufacturing optically transparent cryogels, also suitable for cell incorporation during cyclic treatment, was developed by [11]. With a freeze-thaw cycle consisting of 8 h of freezing at 0 °C and 16 h thawing at 37 °C for 15 cycles, it is thought that the cells would survive this temperature range and could be embedded within the PVA-C matrix for subsequent cell proliferation and protein diffusion for bio-implant applications.

7.6 Conclusions

This is merely a brief introduction into PVA-C manufacturing and biomedical relevance. By optimizing the cryo-treatment of PVA solutions, highly complex anisotropic cryogels can be tailored to meet ever-demanding clinical requirements. Although current biocompatible PVA scaffolds are not biodegradable, it is hypothesized that future research into PVA-C chemistry will improve upon the composition to allow for improved viscoelastic properties while providing the appropriate degradation profile necessitated by many bio-implants.

References

1. Pazos V, Mongrain R, Tardif J-C (2010) Deformable mock stenotic artery with a lipid pool. J Biomech Eng 132(3):034501
2. Mano I, Hitoshi G, Nambu M, Iio M (1986) New polyvinyl alcohol gel material for MRI phantoms. Magn Reson Med 3:921–926
3. Kawase Y, Suzuki Y, Ikeno F, Yoneyama R, Hoshino K, Ly HQ, Lau GT, Hayase M, Yeung AC, Hajjar RJ, Jang I-K (2007) Comparison of nonuniform rotational distortion between mechanical IVUS and OCT using a phantom model. Ultrasound Med Biol 33:67–73
4. Molzahn AE, Starzomski R, McCormick J (2002) The supply of organs for transplantation: issues and challenges. Nephrol Nurs J 30:17–26
5. Humphrey JD (2002) Cardiovascular solid mechanics: cells, tissues, and organs. Springer, New York
6. Ikada Y (2006) Tissue engineering: fundamentals and applications. Elsevier, San Diego
7. Pogue BW, Patterson MS (2006) Review of tissue simulating phantoms for optical spectroscopy, imaging and dosimetry. J Biomed Opt 11:041102
8. Ishihara M, Nakanishi K, Katsuaki O, Masato S, Kikuchi M, Saito Y, Yura H, Matsui T, Hattori H, Uenoyama M, Kurita A (2002) Photocrosslinkable chitosan as a dressing for wound occlusion and accelerator in healing process. Biomaterials 23:833–840

9. Hahn M, Teply BA, Stevens MM, Zeitals SM, Langer R (2005) Collagen composite hydrogels for vocal fold lamina propria restoration. Biomaterials 27:1104–1109
10. Sonntag R, Reinders J, Kretzer JP (2012) What's next? Alternative materials for articulation in total joint replacement. Acta Biomater 7:2434–2441
11. Gupta S, Webster TJ, Sinha A (2011) Evolution of PVA gels prepared without crosslinking agents as a cell adhesive surface. J Mater Sci Mater Med 22:1763–1772
12. Richardson PD (2006) Mechanical properties of atherosclerotic tissues. In: Holzapfel GA, Ogden R (eds) Mechanics of biological tissues. Springer, Berlin
13. Holzapfel G, Gasser T, Ogden R (2000) A new constitutive framework for arterial wall mechanics and a comparative study of material models. J Elast 61:1–48
14. Pazos V, Mongrain R, Tardif J-C (2009) Polyvinyl alcohol cryogel: optimizing the parameters of cryogenic treatment using hyperelastic models. J Mech Behav Biomed Mater 2:542–549
15. Chu B (2010) Characterization of aortic tissue fracture toughness and stiffness under cyclic fatigue loading. Master's thesis, Department of Mechanical Engineering, McGill University. Retrieved from eScholarship@McGill (Publication Number 86772)
16. Gref R, Nguyen QT, Schaetzel P, Néel J (1993) Transport properties of poly(vinyl alcohol) membranes of different degrees of crystallinity. J Appl Polym Sci 49:209–218
17. Pramanick AK, Gupta S, Mishra T, Sinha A (2012) Topographical heterogeneity in transparent PVA hydrogels studied by AFM. Mater Sci Eng 32:222–227
18. Dawson A, Harris P, Gouws G (2010) Anisotropic microstructured poly(vinyl alcohol) tissue-mimicking phantoms. IEEE Trans Ultrason Ferroelectr Freq Control 57:1494–1496
19. Wang BH, Campbell G (2009) Formulations of polyvinyl alcohol cryogel that mimic the biomechanical properties of soft tissues in the natural lumbar intervertebral disc. Spine 34:2745–2753
20. Kehoe S, Zhang XF, Boyd D (2012) FDA approved guidance conduits and wraps for peripheral nerve injury: a review of materials and efficacy. Injury 43:553–572
21. Lozinsky VI, Plieva FM (1997) Poly(vinyl alcohol) cryogels employed as matrices for cell immobilization. 3. Overview of recent research and developments. Enzyme Microb Technol 23:227–242
22. Pazos V, Mongrain R, Tardif J-C (2010) Mechanical characterization of atherosclerotic arteries using finite-element modeling: feasibility study on mock arteries. IEEE Trans Biomed Eng 57:1520–1528

Chapter 8
Dielectric Properties of Polymers at Low Temperatures

Luís Cadillon Costa and François Henry

Contents

Abstract Dielectric spectroscopy is a powerful method that allows the study of the dynamics of polymers in a wide frequency range. The different regimes of the dielectric function can be observed and the dynamics of the primary and secondary relaxations can be found. In fact, to obtain a complete characterization, a large range of frequencies and temperatures must be used. In this work, the investigation was focused in poly(lactid acid), PLA, in two forms, industrial and purified. This polymer is an aliphatic polyester, and one of the most important biocompatible and biodegradable material that has received increasing attention in the last 10 years. The β relaxation was observed between −150 and −30 °C, in frequency domain measurements between 1 Hz and 100 kHz, and was assigned to the secondary relaxation in the glassy state. The changes in the structure, which are connected

L.C. Costa (✉)
I3N and Physics Department, University of Aveiro, 3810-193 Aveiro, Portugal
e-mail: kady@ua.pt

F. Henry
Laboratoire de Recherche en Polymères, CNRS, 94320 Thiais, France
e-mail: henry@glvt-cnrs.fr

S. Kalia and S.-Y. Fu (eds.), *Polymers at Cryogenic Temperatures*,
DOI 10.1007/978-3-642-35335-2_8,

with the water penetration in the polymer, directly affect that relaxation process. Water molecules confined by the polymer chains and in the polymer networks itself play an important role in the degradation of the material. We studied the evolution of that degradation during 4 weeks, in a controlled humidity environment. It is accepted that water preferentially enters in the amorphous zones, but also affects the crystalline regions. It is observed a clear evolution of the relaxation activation energy during the degradation of the polymer. The dielectric relaxation studies are complemented with water permeability measurements during the degradation process with time.

Keywords Poly(lactic acid) • Dielectric relaxation • Impedance spectroscopy • Polymer degradability

8.1 Introduction

The strong industrial interest in electrical properties of polymers reflects the growing use of these materials in electronic devices, optoelectronic switches, fuel cells, batteries and printed board circuits. Dielectric spectroscopy presents good performances that permit to investigate the fundamental aspects of those properties, yielding a wealth of information about the molecular motions and relaxation processes. In the case of polymers, this technique is particularly useful to study the material, and frequently complemented by other measurement methods.

8.1.1 Theoretical Considerations

When an electric field, E, is applied to a material, the resulting polarization of the medium, P, can be expressed by

$$P(\omega) = (\varepsilon^*(\omega) - 1)\varepsilon_0 E(\omega) \tag{8.1}$$

where ε^* is the dielectric permittivity of the material, also known as dielectric response, and ε_0 the permittivity in vacuum [1]. In fact, the permittivity is a complex entity,

$$\varepsilon^*(\omega) = \varepsilon'(\omega) - \mathrm{i}\varepsilon''(\omega) \tag{8.2}$$

where both parts have a physical interpretation. The real part, ε', represents the energy stored in the material, and the imaginary part, ε'', is the energy dissipated per cycle [2]. The complex permittivity, $\varepsilon^*(\omega)$, can be related to the time-dependent dielectric function, $\varepsilon(t)$, using the one-sided Fourier transformation,

$$\varepsilon^*(\omega) = \varepsilon_\infty - \int_0^\infty \frac{\mathrm{d}\varepsilon(t)}{\mathrm{d}t} \mathrm{e}^{-\mathrm{i}\omega t}\,\mathrm{d}t \tag{8.3}$$

where ε_∞ is the dielectric constant at extremely high frequency.

The simplest assumption is consider that, under the application of an electrical field in a material, the change of the polarization is proportional to its actual value [3], that is,

$$\frac{\mathrm{d}P(t)}{\mathrm{d}t} = -\frac{1}{\tau_{\mathrm{d}}} P(t) \tag{8.4}$$

where τ_{d} is the characteristic relaxation time. Solving this equation, we conclude that the current response to the application of a step function potential difference is a decreasing exponential. This behaviour, known as Debye relaxation [4], corresponds to a single relaxation time. In the frequency domain this is expressed in the complex dielectric function as

$$\varepsilon^*(\omega) = \varepsilon_\infty + \frac{\varepsilon_{\mathrm{s}} - \varepsilon_\infty}{1 + \mathrm{i}\omega\tau_{\mathrm{d}}} \tag{8.5}$$

where ε_{s} is the static dielectric constant.

Only in very rare cases this behaviour is observed. In particular, in complex systems, like polymers, the dielectric functions do not present a single relaxation time, that is, the Debye model is not valid. Formally, this non-Debye relaxation can be expressed by a superposition of Debye mechanisms with several relaxation times [5]. From a practical point of view, this can be empirically described by the Kohlrausch-Williams-Watts function in the time domain [6], which in frequency domain can be described by several empirical models [7].

One of the most used is the Cole–Cole model [8], described by

$$\varepsilon^*(\omega) = \varepsilon_\infty + \frac{\varepsilon_{\mathrm{s}} - \varepsilon_\infty}{1 + (\mathrm{i}\omega\tau_{\mathrm{cc}})^{1-\alpha}} \tag{8.6}$$

where τ_{cc} is the characteristic relaxation time, and a α parameter ($0 < \alpha \leq 1$) that leads to a broadening for the relaxation function.

Other possibility, quite used in complex materials, is the Cole-Davidson model [9],

$$\varepsilon^*(\omega) = \varepsilon_\infty + \frac{\varepsilon_{\mathrm{s}} - \varepsilon_\infty}{(1 + \mathrm{i}\omega\tau_{\mathrm{cd}})^{\beta}} \tag{8.7}$$

where again the exponent β reflects the broadening of the relaxation function ($0 < \beta \leq 1$) and τ_{cd} the relaxation time.

A more general model function was introduced by Havriliak and Negami [10], that is a combination of Cole–Cole and Cole-Davidson functions,

$$\varepsilon^*(\omega) = \varepsilon_\infty + \frac{\varepsilon_{\mathrm{s}} - \varepsilon_\infty}{\left(1 + (\mathrm{i}\omega\tau_{\mathrm{hn}})^{1-\alpha}\right)^{\beta}} \tag{8.8}$$

The particular case of $\alpha = 0$ and $\beta = 1$ leads to the Debye relaxation.

8.1.2 Broadband Dielectric Spectroscopy

The dielectric response of a material, $\varepsilon^*(\omega) = \varepsilon'(\omega) - i\varepsilon''(\omega)$, can be measured in the broad band frequency, from a few mHz to hundreds of GHz [11–13]. Nevertheless, to cover a so high frequency range, it is necessary to use different methods, based on several measurement techniques. Low frequency measurements can be made using bridges [14], oscilloscopes methods based on Lissajous figures [15] or phase sensitive detection with lock-in amplifiers [16]. Impedance analysers and transformer ratio arm bridges [17] have applicability at radio frequency frequencies. For the microwave range it can be used network analysers with resonant cavities, based on the small perturbation theory [18]. These methods have advantages and limitations, but there are an answer to the increasing demand of the industrial interest in dielectrics and electrical properties of materials. Systematic studies of systems are reported in several articles [19–25], but the use of broadband dielectric spectroscopy to observe the temporal evolution of structures have relatively recent origin.

In the early years, low frequency measurements were made, particularly in semiconductors [26], using transient current methods. An important improvement was done with the advent of radio frequency electronic devices, in the beginning of twentieth century, increasing the frequency range of measurement up to MHz. Finally, in the 1950s, methods including microwave frequencies measurements were developed [27].

Usually, the electrical measurements are made in a neutral atmosphere using electrodes applied to the sample. The general approach is to apply an electrical stimulus and observe the response of the material. It is then assumed that the properties of the electrode-material system are time invariant, which is an important purpose to the measurement method [28].

Polymeric materials are very complex systems, when compared with low molecular weight compounds. A large number of macromolecular chains are responsible for a great number of conformations with consequences at the level of chain flexibility. Temperature has an important influence in that flexibility, and consequently this behaviour is reflected in the dielectric measurements by broadband dielectric spectroscopy. In fact, this technique allow us to understand the polarization mechanisms present in polymers, that is, the charge migration and that one due to the orientation of permanent dipoles.

From a practical point of view, dielectric spectroscopy can provide the measurement of the complex permittivity, $\varepsilon^*(\omega) = \varepsilon'(\omega) - i\varepsilon''(\omega)$, or derived quantities related to it. The dielectric modulus [29], $M = \varepsilon^{-1}$, a complex quantity $M^*(\omega) = M'(\omega) + iM''(\omega)$, is often used in more conducting materials. The macroscopic properties, impedance $Z^*(\omega) = Z'(\omega) - iZ''(\omega)$, or admittance, $Y = Z^{-1}$, that is, $Y^*(\omega) = Y'(\omega) + iY''(\omega)$ can also express the properties of a material. The interrelations between these quantities are simple when it is known the shape and size of the samples where the measurements are made.

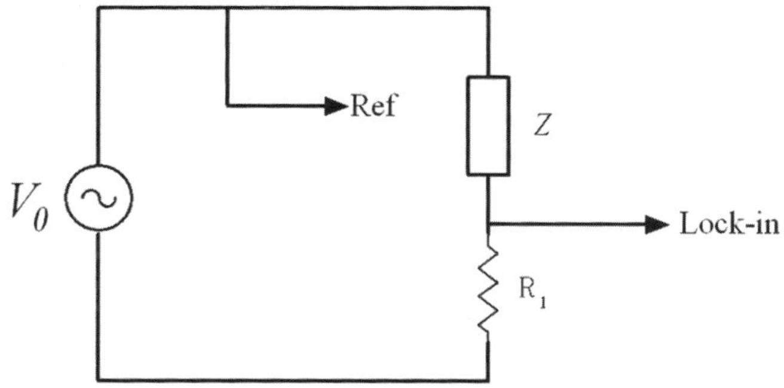

Fig. 8.1 Lock-in amplifier method of measurement

8.2 Measurement Techniques

Different techniques can be used to characterize electrically materials, depending on the type of excitation functions. When using a small amplitude sinusoidal signal and considering the frequency as the independent variable, it is known as frequency domain measurement. Other possibility is to work with time as independent variable, that is, the signal is measured in time domain. In this chapter, only the first technique is considered, presenting three different methods of measurement.

8.2.1 Frequency Domain Methods

8.2.1.1 Lock-in Amplifier Method

This technique can be used when it is necessary to make the detection of signals that are occulted by high levels of noise. Essentially, a lock-in amplifier is a filter with a narrow bandwidth, centred in the frequency of the signal to be measured. Moreover, it can have a signal/noise ratio of the order of 10^9 [30]. It still possesses the possibility to measure a signal with a particular difference of phase to the reference signal, fact that is explored in the implementation of this method.

Figures 8.1 and 8.2 represent schematically the method of measure used for the determination of the complex impedance of the sample [16].

A sinusoidal signal of amplitude V_0, is simultaneously used as reference and applied signal to the association in series, consisting of the sample impedance, Z, and the measuring impedance. This one is formed by a resistance of precision, R_1, in parallel with the input impedance of the amplifier (C_i and R_i). The equivalent circuit of the sample is considered a parallel RC. In a practical point of view, the input impedance of the lock-in amplifier is very high, and then the precision resistance doesn't suffer the charge effect.

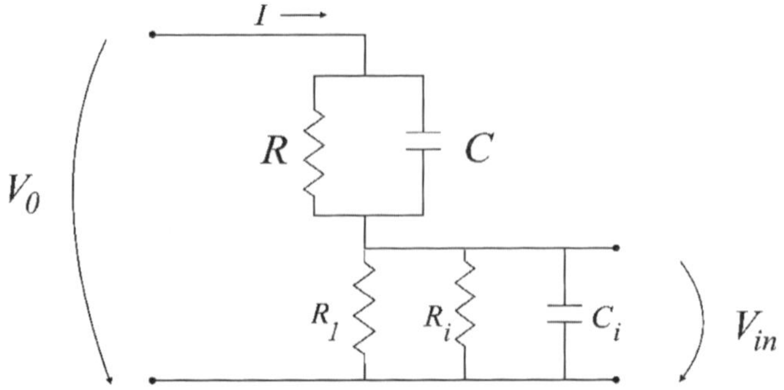

Fig. 8.2 Lock-in amplifier method of measurement. R and C represent the equivalent circuit of the sample, R_l a precision resistance and R_i and C_i the input impedance of the amplifier

Measuring the in-phase, V_f, and the out-of-phase, V_q, components of the sample signal, V_{in}, it is possible to calculate the real and imaginary parts of the impedance, using the expression [31],

$$Z^*(\omega) = \frac{R_i(V_f V_0 - V_f^2 - V_q^2) - \omega C_i^2 R_i^2 V_q V_0}{(1+\omega^2 C_i^2 R_i^2)(V_f^2 + V_q^2)} - i\frac{V_q V_0 R_i + \omega C_i^2 R_i^2 (V_f V_0 - V_f^2 - V_q^2)}{(1+\omega^2 C_i^2 R_i^2)(V_f^2 + V_q^2)} \quad (8.9)$$

where ω is angular frequency. Using the known sample shape and dimensions, it is usually possible to calculate the microscopic dielectric properties of the material.

The accuracy of this technique is highly dependent of the precision of the resistance R_l, and then it must be guaranteed that this is maintained at constant temperature and humidity. Also, it is important to eliminate noise sources, in particular the capacitive and inductive coupling with the detector, ground loops and microphone effects.

8.2.1.2 Impedance Analyser Method

This technique uses the auto balance method, with high range of applicability. The signal is directly applied to the unknown impedance, Z, as shown in Fig. 8.3.

The point L, in the figure, is a virtual ground, and in the equilibrium, the detected current must be null. That is, the gains A and B must be adjusted to obtain this condition and then, as $V_r = AV + iBV$,

$$Z^*(\omega) = R\frac{V}{V_1} = R\frac{V}{AV + iBV} = \frac{R}{A + iB} \quad (8.10)$$

Fig. 8.3 Impedance analyzer method schematic

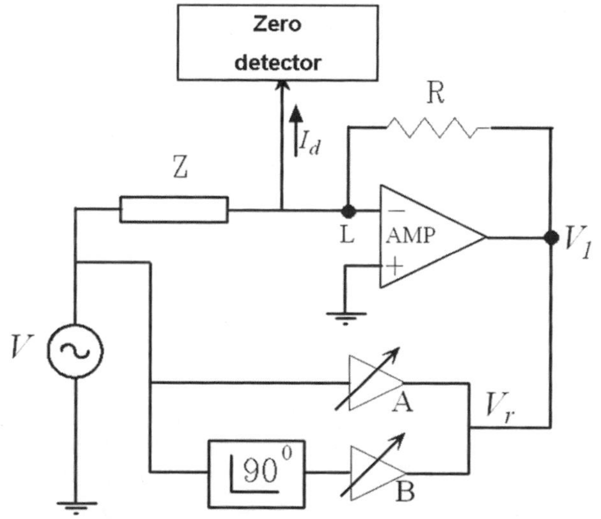

The accuracy of this technique is dependent of several parameters. The most important are the zero detector resolution and the accuracy of the phase shifter.

8.2.1.3 Reflectometry Method

The circuit analysis models that are used at low frequencies are inadequate when the circuit dimensions are of the same order of magnitude of the wavelength. Then, the complex impedance can not be calculated using the complex currents and voltages, but using the complex reflection factor, ρ^*, that can be measured by a reflectometer. This coefficient depends on the location of the measurement along the line. For a line with length L,

$$\rho^*(L) = \rho^*(0)\exp(2l(\alpha_r + i\beta_r)) \tag{8.11}$$

where $\rho^*(0)$ is the reflection coefficient at the beginning of the line, α_r the attenuation and β_r the propagation coefficients.

When a transmission line with characteristic impedance Z_0 is ended by a sample with impedance Z^*, as shown in Fig. 8.4,

$$Z^*(\omega) = Z_0 \frac{1 + \rho(L)}{1 - \rho(L)} \tag{8.12}$$

The accuracy of this technique depends on the quality of the transmission line and to enhance it, sophisticated calibration procedures have to be done. The α_r and β_r parameters must be homogeneous over the line and independent of temperature, as the calibration usually can only be done at room temperature [32]. For the low

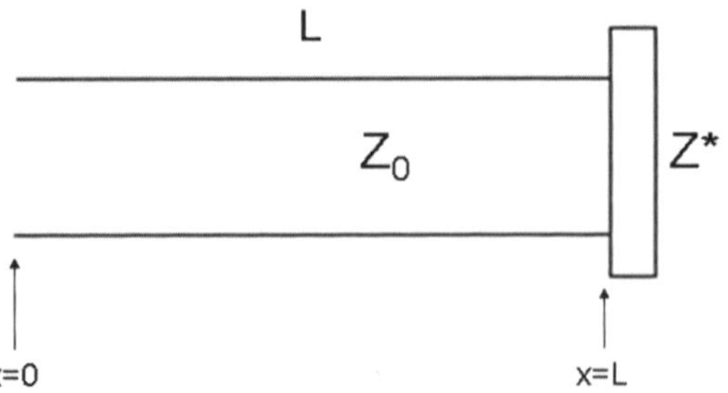

Fig. 8.4 A sample, with complex impedance Z^*, is the termination of a transmission line with characteristic impedance Z_0

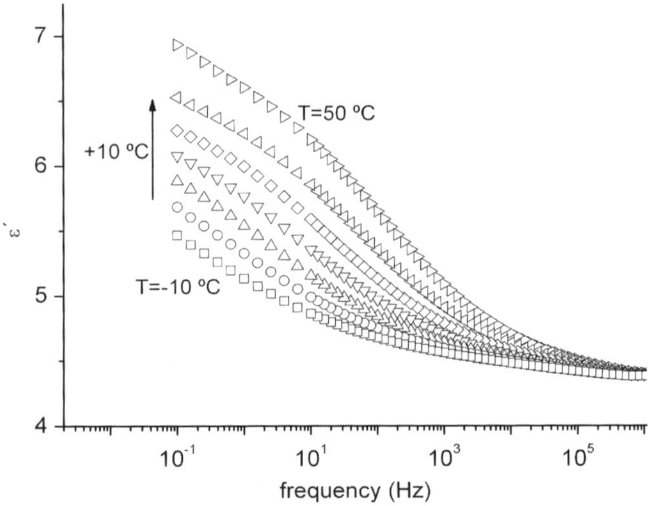

Fig. 8.5 Real part of the complex permittivity for PEMA

temperature measurements, the cryostat must be close to the reflectometer, in order to keep the waveguide as short as possible. Also important is to avoid ambiguous reflections due to mismatched impedances in the measuring system.

8.2.2 Data Analysis

The dielectric response of great part of materials, in particular polymers, shows multiple relaxation behaviour. Each relaxation process is then indicated by a peak in the imaginary part of the complex permittivity, corresponding to an inflexion in the real part. This particular frequency is then the relaxation frequency, and it changes with temperature. The dielectric relaxation behaviour can also be measured isochronally at a fixed frequency vs. temperature [33].

Figures 8.5 and 8.6 shows the real and imaginary parts of the dielectric permittivity for poly-ethyl methacrylate, PEMA, where a relaxation process is visible.

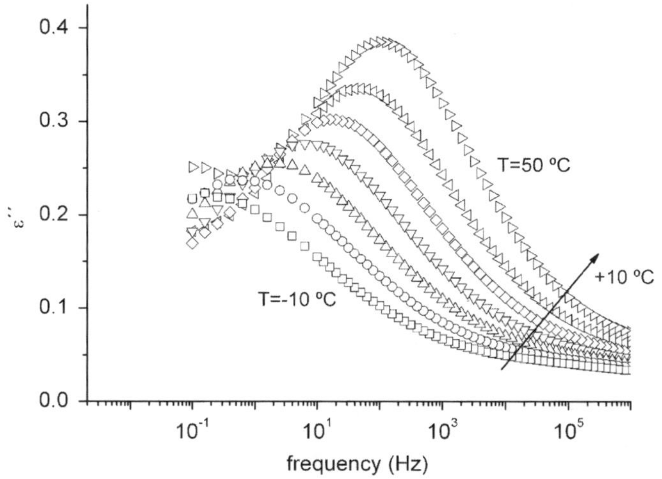

Fig. 8.6 Imaginary part of the complex permittivity for PEMA

This is one way to represent the dielectric data, that is, ε' and ε'', as a function of frequency, for different temperatures. Then, using a particular model, expressed by expressions (8.5), (8.6), (8.7) and (8.8), it is necessary to fit the data. Since a good estimation of relaxation parameters is needed for a correct characterization of a material, complex nonlinear least squares data fitting (CNLS) must be applied. Not only CNLS fitting estimates the parameters of the model, but it also provides estimates of their standard deviations, measures of how well they have been determined by the data fit. These standard deviations are valuable in deciding which parameters are crucial to the model and which are useless, or at least not well determinable from the data [34].

A different way to present the results is the graphical representation, Cole–Cole plot [8]. It consists in the graphic of ε'' vs. ε', where each point corresponds to a measured frequency that increases in the counter-clockwise direction. The relaxation frequency, which corresponds to the maximum of ε'', is easily identified. The data points lie on a semicircle which intersects the ε' axis at ε_∞ and ε_s. For the Debye relaxation this semicircle is centred in the ε' axis, but for the non-Debye models the centre is below that axis. The number of arcs is the number of relaxation processes in the system. Analogous representation can be made using *M*, *Z* or *Y*.

Figure 8.7 show the same results of the precedent figures, using the Cole–Cole plot.

This representation is very useful, but it has a disadvantage to loss the information about the frequency of measurement.

Figure 8.8 show the different models in the Cole–Cole plot. Debye relaxation is represented by a perfect semicircle, centered in the ε' axis. Cole–Cole model presents a symmetric curve, but the centre is below that axis. Cole-Davidson model corresponds to a deformed semicircle. The angles in the intersection are related with the relaxation exponents of the (8.5), (8.6) and (8.7).

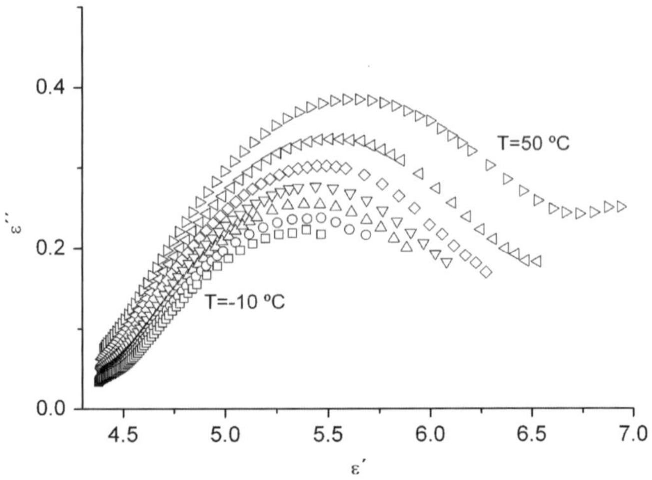

Fig. 8.7 Cole–Cole plot for PEMA. Scales are different in the axis for a better view

Another possibility is to present the complex permittivity sweeping the temperature, for different frequencies. In this representation, low temperatures correspond to high frequencies, whereas high temperatures are related to lower motional frequencies [35]. An example for such representation is shown in Fig. 8.9.

In this figure, we can observe two relaxation processes. That one, at lower temperatures is occulted by that one at high temperatures. In the inset of the graphic it is amplified allowing observing it. Both peaks in the imaginary part of the complex permittivity are moving for higher temperatures, when the frequency increases.

8.3 Following the Ageing of Polyesters Using Thermo-Dielectric Measurements

In order to follow the ageing of poly(lactid acid), dielectric measurements were made. Two relaxations processes were detected, in the frequency and temperature ranges used in this study. The PLA films, industrial and purified, were conditioned at controlled and constant temperature, 25 °C, and humidity, 78 %, during the ageing time, that is, about 2 months. Complementary, other measurement techniques were used to correlate the results.

8.3.1 The α (Cooperative) and β (Noncooperative) Relaxations of PLA

Figure 8.10 shows the imaginary part of the complex permittivity of industrial PLA, as a function of temperature.

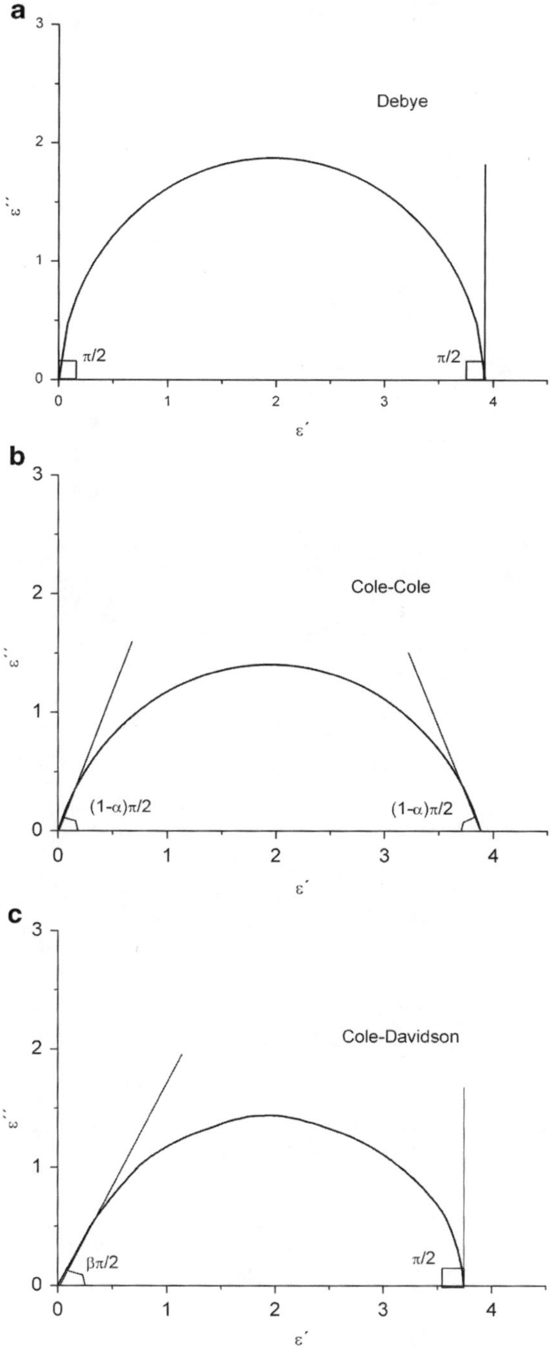

Fig. 8.8 Cole–Cole plot for different relaxation models

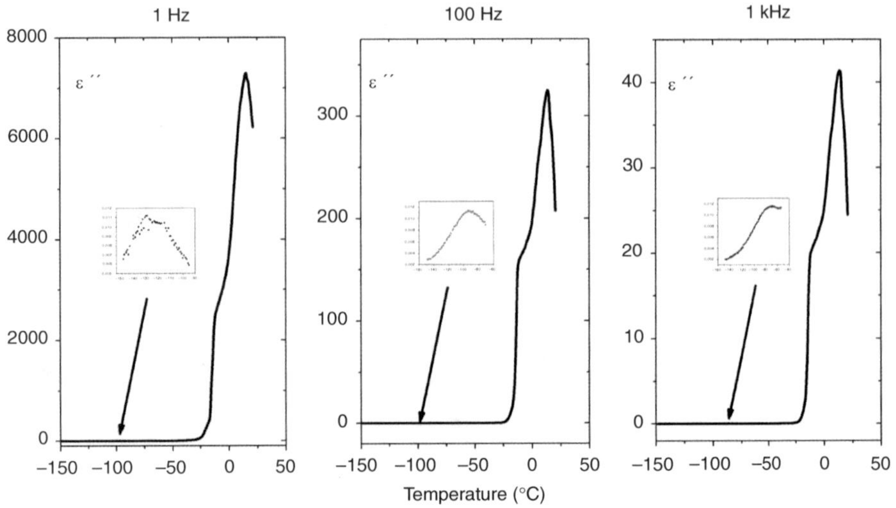

Fig. 8.9 Imaginary part of the complex permittivity, as a function of temperature, for three different frequencies, for poly(lactid acid), PLA

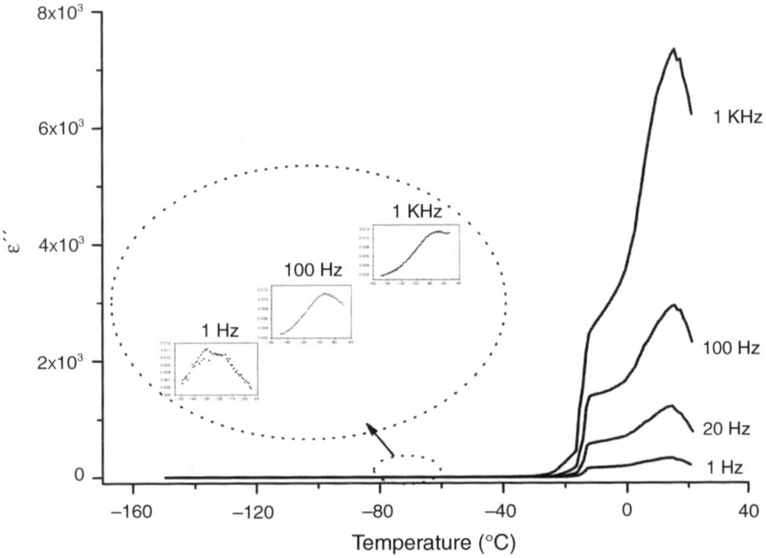

Fig. 8.10 Imaginary part of the complex permittivity, for industrial poly(lactid acid), PLA, showing two relaxation processes

The peak at high temperatures is due to the α relaxation process. This primary relaxation process is associated with the glass transition. It consists of cooperative segmental motion and exhibits a non-Arrhenius temperature dependence.

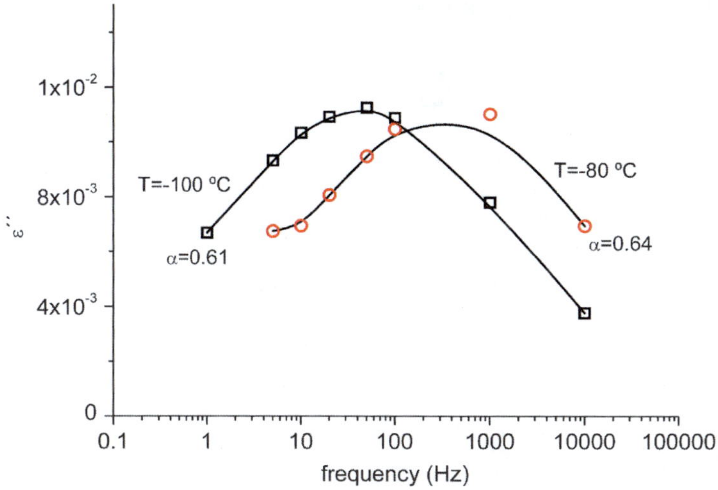

Fig. 8.11 Imaginary part of the complex permittivity, for industrial poly(lactid acid), at two different temperatures. The calculated α, is far from 0, showing that the process is non-Debye

The extent of cooperativity can usually be described by a length scale, but this is a topic of controversial discussion.

At lower temperatures the β relaxation process is observed, which arises from localized rotational fluctuations. That is, fluctuations of localized parts of the main chain or of side groups or parts of them. This relaxation is, in polymers, a non-Debye process. Figure 8.11 shows this behaviour for the industrial PLA, at two different temperatures. The calculated values for the α parameters of (8.6) are about 0.6, that is far from the single relaxation time that corresponds to $\alpha = 0$.

The temperature dependence of this β relaxation is found to be Arrhenius like, that is,

$$f_{\max} = F \exp\left(-\frac{E_{\mathrm{a}}}{kT}\right) \tag{8.13}$$

where $f_{\max}$ is the frequency of the peak in ε'', that is the relaxation frequency, F is the pre-exponential factor, E_{a} the activation energy and k the Boltzmann constant. Then,

$$\ln f_{\max} = \ln F - \frac{E_{\mathrm{a}}}{kT} \tag{8.14}$$

Figure 8.12 shows the imaginary part of the complex permittivity, as a function of temperature, at different frequencies for the industrial PLA. The peak, $f_{\max}$, moves to higher temperatures with the frequency increasing.

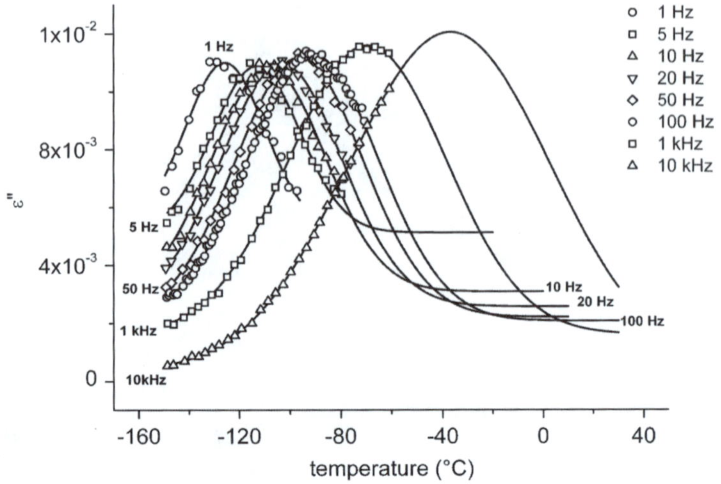

Fig. 8.12 Imaginary part of the complex permittivity, as a function of temperature, at different frequencies, for the industrial PLA [36]

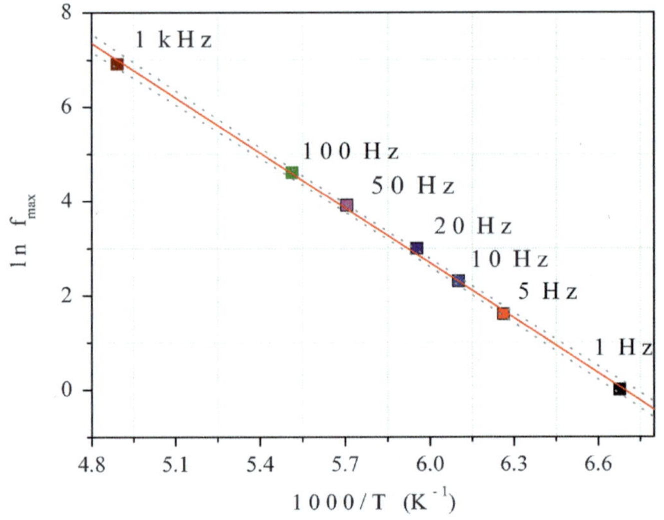

Fig. 8.13 $\ln f_{\mathrm{max}}$ vs. 1,000/T, at the beginning of the ageing process [36]

Figure 8.13 presents the graphic of $\ln f_{\mathrm{max}}$ as a function of 1,000/T, according to (8.14), in order to calculate the activation energy, at the moment $t = 0$, that is, in the beginning of the ageing process.

This methodology was used to obtain the activation energy, for the industrial and purified PLA, during the ageing time. In Fig. 8.14 these data are presented.

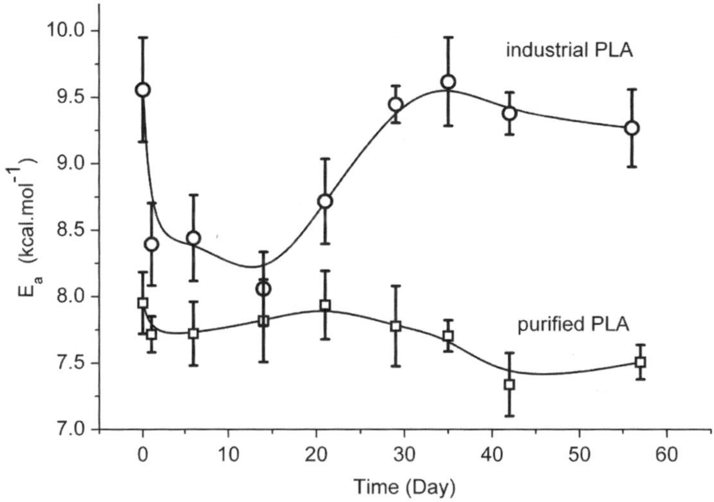

Fig. 8.14 Activation energy as a function of ageing time, for the β relaxation process [36]

For the industrial PLA, it is possible to distinguish three zones. A first one, where the activation energy decreases, very abruptly in the first day, and then more slowly until about 14 days. This behaviour can be associated to the plastification of the polymer by water. A second region presents an increasing of the activation energy, until 35 days, due to the installation of water in the polar sites, corresponding to the anti-plastification, or due to the restructuration of the polymer. The presence of oligomers and water in the material permits that the macromolecular chains can be rearranged. Finally, a small decreasing of the activation energy, which corresponds to the progressive hydration of the polar functions lately formed by hydrolysis.

For the purified PLA, two regions can be distinguished. A first one, with decrease of the activation energy, associated to the plastification of the polymer by water. A second one, with a small oscillating decrease, due to the hydrolysis of the macromolecular chains and to the material restructuration.

The higher activation energy for the industrial PLA can be explained by it higher hydrophobicity.

8.3.2 Correlating Relaxation Parameters with Physical Properties

In order to correlate the dielectric properties results, described in the previous subsection, with other properties, the permeability of the PLA films was measured. Each film seals a cell, maintained at temperature $T = 25$ °C and water vapour constant pressure, $p = 18.6$ mmHg. During the ageing of the material, it is possible to measure the quantity of water vapour that crosses the film. The permeability coefficient K of the membrane can be calculated, as

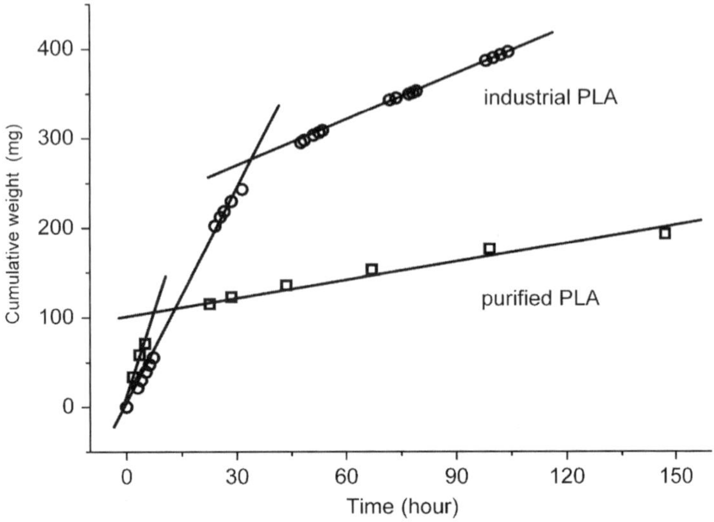

Fig. 8.15 Cumulative weight of water vapour that crosses the films, in the first moments of the ageing process

$$\frac{\partial m}{\partial t} = KA\frac{\Delta p}{d} \tag{8.15}$$

where m is the mass of vapour crossing the film, A the area surface, Δp the gradient of pressure and d the thickness. Figure 8.15 shows the cumulative mass, as a function of the ageing time, for PLA films.

The industrial PLA presents, regarding the break between the two slopes, a higher quantity of water diffusing in a longer amount of time, when compared with the purified PLA. This observation confirms its higher absorbent character, with lower permeability coefficient. Indeed, the lowest permeability is related to the interactions between the polar functions in great number and the water molecules. This confirms the dielectrics results, where the activation energy for β relaxation decreases in the first day.

For longer times, the water vapour crossing the films is placed in the polar sites, provoking gradually the hydrolysis and creating new polar sites by hydrolytic cut, which will retain much more water. This phenomenon is more marked in the industrial PLA, causing a lowering of the water vapour circulation. On the other hand, the purified PLA undergoing less this phenomenon permits that water vapour pass easily. This can be observed in Fig. 8.16, and is directly correlated with the changes of the activation energy for β relaxation, during the ageing process.

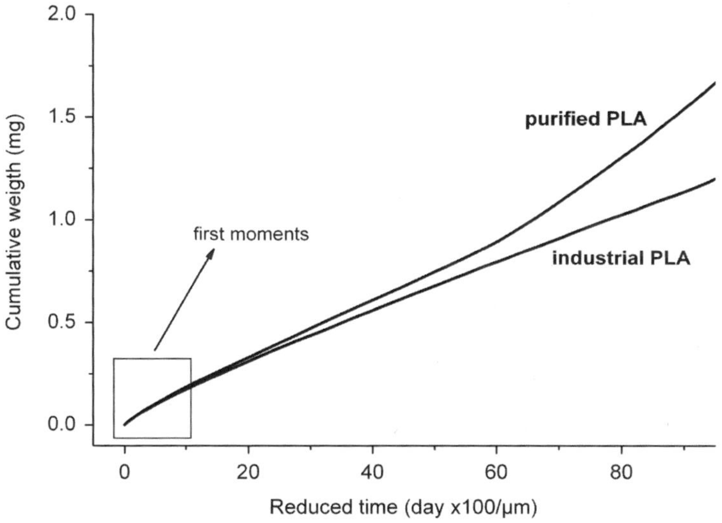

Fig. 8.16 Cumulative weight of water vapour that crosses the films

8.4 Conclusion

Dielectric spectroscopy is a powerful method of measurement, which made a new contribution to the understanding the dynamics of degradation of PLA. Since the capacitance of a dielectric sample is inversely proportional to its thickness, this method is highly suitable for studies of thin films, in contrast with other techniques.

Dielectric relaxation results, in particular the β relaxation observed at low temperatures, were used to follow the ageing of the polymers.

The plasticization of the material by water in the first days of contact with the controlled atmosphere at constant temperature and humidity is indirectly observed using the variations of the activation energy of the β relaxation, which is more evident in the industrial PLA film. After this first period, a restructuring of the material, favoured by the oligomers present in the film and by the progressive penetration of the water in the ester sites, is observed by the increase of the activation energy.

Relaxation studies were complemented by water vapour permeability that confirms the explanations.

References

1. Jonscher AK (1983) Dielectric relaxation in solids. Chelsea Dielectric Press, London
2. Bottcher CJF, Bordewijk P (1978) Theory of electric polarization. Elsevier, Amsterdam
3. Hevig P (1977) Dielectric spectroscopy of polymers. Adam Hilger, Bristol
4. Debye P (1929) Polar molecules. Chemical Catalog, New York

5. Macdonald JR, Johnson WB (2005) Fundamentals of impedance spectroscopy. In: Barsoukov E, Macdonald JR (eds) Impedance spectroscopy, theory, experiment and applications. Wiley, New Jersey
6. Williams G, Watts D (1970) Non-symmetrical dielectric relaxation behavior arising from a simple empirical decay function. Trans Faraday Soc 66:80–85
7. Havriliak S, Havriliak SJ (1997) Dielectric and mechanical relaxation in materials. Hanser, New York
8. Cole KS, Cole RH (1941) Dispersion and absorption in dielectrics. I. Alternating current characteristics. J Chem Phys 9:341–352
9. Davidson DW, Cole RH (1950) Dielectric relaxation in glycerol. J Chem Phys 18:1417–1419
10. Havriliak S, Negami S (1967) A complex plane representation of dielectric and mechanical relaxation processes in some polymers. Polymer 8:161–210
11. Kremer F, Arndt M (1997) Broadband dielectric measurements techniques. In: Runt JP, Fitzgerald JJ (eds) Dielectric spectroscopy of polymeric materials. American Chemical Society, Washington
12. McCrun NG, Read BE, Willians G (1991) Anelastic and dielectric effects in polymeric solids. Wiley, New York
13. Westphal WB, von Hippel AR (1954) Dielectric materials and applications. MIT Press and Wiley, New York
14. Amstrong D, Race WP, Thirsk HR (1968) Determination of electrode impedance over an extended frequency range by alternating current bridge methods. Electrochim Acta 13:215–239
15. Macdonald JR (1992) Impedance spectroscopy. Ann Biomed Eng 20:289–305
16. Costa LC (1995) Propriedades eléctricas de vidros com alguns iões de terras raras. Ph.D. Thesis, Aveiro
17. Calvert R (1948) A new technique in bridge measurements. Electron Eng 20:28–29
18. Henry F (1961) Développement de la métrologie hyperfréquences et application à l′étude de l′hydratation et la diffusion de l′eau dans les matériaux macromoléculaires. Ph.D. Thesis, Paris
19. Karasz FE (1972) Dielectric properties of polymers. Plenum, New York
20. Costa LC (2011) Double relaxation processes in the glass system $x Eu_2O_3.PbO.2B_2O_3$ studied by Broadband Dielectric Spectroscopy. J Non Cryst Sol 357:2178–2181
21. Ku CC, Liepins R (1987) Electrical properties of polymers. Hanser, Munich
22. Adachi K (1997) Dielectric relaxation in polymer solutions. In: Runt JP, Fitzgerald JJ (eds) Dielectric spectroscopy of polymeric materials. American Chemical Society, Washington
23. Riande E, Saiz E (1992) Dipole moments and birefringence of polymers. Prentice Hall, Englewood Cliffs, NJ
24. Fuoss RM, Kirkwood JG (1941) Electrical properties of solids.VIII. Dipole moments in polyvinyl chloride-diphenyl systems. J Am Chem Soc 63:385–394
25. Blythe AR (1979) Electrical properties of polymers. Cambridge University Press, Cambridge
26. Macdonald D (2006) Reflections on the history of electrochemical impedance spectroscopy. Electrochim Acta 51:1376–1388
27. Collin RE (1966) Foundations for microwave engineering. McGraw Hill, New York
28. McKubre MCH, Macdonald DD (2005) Impedance measurement techniques. In: Runt JP, Fitzgerald JJ (eds) Dielectric spectroscopy of polymeric materials. American Chemical Society, Washington
29. Belatar J, Graça MPF, Costa LC, Achour ME, Brosseau C (2010) Electric modulus-based analysis of the dielectric relaxation in carbon black loaded polymer composites. J Appl Phys 107:124111
30. Stanford Research Systems (1995) SR850 Lockin Amplifier Manual
31. El Hasnaoui M, Graça MPFG, Achour ME, Costa LC, Outzourhit A, Oueriagli A, El Harfi A (2010) Effect of temperature on the electrical properties of copolymer/carbon black mixtures. J Non Cryst Sol 356:1536–1541

32. Baker-Jarvis J (1990) Transmission/reflection and short circuit line permittivity measurements. National Institute of Standards and Technology, Colorado
33. Schonhals A, Kremer F (2003) Analysis of dielectric spectra. In: Kremer F, Schonhals A (eds) Broadband dielectric spectroscopy. Springer, Berlin
34. Tsai YT, Whitmore DH (1982) Nonlinear least-squares analyses of complex impedance and admittance data for solid electrolytes. Sol Stat Ion 7:129–139
35. Schonhals A (1998) Dielectric spectroscopy on the dynamics of amorphous polymeric systems. Novocontrol application note dielectrics 1. pp 1–16
36. Henry F, Costa LC, Devassine M (2005) The evolution of poly(lactid acid) degradability by dielectric spectroscopy measurements. Eur Polym J 41:2122–2126

Chapter 9
Influence of Cryogenic Treatment on Mechanical Behavior of Glass Fiber-Reinforced Plastic Composite Laminate

C.G. Sreenivasa and Ajith. G. Joshi

Contents

Abstract Composites are the modern trend materials replacing monolithic metals and alloys in many industrial applications. It has also been widely used in many of engineering applications where materials are subjected to very low temperatures. Hence it finds essential to investigate its mechanical properties after cryogenic treatment of composite materials and compare the results with behavior of materials at ambient conditions. This chapter describes the preparation of glass/epoxy

C.G. Sreenivasa (✉)
Department of Industrial and Production Engineering, University B.D.T College of Engineering, Davangere 577004, Karnataka, India
e-mail: sreenivasacg@gmail.com

A.G. Joshi
Department of Mechanical Engineering, Canara Engineering College, Benjana Padavu, Mangalore 574 219, Karnataka, India

S. Kalia and S.-Y. Fu (eds.), *Polymers at Cryogenic Temperatures*,
DOI 10.1007/978-3-642-35335-2_9,

composites for different volume fraction. The developed specimens were subjected to cryogenic treatment with the aid of liquid nitrogen. Mechanical properties such as tensile, impact, and flexural behavior were presented separately of cryogenic treated specimens and untreated specimens. The obtained results exhibited enhanced mechanical properties.

Keywords Polymer matrix composites • Cryogenic treatment • Mechanical behavior

9.1 Introduction

Cryogenics is the branch of science which deals with methods to obtain temperature well below the room temperature and methods employed to measure it. The developments of cryogenic begin with the Michael Faraday experiments of liquefying gases such as CO_2 and NH_3. Till today liquefaction of different types of gases such as H_2, N_2, O_2, ^3He–^4He, and nuclear cooling stages have been attained. The liquid nitrogen, hydrogen, oxygen, and helium are commonly used cryoliquids. The cryogenic technology is utilized in wide range of applications from conservation of food by freezing to advance scientific computing (quantum computing) [33]. In general, cryogenic treatment of materials results in ductile to brittle transition. Cryogenic treatment of materials indicates that change in material behavior at temperatures well below the room temperature. Thus study of behavior of materials at cryogenic temperature is quite significant. As compared to monolithic metals, the study of behavior of composites is complex due to its heterogeneous nature.

Composites are special class of materials with combination of two or more distinctly recognizable constituent materials. The bamboos reinforced in mud which is used in ancient times as building materials is the best example of composite applications in early days. Modern composites are designed to fulfill the balanced properties requirement of engineering and industrial applications [4]. When compared to single phase metallic materials, composites allow the tailoring of properties which meets the requirements of specific applications. Composites are made up of base matrix and stronger reinforcing materials. Composites are specially utilized in applications where weight saving is an important criteria such as aerospace and marine. They are also used in many structural, automotive applications, and so on. From past few decades, polymers composites are gaining wide scope in many engineering applications such as navel applications, aerospace structures, spaceship structures, gears because of high modulus and high toughness, resistance to corrosion, good tribological properties, and so on [6, 8, 9].

Fiber-reinforced plastics (FRP) are in greater demand since from past few decades because of their superior properties such as higher specific strength and specific modulus, anti-corrosion, good range of operating temperatures, thermal insulation. Some of the advantages are weight reduction, tailorable properties, tolerable thermal expansion, redundant load paths, longer life are the primary advantages of PMCs. Some of the disadvantages are—matrix is weak and low toughness, reuse and disposable may be difficult, cost of raw materials and fabrication is high, difficult to analysis. In recent years, reinforced FRPs have become an integral part of many engineering and commercial applications.

Carbon fibers and E-glass fibers are common type of reinforcements found in many of the engineering applications due to its best characteristics available. As with resin development, the pace of fiber development has accelerated since the 1950s with the introduction of E glass, R glass, S glass, and special acid and alkali-resistant glass. Glass fiber has been a major reinforcement for the FRP industry, but the drive for lighter, stronger, and stiffer structures has seen the introduction of carbon and poly-aramid fibers [10]. In case on thermoset polymeric matrices, epoxy are most widely as structural applications and also in other industrial applications due to its high resistance to corrosion and chemicals, good thermal, mechanical properties [5, 11]. The use of reinforcements in the resin may provide increasing load withstanding capability, improved tribological properties, better thermal properties, and increased mechanical strength [6]. Carbon fibers have good mechanical, thermal, and tribological properties than glass fibers but the principal advantages of glass fibers are the low cost with high strength. The challenge for the use of glass fiber-reinforced polymer composites for primary structures is to avoid the damage propagation during its applications [29]. Among FRPs, Glass fiber-reinforced plastic surfaced the industrial market because of their low cost and excellent properties they enhance. They are also being in demand for cryogenic applications such as cryogenic liquid storage tanks, pressure vessels.

FRP are susceptible to mechanical damage, when subjected to tension, compression, and flexural loads resulting in interlayer delamination. Paiva et al. [29] have reported improving the toughness of the matrix materials can enhance the mechanical properties, which may be due to interface between the reinforcements and matrix. Aramide et al. [3] have obtained the results that increase in fiber volume fraction can lead to increase in elastic modulus and extension to break till certain threshold value and further increase in fiber volume implies in decrease in the same properties. However hardness increases with increase in fiber volume but strain decreases. Al-Hasani [2] has reported that increasing the glass fiber content in the glass/epoxy composites increases its tensile strength and also its hardness. Liu et al. [25] have worked on the laminated epoxy composites and assembled composite plates, and have estimated the perforation resistance offered by those under the impact testing at room temperature. They have suggested that assembled composite plates offer higher impact strength than these laminates. Thus mechanical properties of PMCs depend on factors such as fiber volume fraction, interfacial bonding strength between the reinforcement and matrix, types of fiber, orientation of fiber, operating temperature, and so on. The present chapter is focused on the study of mechanical behavior of GFRP when they are subjected to cryogenic conditions.

9.2 Fabrication of GFRP Composite Laminate

9.2.1 *Matrix Material*

Epoxy resins are widely used type of polymer matrices in GFRP laminates for applications ranging from domestic to engineering applications. They possess several advantages over other types of polymers. The principal advantages are listed below [16]:

- Inherently good adhesion property with variety of reinforcements (especially with different types of fibers) and hence avoids the misalignment of fibers.
- It has low shrink rates, hence it helps to obtain better dimensional accuracy and lowers the tendency of obtaining shear stresses between the matrix and reinforcements.
- Good stiffness, high heat distortion temperature, resistance to water, and better environmental degradation resistance properties.
- Low volatility during curing and minimum volatile by-products, which prevents the voids and bubble formation in cured system.
- It suits good for different FRP processing techniques.

Besides epoxy resin systems represent some of the high performance resin systems, which are available today in market. Generally two types of epoxy systems are employed viz. diglycidyl ether of biphenol (DGEB) and the epoxy novolacs (comprised of glycidyl ethers of cresol novolac or biphenol A novolac). Before dealing with epoxy, an engineer should understand the fundamentals of epoxy resin system. Hence the basic structure, curing temperature, curing time, and properties obtained after curing are further discussed in this chapter. In the discussion, attention is paid on the diglycidyl ether of biphenol A (DGEBA) type of epoxy systems with amine curing. They are principal type of matrices, widely used in FRPs [4].

9.2.1.1 Basic Chemical Structure of Epoxy

The basic chemical structure of epoxy consists of an oxygen atom bonded to two atoms of carbon, which are already bonded. Such a structure is called epoxide group. The molecules containing the epoxide group irrespective of other molecule present in the structure is called epoxide. The epoxide group is represented as shown in Fig. 9.1. The structure is called as "alpha-epoxy or 1,2 epoxy." The basic structure is modified by adding different other molecule for obtaining epoxy in rigid form, i.e., for hardening and for tailoring the properties to suit the desired applications.

9.2.1.2 Curing of Epoxy

Curing is a process of polymerization, which is accomplished by adding curing agents or/and catalysts to neat resin, with or without application of heat and

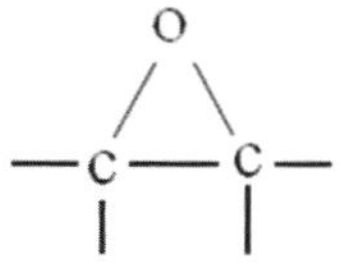

Fig. 9.1 The epoxide group

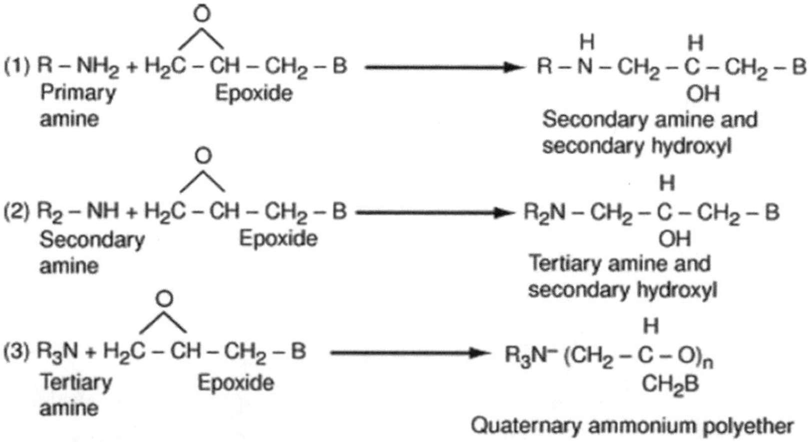

Fig. 9.2 Epoxy/amine reactions

pressure, to obtain the extended network of cross-linked molecules by method of chemical reaction. The curing agents are sometimes also called as "hardeners." The commonly used types of curing agents are aliphatic amines, aromatic amines, and anhydrides. In other type of curing, small amount of catalysts such as Lewis acids or Lewis base are added to achieve polymerization. The type of polymerization occurs in such a case is homopolymerization (self-perpetuating cationic or anionic polymerization. However aliphatic amines are the most often used type of curing agents. In amine curing method, the amine nitrogen has hydrogen, which is utilized to form covalent bond with epoxide group during polymerization. A typical epoxy/amine reaction is shown in Fig. 9.2. As shown there is one hydrogen, each hydrogen bond with each epoxide, when there are two hydrogen available, each hydrogen bonds with different epoxide groups. Thus forms a C–N bond between amine and epoxide group. Hence the environmental resistance of cured epoxy system is good and temperature dependence properties are primarily dependent on the adjacent molecules in the structure [26].

Generally, curing of epoxy resins with amine curing agents can be accomplished with varied temperature range of 25–200 °C. For domestic applications, it is generally cured at ambient temperature, for industrial and elevated temperature applications, elevated temperature is employed for curing. The aliphatic amine systems have low viscosity and require room temperature for curing. Hence it can be used where ambient conditions with low viscosity are desired during curing. The significant

Table 9.1 Typical properties of neat epoxy resin

Property	Values
Ultimate tensile strength	82.74 MPa
Specific gravity	1.1–1.28 g/cc
Young's modulus	3.792 GPa

Table 9.2 Typical properties of HY951 Hardener (1,2-Ethanediamine, N[1]-bis (2-aminoehtyl)-Triethylenetetramine)

Property	Values
Specific gravity	0.97–0.99 g/cc
Flash point	110–120°C
Viscosity at 25 °C	10–20 mPas
Vapor pressure at 25 °C	21.0×10^{-3} mbar
Color	Colorless
Odor	Ammonia
Physical state	Liquid
Solubility in water	Miscible
Decomposition temperature	>200 °C (>392 °F)
pH	13
Percent volatile	Nil

number of aliphatic rings in the chemical structure of aliphatic amine cured system, results in lower T_g value and lower rigidity. The T_g value is significant because the magnitude of the properties above T_g is much lower than properties below T_g. Hence it is considered while designing any commercial applications. The design of commercial applications is nothing but it is expected to be employed at elevated temperatures to low temperatures. However aromatic amine cured systems can be used for elevated temperature applications, but it requires high curing temperature. The typical properties of neat epoxy resin such as ultimate tensile strength, specific gravity, and young's modulus have been presented in Table 9.1. The typical properties of HY951 hardener is also been presented in Table 9.2.

Another important factor considered while curing is gelation time or pot life. Pot life can be defined as the time duration, during which epoxy resin/curing agent formulation remain fluid. Since, once gelation is initiated, abrupt change in the viscosity occurs. The pot lives vary with the type of curing agent used. For aliphatic amine curing, the pot life is some minutes to few hours whereas for aromatic amine curing it is about 24 h. The ASTM (American Society for Testing and Materials) has described the standards as D2393 to measure the viscosity of epoxy resins and related components and D2471 as the method for gel time and peak exothermic temperature of reacting thermosetting resins.

9.2.2 Reinforcement Material

Glass fibers are widely used type of reinforcements from domestic to engineering and industrial applications. The raw materials used for their production are naturally available and virtually renewable materials [23]. They exhibit desirable fiber properties such as stiffness, strength, stability, chemical resistance, inertness, and so on. Glass fibers are most common reinforcing fibers; they occupy the largest volume fraction in a composite laminate and share the major portion of the load acting on a composite structure. Glass fibers are used as reinforcing material due to their principle advantages like low cost and high strength.

Glass fibers are of various types, viz. E-glass fiber, S-glass fibers, C-glass fibers. E-glass fiber was designed for electrical applications. It is used in circuit boards of computers to provide stiffness and electrical resistance. Because of electrical resistance it is suited for applications where radio-signal transparency is desired as in aircraft radomes and antenna. However, it is used for many other purposes now, such as decorations and structural applications. E-glass has good strength properties at low cost. S-glass fibers retain its strength at high temperature compared to E-glass due to its higher content of silica and has higher fatigue strength. It is mainly used in aerospace applications. C-glass fibers are used in chemical environments, such as storage tanks. Composites made from this material exhibit very good electrical and thermal insulation properties. Glass fibers are also transparent to radio frequency radiation and are used in radar antenna applications [4].

9.2.3 Fabrication of GFRP

The different methods of fabrication of composite material are as follows:

- Hand layup method
- Filament winding
- Pultrusion
- Resin transfer molding
- Tape-laying and fiber placement systems
- Autoclave-based method

In this chapter deliberations are made on hand layup method. This is due to the reason that it is the widely used laboratory scale process of fabricating GFRP. Hand layup is the simplest and oldest molding of the composites fabrication process. It is a low volume, labor intensive method suited especially for large components, such as boat hulls. Glass or other reinforcing mat or woven fabric or roving is positioned manually in the open mold, and the resin is poured, brushed, or sprayed over and into the glass piles as shown in Fig. 9.3. Entrapped air is removed manually with squeegees or rollers to complete the laminate structure. Room temperature curing polyester and epoxies are the most commonly used matrix resins. Curing is initiated by a catalyst in the resin system, which hardens the fiber-reinforced composites without external heat for a high quality part surface; pigmented gel coat is first applied to the mold surface [12]. Some of the real time products of FRP are shown in Figs. 9.4, 9.5, and 9.6.

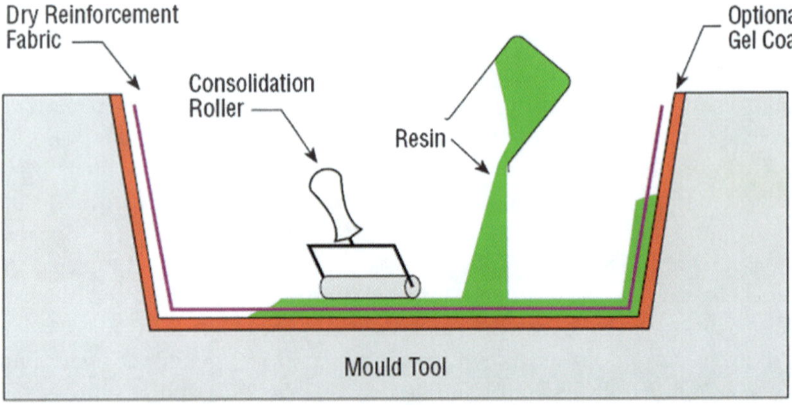

Fig. 9.3 Schematic representation of hand layup technique

Fig. 9.4 Formula student car body developed by AFC in GFRP by hand layup technique

9.3 Cryogenic Treatment

9.3.1 Cryogenic Liquids

9.3.1.1 Liquid Oxygen

Liquid oxygen is a pale blue color liquid with specific gravity of 1.141 g/cc at its boiling point. It freezing point is −222.65 °C, boiling point is −183 °C at 1 atm, and critical temperature is −118.6 °C at 1 atm. (www.scienceclarified.com/Co-Di/Cryogenics.html). In liquid oxygen, small amount of O_4 polymer forms. It requires strict safety measures to handle it. It detonates when it comes in contact of hydrocarbons. Contact of oils, fats, asphalt with liquid oxygen is prohibited.

Fig. 9.5 FRP storage tanks (tradekey.com)

Fig. 9.6 FRP structural towers (Strombergarchitectural.com)

It requires good structural and working materials, when design of materials and development liquid oxygen equipments are considered [18]. However it finds wide engineering and medical applications. Liquid oxygen is used as rocket fuel component in rocket technology, in military aircrafts it is stored on board and lightweight source as breathing gas for pilot [27]. Liquid oxygen was extensively used in making oxyliquit explosives, but nowadays it is rarely used due to high rate of accidents and so on.

9.3.1.2 Liquid Hydrogen

At equilibrium, liquid hydrogen contains 99.8% para hydrogen and orthohydrogen. The density of liquid hydrogen is 70.99 g/L (at 20 K) and has low volumetric energy density but with high specific energy. The use of liquid hydrogen also needs strict safety measures, because of the possibility of detonating mixture (solid oxygen and liquid hydrogen enriched air) formation. Hence attention is paid towards the purity of hydrogen and its composition balance. Liquid hydrogen finds many engineering applications such as rocket fuel element, used in bubble chambers for experiments on elementary particles, used in vehicle-borne electrochemical generators; nowadays it is considered as a promising fuel for transport facilities and has large potential in near future and so on [18].

9.3.1.3 Liquid Helium

Liquid helium is odorless, colorless, inert, noncorrosive and nonflammable liquid, being most interesting and widely used type of cryoliquid. There exists two stable isotopes of helium viz. ^{3}He (boson) and ^{4}He (fermion). The boiling point of ^{3}He is 3.2 K and of ^{4}He is 4.2 K and density at 0 K is 0.0823 and 0.146 g/cc, respectively. Initially it was restricted to the small laboratory scale due to its low availability in nature [24]. The Collins liquefier for the first time allowed for the production separation of liquid helium in large scale from natural gas [13]. The concentration of ^{3}He in natural gases is ~10^{-7} and of ^{4}He is ~10^{-6} of the helium in the atmosphere. The separation is very expensive. Hence ^{3}He is produced as by-product of tritium production in nuclear reactors by beta decay (the half life is 12.26 years). The diffusion process is employed for the separation of ^{3}He. The applications of helium include wide range of engineering and pure science applications. It is used for cooling superconducting devices, for catching infrared radiation of the stars, whose temperature is about 3 K, superconducting magnet applications such as in accelerators (particle physics) and other research facilities, magnetic resonance imaging (MRI) systems in hospitals, and nuclear fusion facilities, and so on.

9.3.1.4 Liquid Nitrogen

Liquid nitrogen has become a standard refrigerant. Its boiling point is 77.4 K at 1 atm [33]. The major advantage is that, it can be obtained in large scale at liquefication and separation plants. In addition, it can be stored and transported at low costs relatively and no toxicity. The leakage of nitrogen as gas also doesn't cause the global warming, since the atmosphere itself contains 70% nitrogen. Hence it has gained wide variety of engineering and medical applications. It is widely used in food preservation, cryopreservation of medical materials such as blood, semen, stem cells and so on, cryosurgery, metallurgical applications, cryogenic machining [31]. To a point (about the temperature of liquid nitrogen), the condensation of material is significantly used in industrial applications. An example is the use of liquid nitrogen in the assembly of some automobile engines. In order to get extremely tight fits when installing valve seats, for instance, the seats are cooled to liquid nitrogen temperatures, whereupon they contract and are easily inserted in the engine head. When they warm up, perfect fit results (www.scienceclarified.com/Co-Di/Cryogenics.html). The major limitation is displacement of oxygen in an enclosed volume. This chapter is focused on the application of liquid nitrogen and its treatment on GFRP composite materials.

9.3.2 Cryogenic Treatment

Cryogenic treatment is a process of treating the work pieces to cryogenic temperatures to modify physical properties such as to enhance mechanical properties by removing residual stresses and enhance tribological properties and so on. Generally work pieces are treated with liquid nitrogen at temperature of 77 K in a chamber. The chamber should be leak proof, since liquid nitrogen boils at 77 K. Hence before the start of process, the chamber should be checked for any leakage. The temperature of the material is slowly decreased to target temperature (77 K) and maintained at that temperature for few hours to day. The time duration again depends on the material and its cross-section (or thickness) to be treated. After maintaining at target temperature for desired duration, temperature is increased slowly to reach room temperature for completing the process (www.cryogenicsociety.org/resources/cryo_central/cryogenic_treatment_of_materials/). This method of cryogenic treatment is called deep cryogenic treatment. Another method of cryogenic treatment is shallow cryogenic treatment (SCT) or subzero cryogenic treatment. In SCT, the materials are placed in a freezer at 193 K and then they are exposed to room temperature.

9.4 Mechanical Properties of GFRP Composite Laminate: Before and After Cryogenic Treatment

In specific applications like marine and structural, composites are exposed to environmental conditions such as freezing, moisture, temperature, solar radiation, and chemical agents, which may cause its degradation and also reduce its service life [1]. In aerospace applications also, the conventional metallic fuel tanks are

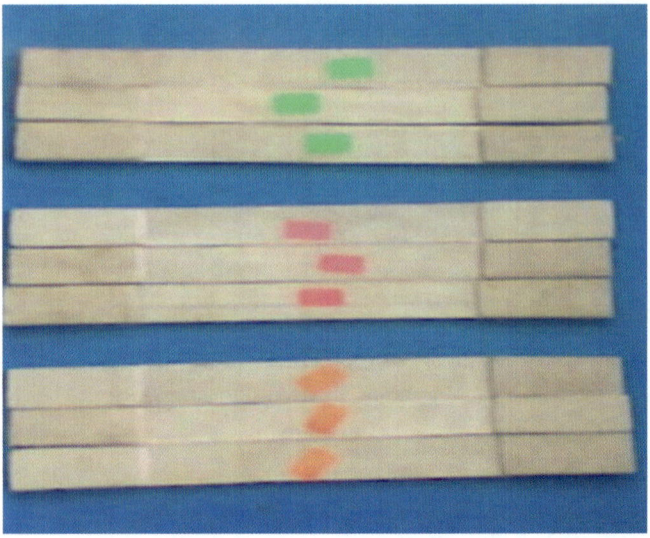

Fig. 9.7 Tensile test specimen made by hand layup technique

replaced with the advanced composite materials. In spacecrafts, the fuels are at cryogenic temperatures like liquid hydrogen. When materials are exposed to cryogenic temperature, the materials behavior becomes brittle [30]. Hence it finds necessary to understand the mechanical behavior of composites at cryogenic temperature. Further a case study on effect of cryogenic treatment on mechanical properties of composites is discussed.

Epoxy resin (araldite LY-556) was used as the matrix material. Hardener used for the epoxy resin is HY 951 as curing agent under ambient conditions. Glass fibers were used as reinforcement, since it has good combination of low cost and high mechanical properties. Laminates of composites are fabricated by hand layup technique as reported elsewhere [7, 38]. After curing under ambient conditions for 24 h, it was cut according to ASTM standards. The specimens were prepared separately for the cryogenic treatment and exposed to liquid nitrogen environment (77 K) for 1 h. To achieve uniform temperature throughout the thickness, it was maintained for 1 h. From the previous studies [1, 14, 17, 19–22, 28, 30], for the specimen thickness considered 1 h is sufficient to achieve the uniform temperature throughout its thickness.

Tensile, impact, and flexural tests were carried out to investigate the mechanical properties of the composites. Tensile test was done on the prepared specimen of untreated and cryogenic treated according to ASTM D3039 standard. Impact Test (Charpy) was done according to ASTM E 23. Flexural test was done in accordance with ASTM D2344. The images of prepared sample specimen are shown in Figs. 9.7, 9.8, and 9.9.

Fig. 9.8 Charpy (impact) test specimen made by hand layup technique

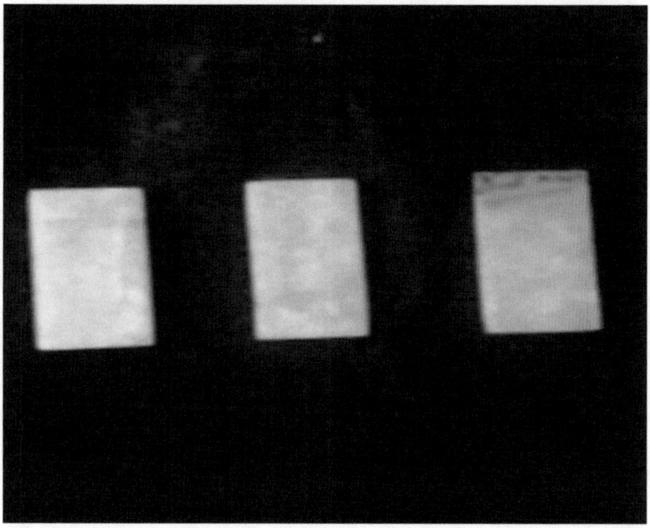

Fig. 9.9 Bending test specimen made by hand layup technique

9.4.1 Tensile Strength

Stress–strain graph for untreated GFRP composites is as shown in Fig. 9.10a [32]. GFRP laminates having a volume fraction of 70:30 have more tensile strength compared to other two volume fractions (65:35, 75:25). The tensile property mainly depends on fiber volume percentage under similar processing and resin conditions.

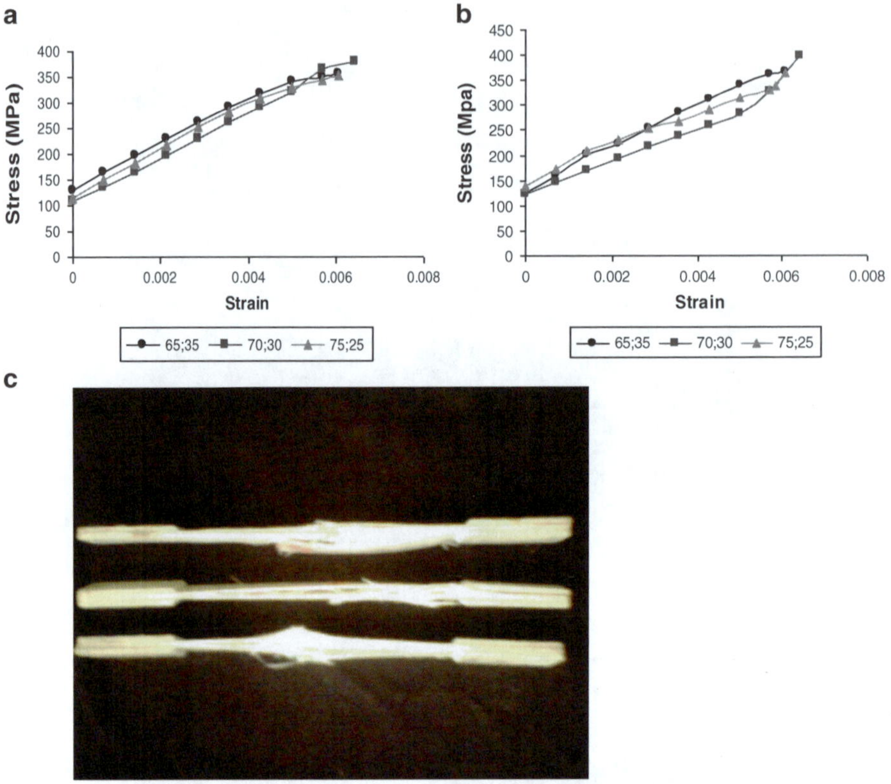

Fig. 9.10 Stress–strain curves of composites for the different volume fraction glass/epoxy and (**a**) before cryogenic treatment and (**b**) after cryogenic treatment [32]. (**c**) Tensile specimen after testing

Figure 9.10b shows the stress–strain graph for the cryogenic treated GFRP composites [32]. The maximum tensile strength was observed for the materials with the volume fraction of 70:30, which is due to the balanced volume of fiber and matrix. Greater order delamination was observed in the tested specimens. The bonding strength between the matrix and fiber was good at the ratio of 70:30 and decreases with the further increase in the volume fiber. Hence the tensile strength decreases. Figure 9.10c shows the specimens after tensile testing.

The influence of cryogenic treatment on tensile strength and corresponding Young's modulus of tested specimens is tabulated in Table 9.3. It is observed that the tensile strength increases for cryogenic treated GFRP specimens when compared with untreated specimens. The enhanced tensile strength of 3% was exhibited by cryogenic treated specimens with volume fraction 65:35 and 75:25 compared to untreated specimens. Whereas for the specimens with volume fraction of 70:30, the increase in tensile strength observed was 5%.

Table 9.3 Tensile properties of GFRP composites (untreated vs. cryogenic treated) [32]

	Ultimate tensile strength (MPa)		Young's modulus (GPa)	
Volume fraction (%) fiber: matrix	Before cryogenic treatment	After cryogenic treatment (77 K)	Before cryogenic treatment	After cryogenic treatment (77 K)
	Before cryogenic treatment	After cryogenic treatment (77 K)	Before cryogenic treatment	After cryogenic treatment (77 K)
65:35	356.5	366.4	58.71	60.35
70:30	378.04	397.6	58.80	61.84
75:25	352.6	362.6	58.07	59.72

The tensile property mainly depends on strength of the fibers and fiber/matrix interface. The cryogenic treatment of laminated fibers reinforced composites increases mainly due to matrix hardening such that disentanglement is almost absent [21]. The cryogenic treatment causes polymer structure to get more densely packed and hence increases the resistance to debonding in GFRP composite materials. The increase in strength at low temperature was also explained on the basis of Weibull scale parameter, which is characterized by the increase in fiber strength and shear strength of the interface at 77 K [15].

In stress–strain curve (Fig. 9.10a, b) it can be observed that initial strain value is zero. The reason for the fact can be explained as follows. The composites are prepared by hand layup technique. In the process, residual stress remain in the composite after curing and in practical conditions residual stress induces during clamping. Hence till stress reaches the total residual stress value which is remained in the composite, displacement obtained is 0. When stress induced due to applied load becomes greater than total residual stress, the displacement is obtained.

9.4.2 Impact Strength

Figure 9.11a indicates increase in impact strength of cryogenic treated GFRP composites. Figure 9.11b shows the specimens after impact testing [32]. The higher impact strength was observed for the specimens with the volume fraction 75:25, when the notch is formed across the laminates. Impact strength increases as the fiber composition increases, due to the decrease of brittle matrix as the fiber composition increases.

The impact strength of the cryogenic treated specimens with volume fraction 70:30 and 75:25 is increased by 0.5 times as compared with the results of untreated specimens, whereas for the specimens with volume fraction 65:35 impact strength was increased by 1.5 times. In the cryogenic conditioning, thermal and residual stresses build up in the matrix material due to contraction and generate higher

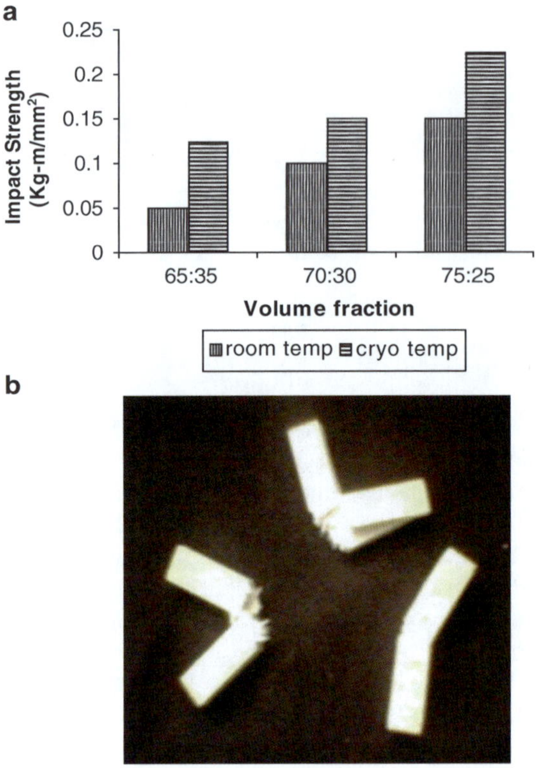

Fig. 9.11 (**a**) Comparison of impact strength of the GFRP composites tested before and after cryogenic treatment. (**b**) Impact test specimen after testing

fracture toughness in the material [22]. Residual stress releases when the load is applied. Therefore, it needs sufficient energy to release and to fracture. When the stress intensity factor exceeds the limit of fracture toughness, material fails. The cryogenic treatment of the composite materials results in the relaxation of the internal stress causing the structural balance throughout the mass of the material. This increases the brittleness, toughness, and durability hence the capacity to absorb energy is increased.

9.4.3 Flexural Strength

Figure 9.12a indicates that there is an increase in flexural strength with the increase in volume fraction for both untreated and cryogenic treated specimens. Figure 9.12b shows the specimens after flexural testing [32]. The volume fraction 75:25 shows the maximum increase in bending strength when compared with other two volume fractions because of the increase of fiber content and hence increase in ductile property of the composites.

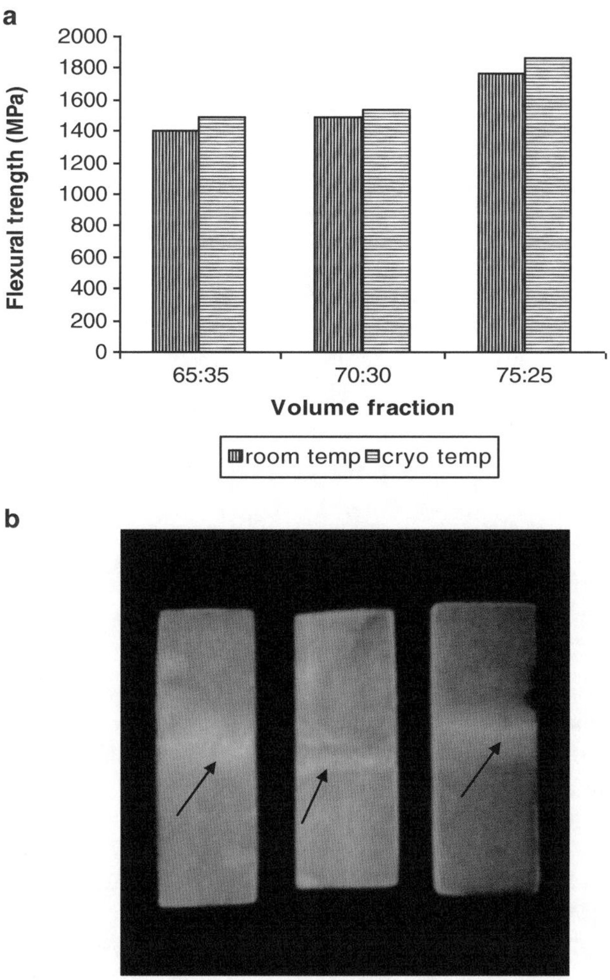

Fig. 9.12 (**a**) Comparison of the flexural strength of GFRP materials tested before cryogenic treatment and after cryogenic treatment. (**b**) Flexural test specimen after testing

After undergoing cryogenic conditioning, the bending strength in GFRP composites increases when compared with the untreated specimens. The comparison between the test results obtained form the cryogenic conditioned specimens with the untreated specimens, shows that there is 6, 3, and 5% increase in bending strength for the laminates with the volume fraction 65:35, 70:30, and 75:25 respectively. The breaking load is more for the cryogenically treated composite materials than the untreated materials, which may be due to the cryogenic hardening of the matrix phase at low temperature. The polymer chains get frozen due to which the deformation process is reduced results in less polymer relaxation that is, it gets hardened.

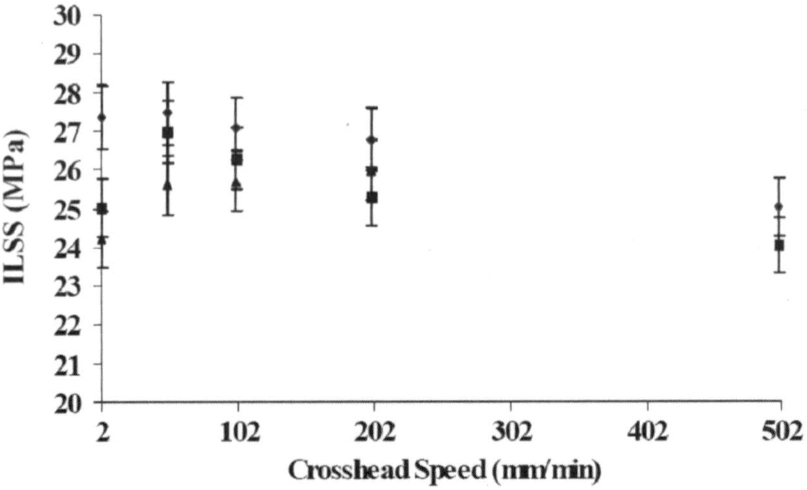

Fig. 9.13 Variation of ILSS of glass-epoxy composites with crosshead speed at cryogenic temperature (*filled diamond*), ambient temperature after cryogenic conditioning (*filled square*) and ambient temperature (*filled triangle*) [20]

9.4.4 Interlaminar Shear Strength

Figure 9.13 illustrates that variation of ILSS of glass-epoxy composites with crosshead speed at cryogenic temperature (filled diamond), ambient temperature after cryogenic treatment (filled square), and ambient temperature (filled triangle). It can be observed for specimen tested at ambient temperature after cryogenic treatment and without cryogenic treatment that, at very low crosshead speed ILSS was found to be lower, increase in crosshead speed above 50 mm/min, the ILSS was also increased. However, furthermore ILSS was decreased with the increase in crosshead speed. The variation of ILSS for specimen at cryogenic conditions was less at lower level of crosshead speed to 102 mm/min. Further ILSS decreased with increase in crosshead. It may be due to the fact that, at low crosshead speed more time is available for the failure leading to more deterioration. Further increase at 102 mm/min very less time is available for failure of material and matrix may not be able to transfer load to the fibers resulting in matrix cracking. Further increase in crosshead speed might have caused in the increased rate of propagation of stress induced cracks, hence ILSS was decreased. In addition, it can also be noticed that, ILSS was found to be higher for specimens at cryogenic condition and cryogenic treated specimen compared to untreated specimen. It may be due to the fact that, matrix gets hardened at cryogenic condition and cryogenic condition causes differential contraction due to different thermal coefficients resulting in increased debonding resistance. It is also evident from the Fig. 9.14 that, due to differential contraction epoxy is trying to squeeze out. During cryogenic treatment, the differential contraction caused due to different thermal coefficients of matrix and fiber

Fig. 9.14 The scanning micrograph shows fiber/matrix debonding due to differential contraction of cryogenically treated specimen [20]

results in the residual stress in the composites. The bonding is mainly influenced by residual stresses remained in the material. Further the generated residual stresses are relieved by viscoelastic flow of matrix. Thus matrix phase was found to be squeezed out. The deformation of matrix may be resulted in the plastic deformation zone near to crack tip and weaker interfacial bonding has caused low ILSS at high loading speed.

9.5 Conclusion

In this chapter, the details of the influence of cryogenic material on mechanical behavior of GFRP composited laminates have been explained. The procedure of fabricating the GFRP specimens and its cryogenic treatment has been presented. The mechanical behavior of developed specimens has been explored for evaluating tensile, impact, flexural, and interlaminar shear strengths. The results of these studies have been compared with before and after cryogenic treatment. From the results, it can be inferred that the fiber volume fraction of 70:30 gives improved ultimate strength with increase in 5% for cryogenic treated specimen when compared at room temperature, than the other volume fraction due to the balanced amount of fiber and matrix composition. The fiber volume fraction of 75:25 gives the improved impact and flexural strength for cryogenic treated specimen when compared at room temperature, than the other volume fraction due to the decrease in the volume of brittle matrix material. The increase in the strengths of the GFRP composite materials exposed to cryogenic temperature is due to the matrix hardening in the materials when it is treated in cryogenic temperature (77 K).

References

1. Aiello MA, Ombres L (2000) Environmental effects on the mechanical properties of glass-FRP and aramid-FRP rebars. Mech Comp Mater 36:395–398
2. Al-Hasani ES (2007) Study of tensile strength and hardness property for epoxy reinforced with glass fiber layers. Eng Technol 25:988–997
3. Aramide FO, Oladele IO, Folorunso DO (2009) Evaluation of the effect of fiber volume fraction on the mechanical properties of a polymer matrix composite. Leonardo El J Pract Technol 14:134–141
4. Daniel BM, Donaldson SL (eds) (2001) ASM handbook, vol 21, Composites. ASM International, Materials Park
5. Bal S (2010) Experimental study of mechanical and electrical properties of carbon nanofiber/epoxy composites. Mater Design 5:2406–2413
6. Barberoi JE, Vivo L (2001) A constitutive model for elastic damage in fiber-reinforced PMC laminate. Int J Damage Mech 10:73–93
7. Basavarajappa S, Joshi AG, Arun KV, Praveen Kumar A, Prasanna Kumar M (2010) Three-body abrasive wear behaviour of polymer matrix composites filled with SiC particles. Polym Plast Technol Eng 49:8–12
8. Bijwe J, Awtade S, Ghosh A (2006) Influence of orientation and volume fraction of Aramid fabric on abrasive wear performance of polyethersulfone composites. Wear 260:401–411
9. Chand N, Naik A, Neogi S (2000) Three-body abrasive wear of short glass fibre polyester composite. Wear 242:38–46
10. Chauhan S, Kumar A, Patnaik A, Satapathy A, Singh I (2009) Mechanical and wear characterization of GF reinforced vinyl ester resin composites with different co-monomers. J Reinf Plast Compos 28:2675–2684
11. Chatys R (2009) Mechanical properties of polymer composites produced by resin injection molding for applications under increased demands for quality and repeatability. Ultragarsas 64:35–39
12. Chawla KK (1998) Composite materials science and engineering. Springer, Berlin
13. Collins SC (1966) Helium refrigerator and liquefier, advances in cryogenic engineering, vol 11. Plenum, New York, pp 11–15
14. Das S, Sahu SK, Ray BC (2007) Effects of loading speed on the failure behaviour of FRP composites. Aircr Eng Aerosp Technol 79:45–62
15. Gong M, Wang XF, Zhao JH (2007) Experimental study on mechanical behaviour of laminates at low temperature. Cryogenics 47:1–7
16. Kaw AK (2006) Mechanics of composite materials. CRC, Boca Raton
17. Kellogg KG, Patil R, Kallmeyer AR, Dutta PK (2005) Effect of load rate on notch toughness of glass FRP subjected to moisture and low temperature. IJOPE 15:54–61
18. Kostionk VV, Rao DB (2003) A textbook of cryogenics. Discovery Publishing House, New Delhi
19. Kumar MS, Ray BC (2007) Mechanical behaviour of FRP composites at low temperature. In: Second international conference on recent advances in composite materials, Feb 20–23, India Habitat Centre, New Delhi, pp. 268–274.
20. Kumar MS, Chawla N, Priyadarsini A, Mishra I, Ray BC (2007) Assessment of microstructural integrity of glass/epoxy composites at liquid nitrogen temperature. J Reinf Plast Compos 26:1083–1089
21. Kumar MS, Sharma N, Ray BC (2009) Microstructural and mechanical aspects of carbon/epoxy composites at liquid nitrogen temperature. J Reinf Plast Compos 28:2013–2023
22. Kumar MS, Sharma N, Ray BC (2008) Mechanical behaviour of glass/epoxy composites at liquid nitrogen temperature. J Reinf Plast Compos 27:937–944
23. Loewenstein KL (1993) The manufacturing technology of continuous glass fibers. Elsevier, Amsterdam

24. Longsworth R (2006) 100th Anniversary of the discovery of helium in natural gas, advances in cryogenic engineering, vol 51A. American Institute of Physics, New York, pp 863–870
25. Liu D, Raju BB, Dang X (2000) Impact perforation resistance of laminated and assembled composite plates. Proceedings of the thirteenth international conference on composite materials, Beijing, pp. 25–29
26. Peters ST (ed) (1998) Handbook of composites. Chapman & Hall, London
27. Mukhopadhyay M (2009) Fundamentals of cryogenic engineering. PHI Learning, New Delhi
28. Murthy VB, Shivakar VS, Prasanna Kumar K (2006) Prediction of engineering properties of GFRP lamina: tensile behaviour. J Inst Eng Electron 86:191–194
29. Paiva JMF, Mayer S, Rezende MC (2005) Evaluation of mechanical properties of four different carbon/epoxy composites used in aeronautical field. Mater Res 8:91–97
30. Ray BC (2006) Effects of changing environment and loading speed on mechanical behaviour of FRP composites. J Reinf Plast Compos 25:1227–1240
31. Scurlock RG, Rizzuto C (2007) Development of low-loss storage of cryogenic liquids over the past 50 years. In: Timmerhaus KD, Feed RR (eds) Cryogenic engineering. Springer Science, Dordrecht
32. Sreenivasa CG, Joshi AG (2011) Influence of cryogenic treatment on mechanical behavior of glass/epoxy composites. Natl J Technol 7(2):20–25
33. Ventura G, Risegari L (2008) The art of cryogenics: low-temperature experimental techniques. Elsevier, Amsterdam
34. www.cryogenicsociety.org/resources/cryo_central/cryogenic_treatment_of_materials/.
35. www.scienceclarified.com/Co-Di/Cryogenics.html.
36. http://strombergarchitectural.com/products/domes/materials/gfrp–glass-fiber-reinforced-polymer–domes
37. http://img.tjskl.org.cn/pic/z1b53740-0x0-/frp_vertical_storage_tank.jpg
38. Suresha B, Chandramohan G, Siddaramaiah, Samapthkumaran P, Seetharamu S (2007) Three-body abrasive wear behaviour of carbon and glass fiber reinforced epoxy composites. Mater Sci Eng A 443:285–91

Chapter 10
Polyurethane and Polyisocyanurate Foams in External Tank Cryogenic Insulation

U. Stirna, I. Beverte, V. Yakushin, and U. Cabulis

Contents

Abstract External tanks of the spacecrafts need not only efficient, but also safe cryogenic insulation materials and the issues of their development are still urgent. At present, polyurethane (PUR) or polyisocyanurate (PIR) foams' cryogenic

U. Stirna • V. Yakushin • U. Cabulis
Latvian State Institute of Wood Chemistry, Dzerbenes 27, 1006 Riga, Latvia

I. Beverte (✉)
Institute of Polymer Mechanics, University of Latvia, Aizkraukles 23, 1006 Riga, Latvia
e-mail: ilzebeverte@edi.lv

S. Kalia and S.-Y. Fu (eds.), *Polymers at Cryogenic Temperatures*,
DOI 10.1007/978-3-642-35335-2_10,

insulation is widely applied in the partially reusable launch systems. Factors influencing the cryogenic resistance of external thermal insulation performed of spray-on PUR or PIR foams are characterised based on a wide literature search. They include chemical structure and macromolecules' architecture of polymeric matrix, physical and mechanical properties of foams, thermal stability and combustibility, cellular structure of foams etc. Experimental data on physical and mechanical properties at room and cryogenic temperatures are presented for IWC developed PUR foams further named as IWC-Cryo. Technological processes of spray-on PUR and PIR foams cryogenic insulation, cryo-pumping and the main defects of external tank insulation are analysed as well as applications of foams' insulation in space technologies (*Space Shuttle*, *Ariane* and *Buran*).

Keywords External tank • Cryogenic insulation • Polyurethane foams • Polyisocyanurate foams • Spray-on technology • Workability • Foams' properties • Thermal stresses and strains • Insulation defects • Cryo-pumping • Space technologies

10.1 Introduction

The use of liquefied hydrogen (LH_2) and oxygen (LO_2) in the field of power industry in the twenty-first century should occupy a prominent place. Achievements in this field are anticipated in aviation [1, 2] and motor transport [3]. Japan is planning to implement a project for LH_2 transport in tankers with the capacity 200,000 m^3 in 2020 and the choice of the appropriate cryogenic material for this large project is crucial [4]. The launching of space ships and satellites in the Earth orbit, at least in the coming years, is unthinkable without the use of LH_2 as the most efficient fuel.

Both oxygen and hydrogen exist as gases at standard temperature and pressure. Since their density in this state is quite low, the amount of these substances required by the e.g. *Space Shuttle* would take up an enormous volume. The only way to carry sufficient propellant in a reasonable amount of space is to increase the density of the gases by cooling and pressurising them until they become liquids. Cryogenics considers very low temperatures: <123 K. LO_2 is cryogenically cooled to 89 K while LH_2 is chilled to 20 K. These liquids must be kept at high pressure and very low temperature or they will boil back to a gaseous state. Cryo-technologies enable a significant increase of space launchers capability, when compared to storable propellant solutions. However handling of cryogenic propellants in the warm environment of launch pad, ascent and sun-exposed space flight requires the use of thermal protections on tanks, lines or other equipment [5].

The cryogenic propellant tanks need an efficient thermal insulation over its outer surface to prevent ice formation and avoid in-flight damage to the ceramic tile thermal protection system (TPS) on the adjacent orbiter. At present, polyurethane (PUR) or polyisocyanurate (PIR) foams' cryogenic insulation is applied in the partially reusable launch system and orbital spacecraft *Space Shuttle* (USA) [6–9] and the expandable launch system *Ariane 5* (European Space Agency) [10]. In the former Soviet Union as well, in the space shuttle *Buran* equipped with the carrier rocket "Energy", the PUR foams *Ripor 2 N* [11], developed by the Latvian State Institute of Wood Chemistry (LSI WC), had been used as a cryogenic insulation material.

External tanks (ET) of the spacecrafts need not only efficient, but also safe cryogenic insulation material, and the issues of its development are still urgent. Minor faults in the properties of insulation materials or in their production technology can cause an accident, e.g. a shiver of the ET cryogenic insulation material may smash the thermal insulation plate of the orbital space station on its way to the Earth [12].

10.2 Workability of External Tank Cryogenic Insulation

The cryogenic resistance or workability of the ET cryogenic insulation performed of spray-on PUR and PIR foams is determined mainly by the physical, mechanical and thermal properties of the applied foams Fig. 10.1 [13]. The complex of these properties is affected by the two basic factors: (1) Chemical structure and macromolecule architecture of the PUR foams' polymeric matrix and (2) Parameters of the foam's cellular structure.

Functions of the layers of ET spray-on cryogenic insulation. Cryogenic resistance of foams can be characterised as ability of material to withstand both thermal stresses and strains in insulations' cold and hot side. The spray-on foams layers of external tank insulation (ETI) have different functions [9, 13].

The ETI SOFI upper layer (the hot side) performs mainly the heat protection functions. Therefore, testing the ability of the ETI to resist heat fluxes, it is preferable to cut out samples from the upper layer. The upper layer ETI is also subjected to the action of in-plane compressive stresses therefore, samples for compressive strength testing should be also cut out from the insulation's upper layer. While ascending the upper layer of ET can warm up to 450 °C in 5 min due to heat fluxes.

The ETI SOFI middle layer during the flight should perform mainly the functions of a good thermal insulation material; it should have a fine-cell structure, a high content of closed cells and should be capable of a considerable resistance to cryopumping processes. For determining the samples' thermal conductivity and the content of closed cells, it would be preferable to cut out samples from the ETI block's middle layer.

The functions of ETI SOFI bottom layer, adjacent to the ET surface (the cold side) are especially important: this layer should be capable of resisting the in-plane tensile thermal stresses and strains. Logically, it would be necessary to evaluate the material's ability to resist the thermal stresses and strains performing tensile and compressive testing of the samples cut out from the foam's layer adjacent to the ET surface.

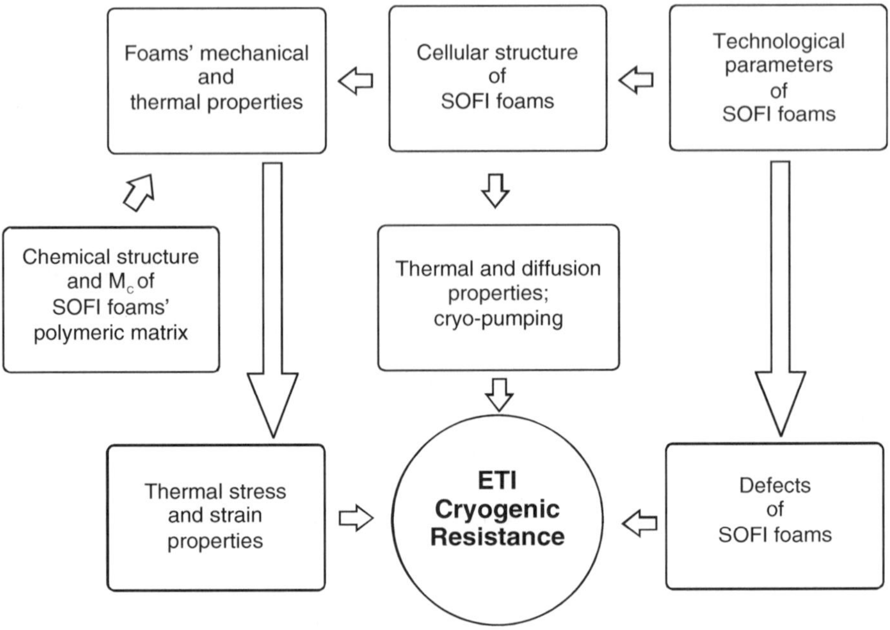

Fig. 10.1 Factors influencing the cryogenic resistance (workability) of external tank spray-on foams' thermal insulation (SOFI)

Comparison of the thermal stresses with the strength values for the tested foams showed that the in-plane stresses are more critical than out-of-plane stresses.

10.3 Influence of the Chemical Structure and Polymeric Matrix Macromolecules' Architecture on Workability of Cryogenic PUR and PIR Foams

To evaluate the effect of the chemical structure of the foams' polymeric matrix on the mechanical and thermal properties of foams, the cohesion energy density (CED) value is often used. This parameter characterises the level of the macromolecules interaction. It is not possible to vary the CED of the PUR polymeric matrix in a wide range, but its growth slightly increases the PUR foams' tensile and compressive strength properties. The growth of the CED values improves the thermal strength of PUR, increases its glass transition temperature and decreases the linear thermal expansion coefficient and tensile elongation [13, 14].

The CED itself is defined as the ratio of the energy of vaporisation to the molar volume. The CED was calculated by the Fedors' group contribution method [15], Table 10.1. It can be concluded that high values of CED are characteristic for urethane, ester and ether groups.

Table 10.1 Cohesion energy density (CED) of groups in PUR structure

Group	Structure	*M* (g/mol)	CED (J/cm^3)
Urethane	–NH–C(O)O–	59.02	1,172
Ester	–C(O)O	28.01	1,000
Ether	–O–	16.00	882
Aromatic ring	–C_6H_4–	76.09	636
Hydrocarbon	–CH_2–	14.03	307

Effect of the cohesion energy of the polymeric matrix of PUR and PIR foams on their cryogenic resistance. The strength and deformation properties of linear heterocyclic polymers have been studied extensively enough [16]. In particular, it has been found that the best flexibility at cryogenic temperatures is typical for polymers, containing aromatic rings and groups capable of forming hydrogen bonds, e.g. polyamides, polycarbonates, polyamidoimides, polysulphones, in the chain. Such polymers, upon elongation in the temperature range 4.2–77 K, do not obey the Hooke's law, i.e. feature a forced flexibility [17, 18]. It is prospective that PUR and PIR foams having the polymer matrix with a high CED would have a high cryogenic resistance.

The cyclic structures of the groups of polar nature are characterised by high CED, and as these groups are purposefully incorporated of polyurethane main chain, it is possible to obtain foams with a high tensile strength. It has been found that it is desirable to incorporate more groups with high CED values in the polymeric matrix structure of the PUR foams designed for cryogenic insulation, than in conventional PUR foams [19].

Effect of molecular weight per branching unit on the PUR foams' cryogenic resistance. The macromolecules' architecture can be characterised by the molecular weight per branching unit (M_c) values [13, 14]. The index M_c shows, how high the molecular weight for one branching unit of the polymeric matrix of the PUR or PIR foams is. Changes of this parameter are among the main instruments to regulate the physical and mechanical properties of PUR or PIR foams. There is a wide range of investigations of the effect of M_c on the glass transition temperature of PUR foams, as well as on the combined contribution of the chemical structure parameters to this property [20, 21]. However, the available evidence of the influence of the parameters of the chemical structure of the polymeric matrix as well as M_c on the physical and mechanical properties of the polymeric matrix, determined by these factors, is not sufficient. Therefore, an empirical approach is mainly employed when compiling the formulations of PUR foams.

Conventional PUR or PIR foams are not suitable for the cryogenic insulation of LH_2 tanks, because they have low safety coefficient indices. The complex of mechanical properties for these foams is commonly achieved by PUR or PIR foams' compositions having polymeric matrix M_c indices in the range 300–600. Such foams have high Young's modulus and relatively low tensile elongation indices [22].

Appropriate physical and mechanical properties for PUR foams designed for cryogenic insulation at temperatures of 77 K and lower are achieved, if the M_c

values of the polymeric matrix are higher than 600. Such parameters of the polymeric matrix are reached, using low functionality polyols [13].

Research carried out at the Latvian State Institute of Wood Chemistry [13] confirmed the above-mentioned results: at the M_c values 600–900 of the polymeric matrix, the obtained foams are characterised by high tensile strength and ultimate tensile strains at 77 K. The ultimate tensile strain of the obtained PUR foams reaches 4.0–5.9 % at 77 K, while values of the index K_{S1}, (10.4), are in the range 3.40–5.41. If the PUR or PIR foams' polymeric matrix has a high cross-link density, then it hampers the intermolecular action, respectively, the formation of physical bonds and such foams have a worse complex of cryogenic properties.

Analysing the data on the cryogenic insulation of the *Space Shuttle* [7–9], it may be concluded that no data on the parameters of the polymeric matrix of the PUR and PIR foams is presented, although the fact that PIR foams are used permits to conclude that the M_c values of the NCFI 24-124 (*North Carolina Industries*) polymeric matrix are not high. The ultimate strain indices of the used BX-265 closeout PUR foams are not high at 72 °F (22.2 °C): 0.132 in./in. and 0.028 in./in. at −320 °F (−196 °C). It is pointed out that high functionality polyols should be used for production of good cryogenic insulation materials.

The cryogenic resistance can be characterised as the ability of an insulation material to withstand both thermal stresses and strains in the insulation's cold and hot sides. This important characteristic is influenced by the polymer matrix parameters and the foams' cellular structure. In its turn, the cell structure is influenced by the technological parameters of SOFI production. The workability of the ETI is influenced by several other parameters, considered in [7–9].

The foams' suitability for cryogenic insulation can be estimated from the elementary work dW to be exerted for its tensile rupture $dW = \sigma_t(\varepsilon)d\varepsilon$. As the majority of the stress—strain curves of PUR foams at 77 K follow to the Hooke's law, work W can be calculated from the (10.1):

$$W = \sigma^{77}{}_{\mathrm{T}} \times \varepsilon^{77}{}_{\mathrm{T}}/2, \tag{10.1}$$

where $\sigma^{77}{}_{\mathrm{T}}$—foams' tensile stress; $\varepsilon^{77}{}_{\mathrm{T}}$—the corresponding strain.

However, it is difficult to apply these methods with an aim to increase the safety coefficient of the cryogenic insulation, and this follows from the Barker's equation, (10.2), [23]:

$$E \times \alpha^{2\sim}15N/(\mathrm{m}^2\ \mathrm{K}^2), \tag{10.2}$$

where E—modulus of elasticity and α—CTE. Barker's equation is valid in respect to the linear polymers and shows that there is a correlation between the majority of the elastic properties of polymers and their CTE. Equation (10.2) indicates that polymers with a low CTE have high Young's modulus values and vice versa.

10.4 Technological Processes of Spray-On PUR and PIR Foams' Cryogenic Insulation Materials and Foams' Cellular Structure

The insulation of ET by PUR or PIR foams can be performed consuming two basic technologies: (a) Gluing of preliminary produced foams' blocks on the ET walls surface and (b) spray application of foams on the ET walls surface. PUR and PIR foams blocks or shells are widely used in cryogenic technologies for requirements of liquefied natural gas (LNG) or liquefied nitrogen (LN_2) storage or transportation systems.

Spray-on of SOFI on substrate surface. While spraying of the PUR or PIR foams to the metal surface, the foaming process in different parts occurs in different conditions. At the substrate surface, the PUR and PIR foams foaming occurs in nonadiabatic (With heat losses) conditions. These conditions are more pronounced, if the substrate heat capacity grows, and the environment and substrate temperature is low. The SOFI material formation is a *chemically-technological* process, and 84 KJ/mol of heat is released as the OH-group reacts with the NCO-group, but cyclotrimerisation of isocyanate groups occurs with a smaller heat release—66.5 J/mol [24]. If a part of this heat is lost before the foams manage to rise (within 3–20 s), the PUR foams polymer matrix formation is hampered at the substrate surface. If the foams' formation at the substrate surface occurs in pronounced nonadiabatic conditions, it can be anticipated that the foams density will grow at the substrate surface and the conversion degree of the functional groups forming the polymeric matrix will decrease. The latter circumstance can be the reason to the formation of the foams with considerably impaired mechanical properties and a poor adhesion to the substrate surface.

While foaming and hardening the core part of the foams layer is formed in adiabatic conditions that facilitate a high conversion degree of polyols' OH groups and NCO groups of isocyanate. Thus appropriate physical and mechanical properties of foams' are ensured. The upper part is formed in quasi-adiabatic conditions since the heat created by the reaction forming the urethane groups, is partly channelled into the surrounding environment. Table 10.2 presents formulation of the patented [25] SOFI material—PUR foams IWC-Cryo [13]. Data on influence of the technological parameters on the cellular structure and properties of these foams is presented further.

Manufacturing SOFI material samples. SOFI material IWC-Cryo were produced by a high pressure spraying device *Glass Craft*. The following parameters were controlled: A-component (polyol) temperature, B-component (polyisocyanate) temperature, A and B components' volume ratios, pressure A and B components' supply lines and room temperature, relative air moisture, cream time, overlap time after that the next layer was sprayed and metal sheet temperature. The main technological parameters for the SOFI material IWC-Cryo production are listed in Table 10.3 [13].

Influence of metal substrate temperature on cellular structure of SOFI (monolayer) PUR foams IWC-Cryo. The following factors were found to influence the

Table 10.2 SOFI material IWC-Cryo formulation

Ingredients	pbw
Polyols	
Polyether polyols	25
Polyester polyols	30
Reactive flame retardant	20
Chain extender	25
Additives	
Blowing agent *Solkane* 365mfc/227ea	25–27
Silicone surfactant	1.5
Water	0.5
Catalyst package	3–5
Additive flame retardant	15
B-component	
Crude MDI	150

Table 10.3 Technological parameters of spray-on PUR foams IWC-Cryo

Parameters	IWC-Cryo
Characteristics of a component	
Density—A-component (kg/m^3)	1,230
Density—B-component (kg/m^3)	1,230
Viscosity—A-component (mPa s)	133
Hand mixing	
Cream time (s)	4
Gel time (s)	9
Tack free time (s)	11
Spray-on process	
Mixing A/B volume ratio	1.00/1.08–1.16
A-component temperature (°C)	40–50
A-component mixing pressure (bar)	110–125
B-component mixing pressure (bar)	125–140
B-component temperature (°C)	40–50
Spray-on foams consumption (kg/m^2)	~2.7
Substrate temperature (°C)	20; 27; 35
Cream time (s)	3–4
Room temperature (°C)	18–20
Relative air moisture (%)	30–50

cellular structure of SOFI material [13]: (a) The temperature of the surface to be insulated and A and B components of PUR foams formulation; (b) The thermal capacity of the surface subjected to insulation; (c) The reactivity of PUR foams composition in foaming process; (d) The ratio and concentration of blow and gel catalysts and (e) The thickness of foaming layer.

The LSI WC developed spray-on PUR foams IWC-Cryo were investigated at the environmental temperature 18–20 °C and the relative air humidity 30–50 %. The effect of the temperature of aluminium sheets on the distribution of the density of

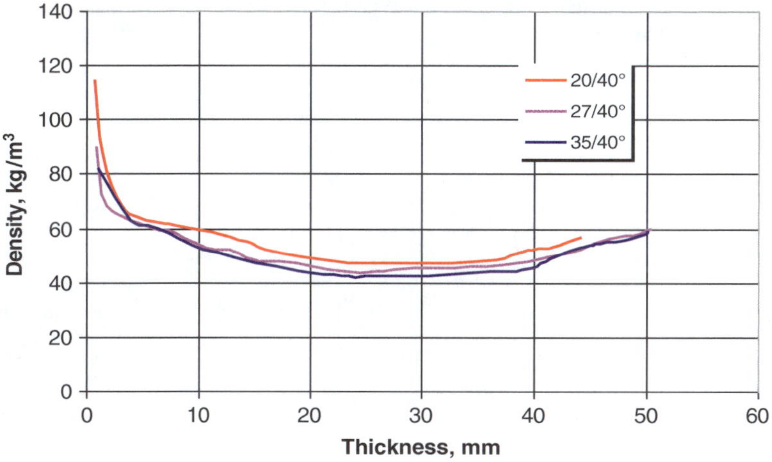

Fig. 10.2 Density of the monolayer SOFI material IWC-Cryo in dependence of layer's thickness at different temperatures of aluminium sheet having thickness 4 mm, and components in foams' formulation [13]

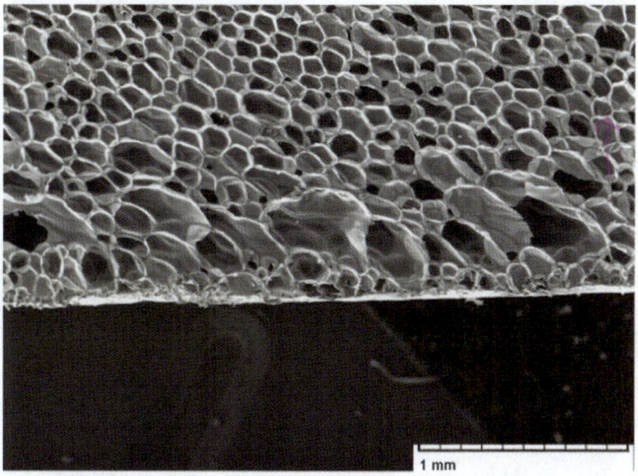

Fig. 10.3 SEM image of the SOFI material IWC-Cryo, aluminium sheet temperature 20 °C

foams with different temperature of the components is shown in Fig. 10.2. As can be concluded the increase of the metal sheet temperature from 20 to 35 °C decreases the core density and the density of bottom layer as well.

The cellular structure of the SOFI material adjacent to the aluminium sheet has a very fine structure, Figs. 10.3 and 10.4, and the cell sizes are similar both in the foams' rise direction and perpendicular to it. The foams' structure has no large or extended cells. SEM images for the monolayer SOFI material sample, spray-applied in the thickness 60 mm on the metal sheet surface having temperature 27 °C are presented in Figs. 10.5 and 10.6.

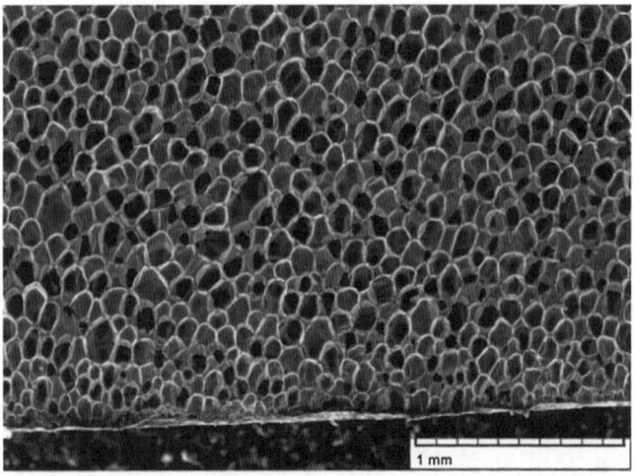

Fig. 10.4 SEM image of the SOFI material IWC-Cryo, aluminium sheet temperature 35 °C

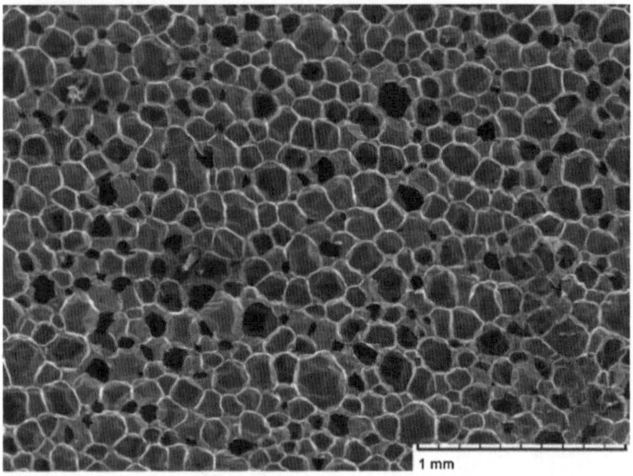

Fig. 10.5 SEM image of the structure of the SOFI material IWC-Cryo in direction parallel to foams' rise, distance 1 cm from the metal sheet surface

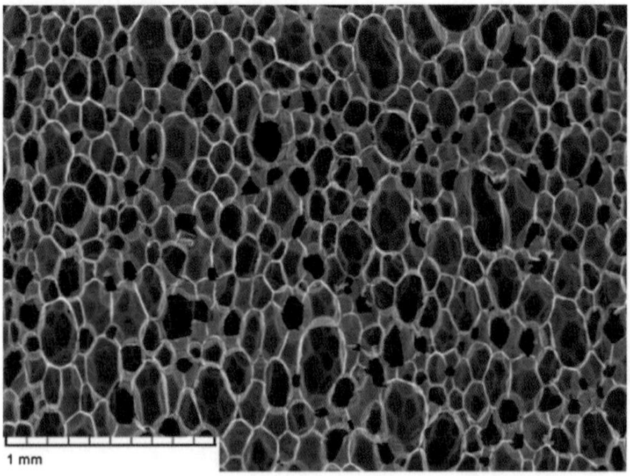

Fig. 10.6 SEM image of the structure of the SOFI material IWC-Cryo in direction parallel to foams' rise, distance 3 cm from the metal sheet surface

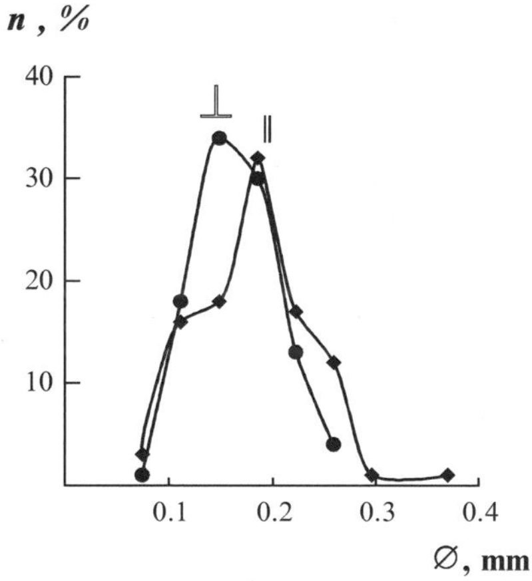

Fig. 10.7 Frequency distribution n of the cells' diameters Ø for the foams adjacent to the metal sheet surface at a distance 1 cm, parallel (//) and perpendicular (⊥) to foams' rise direction

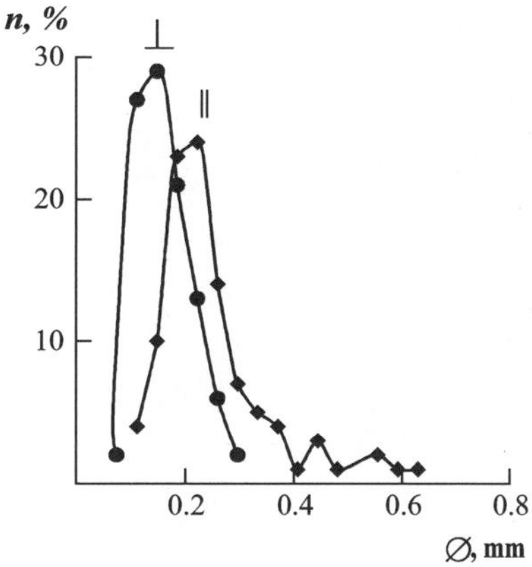

Fig. 10.8 Frequency distribution n of the cells' diameters Ø for the foams in the middle part of the block, parallel (//) and perpendicular (⊥) to foams' rise direction

SEM images reveal that the cellular structure of the SOFI material at the distance about 1 cm from the metal surface can be characterised as isotropic; anisotropy coefficient $A \approx 1.10$. The cell size distribution, calculated from the SEM images for the foams at the distance 1 cm from the metal surface and in the middle part of the block is shown in Figs. 10.7 and 10.8. This cellular structure is fine-porous; the average cell diameter in directions parallel and perpendicular to foams' rise is 0.148 and 0.134 mm, respectively. The 60 mm thick foams' block middle layer cells have a transversally isotropic structure. The foams cell anisotropy coefficient in this case is high $A \approx 1.51$; the cell average sizes in directions parallel and perpendicular to foams' rise are 0.203 mm and 0.134 mm, respectively. The content of large cells in direction parallel to foams' rise is moderate—only several percents of the cells with the size greater than 0.5 mm. With metal sheet surface temperature increasing from 20 to 35 °C, the cellular anisotropy coefficient of the sample's middle layer SOFI material grows from $A = 1.30$ to $A = 1.80$. Thus, the SOFI production technology influences the cellular structure essentially and the structure can be highly different in different layers.

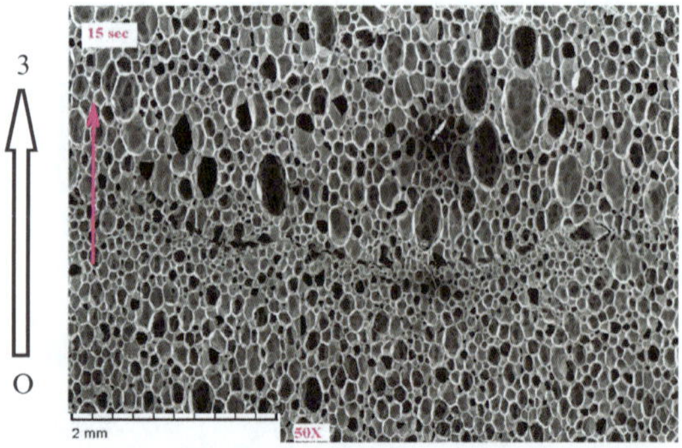

Fig. 10.9 SEM image of the cell structure of the two-layer SOFI foams IWC-Cryo, overlap time 15 s, 50×, foams' rise direction O3

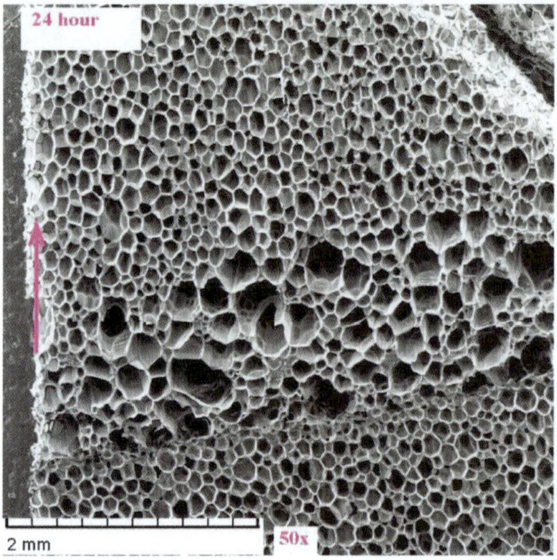

Fig. 10.10 SEM image of the cell structure of the two-layer SOFI material IWC-Cryo with the overlap time 24 h, 50×, foams' rise direction O3

Cellular structure of the two-layer SOFI material IWC-Cryo. An estimate of the effect of the overlap time on the cell structure of the two-layer SOFI material can be made by analysis of the knit line zone from the SEM images.

At the overlap time 15 s, nearly equal to the SOFI material's tack time, the knit line of the material IWC-Cryo is without any considerable defects, Fig. 10.9. Above the knit line, individual extended cells are formed, although their size does not exceed 0.8 mm.

If the overlap time is 24 h, Fig. 10.10, in the ~2 mm thick layer above the knit line, there are mainly relatively large cells, with the diameters parallel to the foams' rise direction ~0.5 mm. Specific cellular structures are formed in the SOFI layer, which is spray-on to the surface of the same mechanically treated PUR foams after 72 h, Fig. 10.11. Because the air is adsorbed at the surface of the treated foams, the formation of a structure of two different types is observed as the second layer is spray-on. Above the knit line, the foams have fine cells, with a diameter of about 0.10 mm and large extended cell as well, which could be formed by the air adsorbed at the surface during the foaming process. This transition zone is about 2 mm thick,

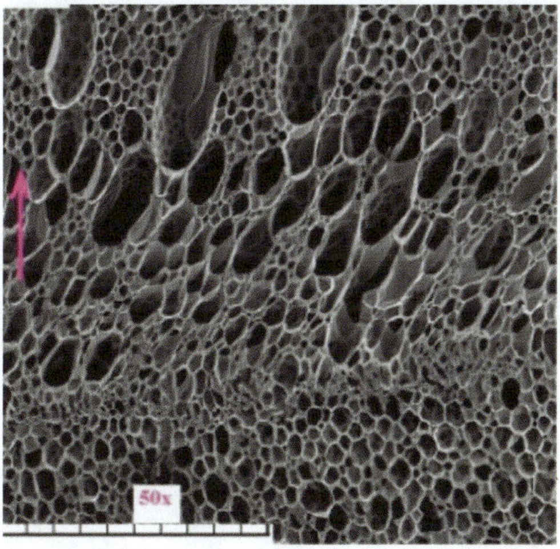

Fig. 10.11 SEM image of the cellular structure of the SOFI material IWC-Cryo, the overlap time 72 h, 50×, foams' rise direction O3 (the bottom layer was mechanically treated)

and the large cells have sizes 0.7–1.2 mm in direction parallel to foams' rise. In direction perpendicular to foams' rise the cells' sizes do not exceed 0.5 mm.

The obtained data permit to choose the optimal technological regime in the cases of mono- and multilayer spray-on technologies. The investigation performed reveals the considerable influence of the technological parameters on the PUR class SOFI structure and properties that in turn influence the workability of the cryogenic insulation.

10.5 Influence of Spray-On PUR and PIR Foams' Cellular Structure on Foams' Properties

The main parameters of the foam's cellular structure include the foam's cell size, content of open or closed cells, cell orientation and anisotropy coefficient A defined as ratio of the cell's dimension along foams' rise direction to the dimension along direction perpendicular to rise direction or as ratio $A_{31}{}^{\sigma}$ of strength values in corresponding directions. With cell anisotropy coefficient of rigid PUR foams' increasing from 1.0 to 2.0, the compressive strength along the cell's elongation direction increases approximately twice, and the linear thermal expansion coefficient decreases considerably [26]. In most cases, the anisotropy coefficient of conventional spray-on foams insulations is approximately $A \approx 1.5$. The anisotropy coefficient is influenced by both the properties of PUR foams compositions and the parameters of the technological process for production of the foams. The purposeful regulation of PUR foams' anisotropy is quite a complicated process, and there is a few data on this topic in scientific information sources.

The ET cryogenic insulation is subjected to the following main effects during the flight or when filling/discharging the cryogenic agent into the ET: (a) thermal

Table 10.4 Effect of the anisotropy coefficient of the cellular structure, the polymeric matrix' M_c and CED of PUR foams on the mechanical, diffusion and thermal properties of ETI [13] at temperatures 77 and 296 K

Properties[a]	M_c (300–1,150) (g/mol)	CED (580–650) (J/cm^3)	Anisotropy coefficient A	
			>1.0	<1.0
Mechanical properties, compression				
$\sigma_{33}{}^{296}{}_{C}$	↓↓	↑	↑↑	↓↓
$\sigma_{33}{}^{77}{}_{C}$	↑	↑	↑↑	↓↓
$\sigma_{11}{}^{296}{}_{C}$	↓↓	↑	↓↓	↑↑
$\sigma_{11}{}^{77}{}_{C}$	↑	↑	↓↓	↑↑
$E_1{}^{296}{}_{C}$	↓↓	↑	↓↓	↑↑
$E_1{}^{77}{}_{C}$	↓↓	↑	↓↓	↑↑
Mechanical properties, tension				
$\sigma_{11}{}^{296}{}_{T}$	↑↑	↑	↓↓	↑↑
$\sigma_{11}{}^{77}{}_{T}$	↑↑	↑	↓↓	↑↑
$E_1{}^{296}{}_{T}$	↓↓	↑	↓↓	↑↑
$E_1{}^{77}{}_{T}$	↓↓	↑	↓↓	↑↑
$\varepsilon_{11}{}^{296}{}_{T}$	↑↑	↓	↑↑	↓↓
$\varepsilon_{11}{}^{77}{}_{T}$	↑↑	↓	↑↑	↓↓
Diffusion properties				
Water absorbance	↑	↑	↑	↓
Cryo-pumping	↑	↑	↑	↓
Thermal properties				
Resistance against action of heat fluxes	↓	↑	↓	↑
Glass transition temperature	↓	↑	Constant	Constant
Coefficient of thermal expansion	↑	↓	↑	↓

[a] *Arrows* designate the parameters which influence the properties of the ETI cryogenic insulation: (a) grey—improve; (b) black—impair and (c) double—influence is the most significant; direction of arrow corresponds to the increase or decrease of the numerical value of the corresponding property

stresses of tensile and compressive character, (b) heat fluxes and (c) cryo-pumping effects. Table 10.4 analyses the effect of the CED of the polymeric matrix of PUR foams M_c and the foams' cellular structure anisotropy on the mechanical, diffusion and thermal properties of the external cryogenic insulation, where σ_{ii}—strength, E_i—modulus of elasticity, ε_{ii}—ultimate strain, $i = 1$, 2 and 3; O3—foams' rise direction; *C*—compression and T—tension.

Considering PUR foams with the isotropic cell structure or having the anisotropy coefficient $A \sim 1.0$ it can be concluded that the foams: (a) will be less subjected to the material's erosion caused by heat fluxes, as well as diffusion and cryo-pumping processes and (b) could have better resistance against thermal stresses both in the ETI cold side (tension stresses) and in the hot side (compressive stresses).

Table 10.5 Influence of technological parameters on the main properties of the IWC-Cryo foams; monolayer block thickness 50–60 mm

Level of SOFI block	T_{met} (°C)	T_{comp} (°C)	ρ (kg/m^3)	$\alpha \times 10^{-5}$ (K^{-1})	$\varepsilon_{11}{}^{77}{}_{T}$ (%)	$\sigma_{33}{}^{296}{}_{C}$ (MPa)	$\sigma_{11}{}^{296}{}_{C}$ (MPa)	K_{S1}
Bottom	20	40	65.8	4.75	4.00	0.20	0.26	3.90
Middle	20	40	42.2	5.81	4.66	0.24	0.19	3.72
Bottom	27	40	62.8	4.38	4.15	0.20	0.25	4.39
Middle	27	40	42.2	6.35	4.56	0.25	0.16	3.33
Bottom	27	50	62.8	5.32	3.81	0.22	0.25	3.31
Middle	27	50	40.8	6.61	5.81	0.28	0.16	4.07
Bottom	35	40	61.4	5.83	3.79	0.20	0.22	3.01
Middle	35	40	41.9	6.70	5.81	0.27	0.15	4.02
Bottom	20	50	67.4	5.32	3.59	0.22	0.26	3.12
Middle	20	50	47.2	5.56	5.59	0.28	0.16	4.49

With temperature of the aluminium sheet surface T_{met} increasing from 20 to 35 °C, the anisotropy coefficient of the SOFI middle layer increases from $A = 1.26$ to $A = 1.80$, Table 10.5, where T_{met}—temperature of the aluminium sheet, T_{comp}—temperature of components in foams' formulation.

These results agree with the results described earlier on the cell structure of these foams. With metal sheet temperature increasing from 20 to 35 °C, the density of the samples' middle layer slightly changes and is in the range 40.8–47.2 kg/m^3. The density of the foams' samples, cut out at the distance 1 cm from the metal sheet surface, is in the range 61.4–67.4 kg/m^3 and tends to increase at lower temperatures of the metal sheet. The CTE (Determined perpendicular to foams' rise direction) is smaller for the layer, cut out from the bottom zone. That has two reasons: the higher foams' density and the isotropic/orthotropic structure of the foams. Because the samples middle part foams have a transversally isotropic structure, CTE has relatively high values. The K_{S1} value is higher for the samples, cut out from the middle part of a block. The K_{S1}, (10.4), for the samples adjacent to the metal surface is in the range 3.0–4.5.

Properties of the monolayer SOFI material IWC-Cryo. To perform the studies, SOFI materials' samples with the thickness from 35 to 80 mm were produced. All samples for mechanical testing were cut out from the middle part of the blocks (core). The control of apparent density was obligatory for each sample. The PUR foams' properties in compression were determined in directions parallel and perpendicular to foams' rise direction, the tensile properties were determined in direction perpendicular to foams' rise direction. Cylindrical samples 20 mm in diameter and 22 mm in height were used for compression properties determination. Ring samples 13–14 mm in width with the inner diameter 43 mm and the outer diameter 53 mm were used for tensile characteristics determination. Testing speed of all samples was 10 %/min. Mechanical testing was performed on testing machines *Zwick/Roell 500 N* and *Zwick/Roell Z100* (sensor 1000 N). A compact cryostat with mechanical testing appliances was used to carry out the low

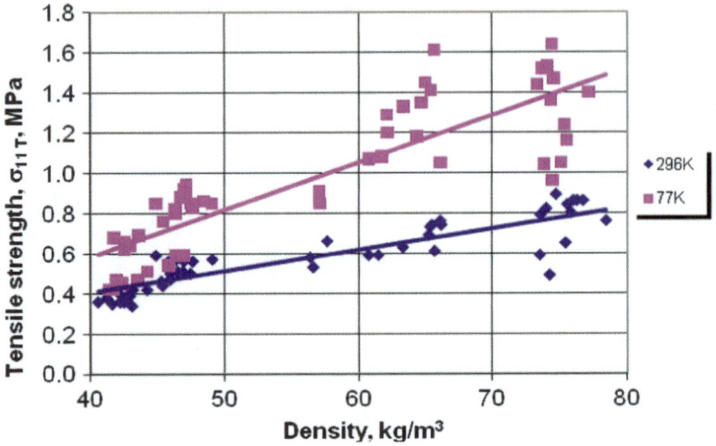

Fig. 10.12 Tensile strength $\sigma_{11\ \mathrm{T}}$ vs. foams' density for the SOFI material IWC-Cryo [13]

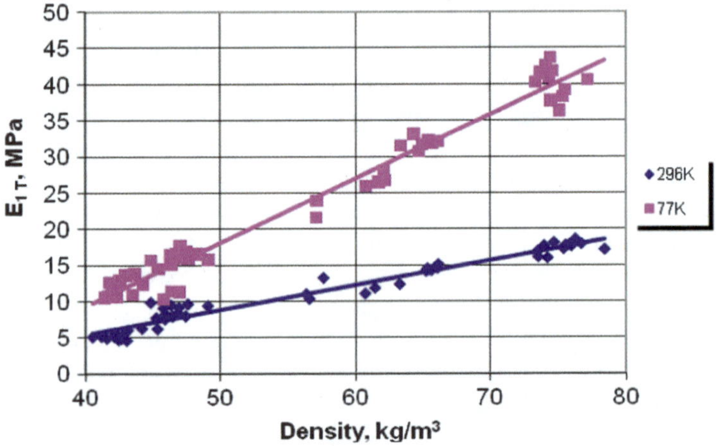

Fig. 10.13 Young modulus in tension $E_{1\ \mathrm{T}}$ vs. foams' density for the SOFI material IWC-Cryo

temperature tests. Prior to testing, the samples were held in the cryostat at the specified temperature for 15–20 min. As a rule, six samples on each experimental point were tested. In the majority of cases, the coefficient of variation of foams characteristics was within 6–15 %.

IWC-Cryo foams exhibit tensile properties acceptable for cryogenic applications both at 296 and 77 K, Fig. 10.12. Samples for testing were cut out from different zones of the block that results in their different densities, e.g. samples with the density ~75 kg/m^3 were cut out from the zone adjacent to the substrate surface. SOFI material IWC-Cryo has a low modulus of elasticity at 77 K when the foam's density is below 50 kg/m^3, Fig. 10.13. Elongation at break, both at 296 and 77 K

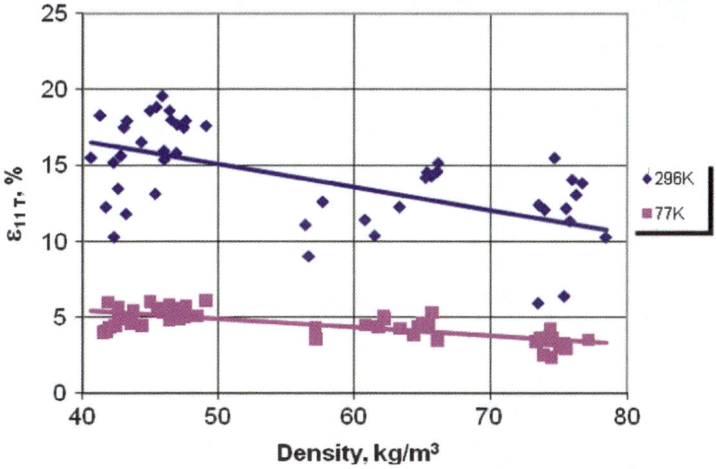

Fig. 10.14 Ultimate strain $\varepsilon_{11\ T}$ at tensile break vs. foams' density for the SOFI material IWC-Cryo

Table 10.6 Effect of the thickness *h* of the spray-on monolayer SOFI material IWC-Cryo on the foams' properties in compression at 296 K

h (mm)	Core density (kg/m^3)	Sample's cut out level	$\sigma_{33\ C}$ (MPa)	$\sigma_{11\ C}$ (MPa)	$E_{3\ C}$ (MPa)	$E_{1\ C}$ (MPa)	$\sigma_{33\ C}/\sigma_{11\ C}$
35	50	Middle	0.21	0.23	4.99	5.35	0.87
40	53	Middle	0.20	0.23	4.85	5.68	0.91
40	52	Middle	0.24	0.24	6.35	5.69	0.87
50	52	Middle	0.25	0.30	5.20	6.83	0.71
65	47	Bottom[a]	0.24	0.25	6.53	5.83	0.96
65	48	Bottom[a]	0.25	0.20	6.95	4.97	1.25
70	51	Bottom[a]	0.18	0.26	3.88	6.64	0.69
70	45	Upper	0.26	0.18	8.12	4.31	1.44
75	45	Bottom[a]	0.26	0.22	7.20	5.07	1.18
80	49	Bottom[a]	0.20	0.24	4.57	6.34	0.83
80	40	Middle	0.24	0.11	6.89	2.48	2.18
80	46	Upper	0.24	0.17	7.36	3.80	1.41

[a]Samples cut out at a distance 1 cm from the metal sheet surface

(Fig. 10.14), decreases with increasing samples' density, and this agrees with the generally accepted notion about the effect of the foams' density on the tensile properties.

Table 10.6 presents data on the mechanical properties of the monolayer SOFI material IWC-Cryo samples with the thickness 35–80 mm. The spray-on of the foams was carried out using a *Glass Craft* machine at the volume ratio of components A/B = 0.93. The temperature of components A and B was 40 °C in all cases.

Table 10.7 Effect of the thickness of the spray-on layer *h* of the monolayer SOFI material IWC-Cryo on the foams' properties in tension

		Samples' cut out	$\sigma_{11\ T}$ (MPa)		$E_{1\ T}$ (MPa)		$\varepsilon_{11\ T}$ (%)	
h (mm)	Density ρ (kg/m^3)	level	296 K	77 K	296 K	77 K	296 K	77 K
40	60	Middle	0.60	1.07	11.6	25.4	11.1	4.44
40	75	At the metal	0.78	1.20	17.3	38.8	11.4	3.08
50	47	Middle	0.57	0.71	9.4	13.4	18.3	5.66
50	54	Middle	0.56	–	9.4	–	18.3	–
65	65	At the metal	0.71	1.34	14.5	31.9	14.5	4.31
65	47	Middle	0.49	0.88	8.0	16.8	15.9	5.13
65	43	Upper	0.40	0.69	5.7	13.6	17.4	5.46
65	43	Upper	0.39	–	5.8	–	17.4	–
65	45	Middle	0.48	–	8.0	–	15.9	–
65	65	At the metal	0.68	–	14.5	–	14.1	–
80	39	Middle	0.39	–	5.7	–	17.9	–
85	75	At the metal	0.75	1.44	10.4	34.9	17,6	–
85	43	Upper	0.36	–	13.0	–	–	–

With thickness of the spray-on layer increasing from 35 to 80 mm, the core density of the foams' middle layer decreases from 53 to 40 kg/m^3. For the SOFI material, spray-on on the metal sheet and having 35–50 mm in thickness, anisotropy coefficient $A_{31}{}^{\sigma}$ is in the range 0.87–0.91. If the thickness of the spray-on layer is approximately 80 mm, then the middle part of the foams block is characterised by very high anisotropy coefficients. For all the foams samples, located at the distance about 1 cm from the metal sheet, $A_{31}{}^{\sigma}$ is in the range 0.69–1.25.

The results summarised in Table 10.7 testify that it is not appropriate to carry out the ETI spray application in one layer with the thickness more than 60 mm, because it favours the formation of a material with a very high foams anisotropy degree in the block's middle layer. Samples, cut out from the block's middle layer, have a high $\varepsilon_{11\ T}$ at 77 K: in the range 4.44–5.66 %. For the samples cut out from the layer adjacent to the metal sheet surface, $\varepsilon_{11\ T}$ at 77 K is lower: 3.08–4.31 %. This could be explained by the foams' greater density for the layer formed in nonadiabatic conditions as well as with the cell orthotropic structure. The closed cell content for all monolayer SOFI material samples was high: 93–97 %. The SOFI material's contraction $\Delta l_{x\ 77\text{–}296}$ for the layer adjacent to the metal surface was in the range 1.112–1.136 %, but for the middle layer in the range 1.332–1.429 %. The average strength of the samples of the foams IWC-Cryo with the shear strength 51.3 kg/m^3 was 0.265 MPa.

The chemical structure of the SOFI material is evaluated from the FTIR spectra for the samples cut out from the block's middle layer and the zone adjacent to the metal sheet. The absorption band corresponding to the H-bonded groups NH_b is observed at 3,317 cm^{-1}. There is another absorption band at 3,450 cm^{-1} that is typical of PUs, containing free (nonassociated) groups NH_{fr}. The degree of the formation of NH_b groups was estimated from the value *R*, determined from the ratio of absorption intensities NH_b/NH_{fr} [19].

Table 10.8 Effect of the overlap time of the SOFI material IWC-Cryo on the mechanical properties in compression and anisotropy of the foams in second layer, T_{met} = 27 °C

N	Overlap time	Density[a] ρ_1 (kg/m^3)	Density ρ_3 (kg/m^3)	$\sigma_{33\ C}$ (MPa)	$\sigma_{11\ C}$ (MPa)	$E_{3\ C}$ (MPa)	$E_{1\ C}$ (MPa)	$A_{31}{}^{\sigma}$
1	15 s	46	45	0.27	0.17	6.49	3.60	1.59
2	30 s	47	46	0.28	0.19	7.21	3.87	1.46
3	45 s	44	44	0.30	0.14	8.48	2.76	2.17
4	120 s	46	43	0.27	0.17	7.42	3.44	1.63
5	300 s	48	49	0.32	0.18	8.30	3.86	1.74
6	60 min	47	46	0.27	0.22	6.31	4.92	1.23
7	24 h	48	46	0.25	0.22	6.10	5.05	1.15
8	72 h treatment	53	50	0.25	0.23	6.50	5.41	1.13

[a]ρ_1 and ρ_3—densities of samples for loading perpendicular (O1) and parallel to foams' rise direction (O3)

It is calculated from the FTIR spectra that the R value for the sample cut out from the block's middle layer is 1.92, but that for the sample from the outer part, this index is smaller, namely, 1.54. The results testify that SOFI material's spray application in nonadiabatic conditions diminishes the possibility of the formation of H-bonds. As a consequence, it has a decreased intermolecular interaction of polymer chains in the foams' polymeric matrix and, respectively, also smaller strength indices. If the SOFI material's foaming process occurs in nonadiabatic conditions (farther than 1 cm from the metal sheet), then the flexibility of the macromolecule chains during the process is sufficiently high, and the polymeric matrix has a higher possibility of H-bonds formation.

Properties of the two-layer SOFI material IWC-Cryo. There are applications that require multilayered foams as well. It is not always possible to spray PUR or PIR foams in the necessary thickness applying the monolayer technology, two- and multilayer technologies are often used. In this case it is necessary to determine the influence of overlap time on the properties of SOFI materials. It was determined: if the overlap time is 15–120 s interlayers breaking strength is in the limits 0.52–0.55 MPa and it is even bigger than the tensile strength of the core foams. When the overlap time is 1 h the breaking strength is 0.23 MPa and at overlap time 24 h the breaking strength is only 0.14 MPa. The optimal overlap time is different for each PUR and PIR foams formulation thus the recommended overlap time for PIR foams NCFI 24-124 is 7–28 s [9].

As can be seen from Table 10.8, the density of the middle part of the second layer foams is in the range 43–48 kg/m^3, and the material has high compressive strength values. The second layer foams have variable structural anisotropy indices, and a material with the anisotropy coefficient $A_{31}{}^{\sigma}$ close to 1.0 is obtained only at the overlap time 1 h or longer; $A_{31}{}^{\sigma}$ defined as compressive strength ratio parallel and perpendicular to foams' rise direction O3:

$$A_{31}{}^{\sigma} = \sigma_{33\ C}/\sigma_{11\ C}, \quad A_{31}{}^{\sigma} \approx A_{32}{}^{\sigma}. \tag{10.3}$$

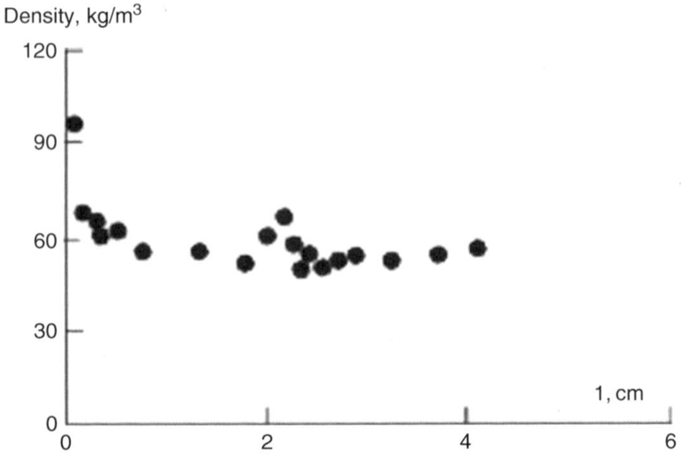

Fig. 10.15 Density of the SOFI material IWC-Cryo (sprayed in two layers) in dependence of height coordinate l perpendicular to interlayer surface; sample's thickness 42 mm. Temperature of aluminium sheet 20 °C

If the spray-on application is performed on the surface of the mechanically treated foams, the anisotropy coefficient of the second layer foams is close to 1.0, too.

The changes in the density of the two-layer SOFI material IWC-Cryo in dependence on the distance of the foams' layer from the surface of the aluminium sheet are depicted in Fig. 10.15. The density of the foams layer at the distance 1 mm from the metal sheet surface is about 90 kg/m^3 and the density in the sample's middle layer decreases considerably to 45–50 kg/m^3. Close to the interlayer knit line, the foams' density is higher than in the block's middle layer and reaches 55–62 kg/m^3.

10.6 Thermal and Mechanical Properties of External Tank Insulation PUR and PIR Foams at Room and 77 K Temperatures

PUR foams specifications are regulated by the standard ISO 8873 "Rigid cellular plastics – Spray-on polyurethane foams for thermal insulation" [27–29] in the European Union. The first part of ISO 8873 "Material specifications" specifies requirements and test methods for spray-on polyurethane rigid cellular plastic, used as thermal insulation. The application of PUR or PIR foams is regulated by the second part of ISO 8873-2 "Application". It specifies requirements for physical properties of rigid cellular plastic spray-applied polyurethane foams and lists the test methods to be used. The designer has the responsibility for confirming that the

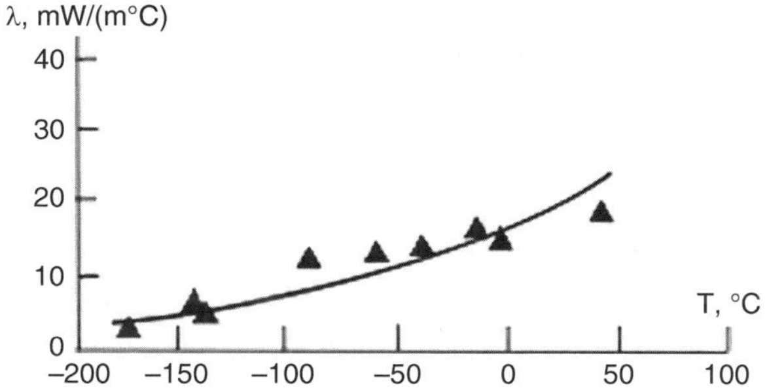

Fig. 10.16 PUR foams' IWC-Cryo thermal conductivity λ in dependence of temperature

physical properties provided by the material manufactured to this part of ISO 8873 will provide the requirements needed for a specific application. The third part of ISO 8873 "Tests methods" specifies the test procedures that are to be used when testing spray-on polyurethane foams to verify that they meet the requirements given in ISO 8873-1.

The standard ASTM C591 - 12a "Standard Specification for Unfaced Preformed Rigid Cellular Polyisocyanurate Thermal Insulation" [30] covers the types, physical properties, and dimensions of unfaced, preformed rigid cellular polyurethane modified PIR plastic material intended for use as thermal insulation on surfaces. Different test methods shall be performed in order to determine the thermal insulation's following properties: density, compressive resistance, apparent thermal conductivity, hot-surface performance, water absorption, water vapour permeability, dimensional stability, closed-cell content, surface bearing characteristics, tensile strength and leachable chloride, fluoride, silicate and sodium ions.

These standard specifications provide a point of reference for further analytical comparisons with both ambient test data and cryostat data obtained under large temperature differential cryogenic conditions [31].

Thermal conductivity of PUR and PIR foams. Thermal conductivity (λ) is a specific material property. It represents the heat flow in watts (W) through a 1 m^2 surface and 1 m thick flat layer of a material when the temperature difference between the two surfaces in the direction of heat flow amounts to 1 Kelvin (K). The unit of measurement for thermal conductivity (λ) is W/(m K). The thermal conductivity of rigid PUR and PIR foams is dependent on the cell gas used density, temperature, behaviour in the presence of water and moisture time of measurement.

The study Fesmire et al. [31] provides data on the cryogenic thermal performance of SOFI under large temperature differentials and basic information to apply in the design and optimization of future launch vehicles and other cryogenic systems. The test measurements were made at the full temperature difference (78–293 K) and included a full vacuum pressure range. The results are reported in terms of the apparent thermal conductivity λ and the mean heat flux.

Thermal conductivity data for the cryogenic PUR foams IWC-Cryo is presented in Fig. 10.16. The low values of λ are caused by the high amount of closed cells, the high porosity, very small cells; thin cell walls as well as anisotropy of the material. In addition formation of water in cells during foaming and residuals of the blowing agent is reduced to a minimum.

The blowing agent *Solkane 365mfc/227* used to produce IWC-Cryo type SOFI materials is a mixture of pentafluoro butane and heptafluoro propane having a boiling point at 24 °C. The obtained IWC-Cryo foams are characterised by a high content of closed cells: 93–97 % and these values largely determine the low thermal conductivity. As can be seen from Fig. 10.16 the coefficient of thermal conductivity of IWC-Cryo at 303 K is 19 mW/(m K) and only 4 mW/(m K) at 103 K. It is known that thermal conductivity is influenced by the foaming agents used in the blowing process. Since the water content in the polyol system is low, the calculation testifies that the blowing gasses ratio in the ready foams is CO_2-14 vol.% and *Solkane 365mfc/227*-86 vol.%.

Thermal expansion of polyurethane foams. The major advantages of foams insulations are their ease of fabrication, relatively low cost and self-supporting structure. However, one of the major disadvantages of foams is their large thermal expansion. In general, the thermal expansion coefficient of the foams considered is two to five times that of aluminum and four to ten times that of steel. Because of this large thermal expansion, it is necessary that the foam being developed accommodate this difference. The foams shrink more than the metal when the foams are in contact with the cold surface, creating cracks that open passageways for the admission of both water vapour and air, which considerably increases the thermal conductivity of the insulation [32]. Several solutions to this problem of thermal expansion have been investigated, including the addition of glass fibers, reinforcing threads, expansion joints and the development of vacuum-panel insulation. Another problem of concern with foams is the removal of the expansion gas from the interstices of the cellular foams [33].

There is relatively scarce information about the influence of the polymeric matrix of PUR or PIR foams on the linear expansion coefficient (CTE) of these materials. The companies offering cryogenic insulation materials for insulating LN_2 or LNG storage systems commonly indicate α values in a wide range of $30–70 \times 10^{-6}\ K^{-1}$ [34, 35]. The investigation [34] focuses on the thermal expansion tests and the response of the microstructure. A novel optical method is described, which is appropriate for measuring thermal expansion at high temperatures without influencing the thermal expansion measurement. A detailed micro structural investigation is described, which show cell expansion as a function of temperature. Finally, a phenomenological model on thermal expansion is presented. The thermal expansion (contraction) of the *Space Shuttle* cryogenic insulation foams is analysed in [36].

Both thermal expansion and conductivity are dependent on the determination direction with regard to the foams' rise direction, too. Thermal property testing is done to verify that the process environment would not affect the thermal properties. The main thermal property tests comprise thermal-vacuum testing, hot gas panel testing, thermal conductivity testing, specific heat etc.

Table 10.9 A typical PUR foams' formulation

Component	pbw
Chain extender	10–70
Polyester polyols	10–80
Polyether polyols	0–70
Blowing agent *Solkane 365mfc/227*	15–20
Catalyst—dimethylethanolamine or *Polycat 5*	1.0–2.0
Gel catalyst *Kosmos 19*	0–0.2
Water	0–1.0
Silicone surfactant *L6915*	1.0–2.0
Polymeric MDI	With isocyanate index 110

Physical and mechanical properties. Extended studies on the mechanical properties of PUR foams at low temperatures are reported in Reed et al. [37]. Results on the tensile strength and the Young s modulus indices of PUR and polystyrene foams with different density at 76, 195 and 300 K have been analysed. Properties of PUR foams and their applications in cryogenic technology are characterised in Sparks and Arvidson [38]. The mechanical properties and applicability of PUR foams at the temperatures from 77 to 403 K have been described in Demharter [32]. In Nadeau et al. [39] the properties of PIR foams and their application potentialities in cryogenic technology are characterised. The physical and mechanical characteristics of spray-on rigid polyurethane foams at normal and cryogenic temperatures have been analysed in Yakushin et al. [40].

Different test methods for physical and mechanical characteristics of cryogenic insulation foams comprise plug pull testing, lap shear tests, flexure test, bond tension tests, monostrain testing, gradient cryo-flex testing, compression and density determination etc. [7–9]. A wide range of physical and mechanical properties of spray-on and pour rigid polyurethane foams at room and cryogenic temperatures has been investigated at the LSI WC and Institute of Polymer Mechanics (University of Latvia) [19]. Mechanical testing was performed on testing machines *Zwick/Roell 500 N* and *Zwick/Roell Z100* (sensor 1000 N). A compact cryostat (Cryogenic agent liquid nitrogen LN_2) with mechanical testing appliances was used to ensure low temperature during tests. Prior to testing, the samples were held in the cryostat at the specified temperature for 15–20 min. As a rule, five samples were tested for each experimental point. In the majority of cases, the coefficient of variation of foams' characteristics was within limits 6–15 %.

In order to investigate influence of polymeric matrix' molecular weight per branching unit on mechanical properties of foams, the foams' formulation presented in Table 10.9 was used. The content of polyether and polyester polyols as well as content of chain extenders was varied in the formulation.

As can be seen from Fig. 10.17, with increasing polymeric matrix' M_c from 366 to 1,150, the foams' tensile strength perpendicular to rise direction and at 296 K $\sigma_{22\ T}$ grows from 0.66 to 0.95 MPa, while at 77 K its values grow from 0.72 to 1.85 MPa, respectively. Relatively low tensile strength $\sigma_{22\ T}$ indices at 77 K and low M_c values (the first group PUR foams) could be connected with the fact that the cross-links of

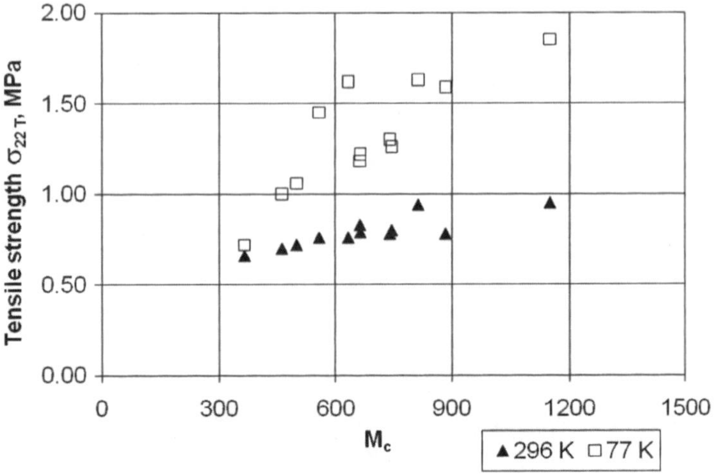

Fig. 10.17 Tensile strength $\sigma_{22\ \mathrm{T}}$ of PUR foams at 77 (*open square*) and 296 K (*filled triangle*) in dependence of polymeric matrix' M_c

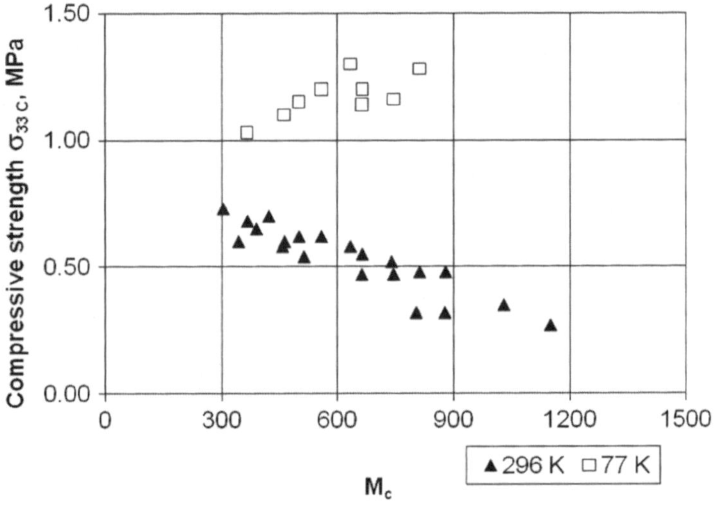

Fig. 10.18 Compressive strength $\sigma_{33\ \mathrm{C}}$ of PUR foams at 77 (*open square*) and 296 K (*filled triangle*) in dependence of the polymeric matrix' M_c

covalent nature hamper the intermolecular action of the polymeric chains, hence, also the formation of H-bonds.

The compressive strength of pour PUR foams was tested in the direction parallel to foams' rise direction O3. As can be seen from Fig. 10.18, with increasing M_c from 300 to 1,150, compressive strength $\sigma_{33\ \mathrm{C}}$ decreases from 0.72 to 0.27 MPa,

Table 10.10 Experimental data on spray-on and pour PUR foams' properties parallel (//) and perpendicular (⊥) to foams' rise direction

N	Characteristics	Units	Values		
1	Production mode		Spray-on	Pour	
2	Molecular weight per branching unit M_c	g/mol	740	450	740
3	Apparent core density ρ	kg/m^3	48	65–70	65–70
4	Closed cell content	%	96.9	93.0	93.0
5	Anisotropy coefficient		1.00–1.05	1.20–1.30	1.00–1.05
	Temperature 296 K				
6	Compressive strength // $\sigma_{33\ C}$	MPa	0.16	0.60	0.50
7	Compressive strength ⊥ $\sigma_{22\ C}$	MPa	0.17	–	–
8	Modulus in compression // $E_{3\ C}$	MPa	4.21	12.00	10.00
9	Modulus in compression ⊥ $E_{2\ C}$	MPa	4.61	–	–
10	Poisson's coefficient in compression // $\nu_{32\ C}$		0.15	–	–
11	Poisson's coefficient in compression ⊥ $\nu_{23\ C}$		0.29	–	–
12	Tensile strength ⊥ $\sigma_{22\ T}$	MPa	0.41	0.65	0.75
13	Modulus in tension ⊥ $E_{2\ T}$	MPa	7.41	13.00	10.00
14	Elongation at break ⊥ $\varepsilon_{22\ T}$	%	23.1	12.00	12.00
15	Interlayer adhesion	MPa	0.38	–	–
	Temperature 77 K				
16	Compressive strength // $\sigma_{33\ C}$	MPa	0.61	1.10	1.20
17	Compressive strength ⊥ $\sigma_{22\ C}$	MPa	0.62	–	–
18	Modulus in compression // $E_{3\ C}$	MPa	6.37	25.00	22.00
19	Modulus in compression ⊥ $E_{2\ C}$	MPa	9.76	–	–
20	Tensile strength ⊥ $\sigma_{22\ T}$	MPa	1.03	1.00	1.30
21	Modulus in tension ⊥ $E_{2\ T}$	MPa	20.49	28.00	21.00
22	Elongation at break ⊥ $\varepsilon_{22\ T}$	%	5.33	3.00	4.30
23	Ratio $E_2{}^{77}{}_T/E_2{}^{296}{}_T$		2.76	1.95	2.10
24	Ratio $E_3{}^{77}{}_C/E_3{}^{296}{}_C$		1.51	2.00	2.20

while, in its turn, at 77 K, with increasing M_c from 360 to 813, $\sigma_{33\ C}$ grows from 1.03 to 1.28 MPa.

The main results of Stirna et al. [19] are presented in Table 10.10. At 296 K both compressive moduli and strength of spray-on foams are nearly equal for loading parallel and perpendicular to foams' rise direction: $E_{3\ C} \approx E_{2\ C}$ and $\sigma_{33\ C}{}^{10\ \%} \approx \sigma_{22\ C}$ due to nearly isotropic character of structure. The slightly higher values in compression perpendicular to rise direction could be explained by higher mechanical resistance of the internal skin and near-by foams, that act as a reinforcement. The relationship between Poisson's coefficients $\nu_{23} > \nu_{32}$ could be explained by different action of internal skin and surrounding foams in compression parallel and perpendicular to rise direction, as well as mode of measuring the transversal displacement.

The following was concluded: (1) With M_c values increasing from 360 to 1,150, the PUR foams' tensile strength perpendicular to rise direction and the corresponding elongation at break increases, while modulus in tension perpendicular to rise direction decreases both at temperatures 296 and 77 K. For PUR foams, core density 65–70 kg/m^3, tensile strength perpendicular to rise direction and corresponding elongation at break at can be increased for 30–40 % by increasing polymer matrix' molecular weight per branching unit M_c value from 450 to 740, (2) With polymeric matrix' M_c values increasing from 300 to 1,150, the PUR foams' compressive strength parallel to rise direction at 296 K decreases, but grows at 77 K, (3) With PUR foams' polymeric matrix' M_c increasing, the intermolecular interaction increases, which is testified by the growth in the concentration of the urethane groups connected with H-bonds. This factor promotes the increase in the tensile strength perpendicular to rise direction of PUR foams and (4) Layered spray-on PUR foams with core density 48 kg/m^3 and $M_c = 740$ exhibit competitive tensile strength and elongation at break at cryogenic temperature [19].

The complex of the physical and mechanical properties of the ET cryogenic insulation foams is determined to a great extent by the parameters of the PUR or PIR foams' polymeric matrix: (a) content of groups with high CED, (b) the presence of side chains in the polymeric structure and (c) the architecture of macromolecules determined by the cross-link density of the polymeric matrix.

The mechanical and thermal properties of PUR and PIR foams are influenced as well by the flame retardants used in the foams formulation.

10.7 Thermal Stability and Combustibility

The thermal stability and combustibility of PUR and PIR foams are dependent on the chemical structure of the groups present in polymeric matrix. It is known that polyurethanes, comprising rings of aromatic structure, isocyanurate or carbodiimide groups, exhibit a high thermal stability. Therefore, PIR foams are characterised by a high thermal resistance; too, although decomposition at the first stage for these foams begins at 200–300 °C, since their polymeric matrix possesses thermally not durable urethane groups [41], too. Data on the thermal resistance of the groups present in the structure of PUR and PIR foams is summarised in Table 10.11. The thermal stability of the groups incorporated in the PUR foams' polymeric matrix macromolecules can be aligned in the growth direction in the following order: aromatic rings > isocyanurate > carbodiimide > urea > urethane > alophonate > biuret.

It can be concluded: when forming cryogenic insulation PUR or PIR foams compositions, compromise solutions should be chosen, since, the urethane group is characterised by high cohesion energy, although the thermal stability of this group is relatively low. It is not desirable to incorporate allophanate or biuret groups in the structure of the polymeric matrix, Table 10.11. Hence, the designing of the polymeric matrix of cryogenic insulation is a highly complicated task.

Combustibility and flame retardants. In order to reduce the PUR and PIR foams' combustibility flame retardants comprising phosphorus, chlorine and bromium are widely applied. They can be divided in two groups: reactive and additive. The reactive flame retardants comprise OH groups, capable of reaction with the isocyanates but the additive retardants do not react with the isocyanate component

Table 10.11 Thermal stability of the groups comprised in the structure of PUR or PIR foams

Name	Groups	Temperature range (°C)
Isocyanurate	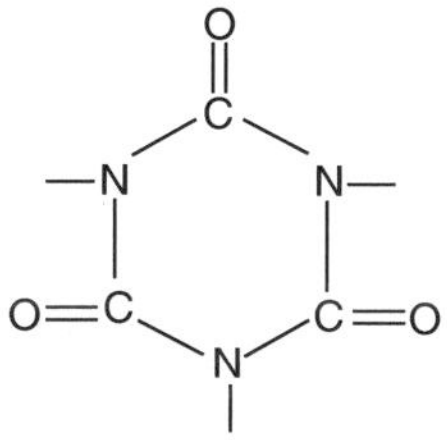	>270
Carbodiimide	—N=C=N—	>270
Urea	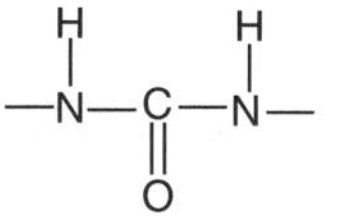	180–250
Urethane	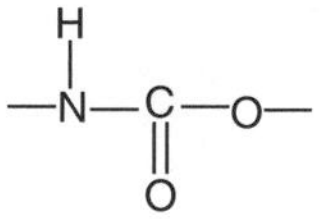	150–250
Allophanate	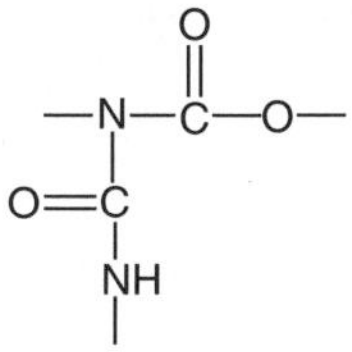	120–180
Biuret	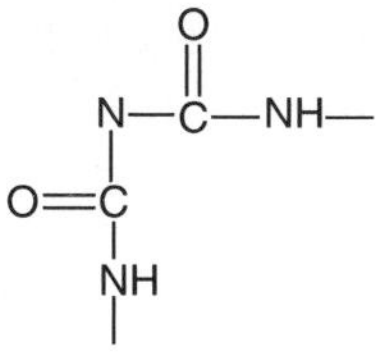	120–180

of foams formulation. The additive flame retardants, e.g. trichlorethylphosphate (TCEP) and trichlorpropylphosphate (TCPP) are used in formulations of PUR and PIR cryogenic foams. The additive retardants influence the thermal and mechanical properties of the foams considerably.

A study on the effect of the flame retardant on the strength, deformation and thermal properties was performed on the basis of the formulation of the PUR foams [42]. In PUR foams compositions, TCEP acts not only as a flame retardant, but also as a plasticizer [43]. In contrast to the plasticization of linear polymers, the plasticization of cross-linked polyesterurethane (PEU) has an essential distinction to polymerisation, namely, it is mixed together with the initial compounds, i.e. polyol and isocyanate, while plasticization is realised in the polymer formation process. The plasticized cross-linked PEU should be regarded as a polar polymer—polar solvent system.

Since the molecules of TCEP fill only a part of the fluctuation free volume, it could have been expected that this would favour the intermolecular interaction due to the formation of solvates with TCEP molecules by urethane groups. This hypothesis has been confirmed by IR-spectra. The increase in the relative intensity of the 3,320 cm^{-1} band is connected mainly with the formation of solvates of the PEU…TCEP…PEU type [43]. It was determined that at TCEP concentrations higher than 0.06 pbw the plasticisation of PEU networks happens. This flame retardant had an adverse action on the safety coefficient value of the cryogenic insulation material.

The fact that the TCEP in the composition of the polymeric matrix of PUR foams causes anti-plasticization is confirmed by the data of shear storage modulus depending on temperature. A study of the dynamic mechanical properties of the polymeric matrix of PUR foams containing from 0.03 to 0.196 parts by weight of TCEP in the temperature region 77–450 K, revealed that, with a growth of the content of TCEP, the β-relaxation is inhibited, which testifies the anti-plasticization effect [42, 44]. It should be mentioned that the anti-plasticization in the region of cryogenic temperatures is an undesirable effect.

TCEP as a flame retardant is currently seldom used—instead of it, TCPP is applied, which is characterised by a low corrosion activity. In the presence of this flame retardant, polyol systems can retain their activity up to 6 months. There is evidence that TCPP is used also in the cryogenic insulation material NCFI 24-124, although it exerts an adverse action on the properties of the PUR and PIR foams cryogenic insulation [1].

Figure 10.19 presents the SOFI material IWC-Cryo samples after the combustibility test. The length of the burned up parts of this material for the sample obtained at the components' ratio A/B = 0.86 is 13.2 cm and for the sample obtained at the ratio A/B = 0.90 is 12.8 cm. These indices are smaller than the length of the permitted burned up part 15 cm, and the material, in compliance with EN 13501-1, corresponds to class E.

Thermogravimetric analysis. The thermal resistance of foams can be estimated by thermogravimetric (TG) analysis. Samples of SOFI material IWC-Cryo were tested at the heating rate 10 °C/min in the air and the N_2 atmosphere [13].

Figure 10.20 presents the IWC-Cryo sample's decomposition in the N_2 atmosphere having a three-stage character. The first decomposition stage (DTG_{max} at about 190 °C) is connected with the TCPP as well as catalyst evaporation, and the urethane groups decomposition losses are about 10 mass%. The second stage is connected mainly to the decomposition of the polyol structure fragments. The coke formation process in N_2 starts at 350 °C, but in the air at 300 °C. Table 10.12 summarises the data on the progress of the decomposition of the SOFI material IWC-Cryo in the N_2 and air atmosphere.

Fig. 10.19 IWC-Cryo foams' after combustibility test in compliance with the Standard EN 13501-1

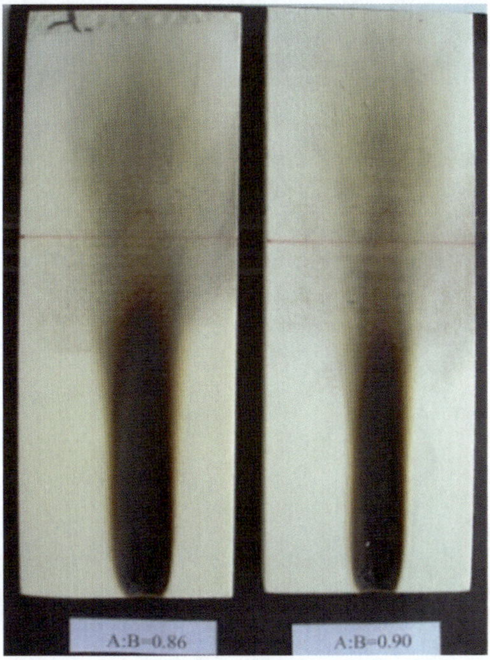

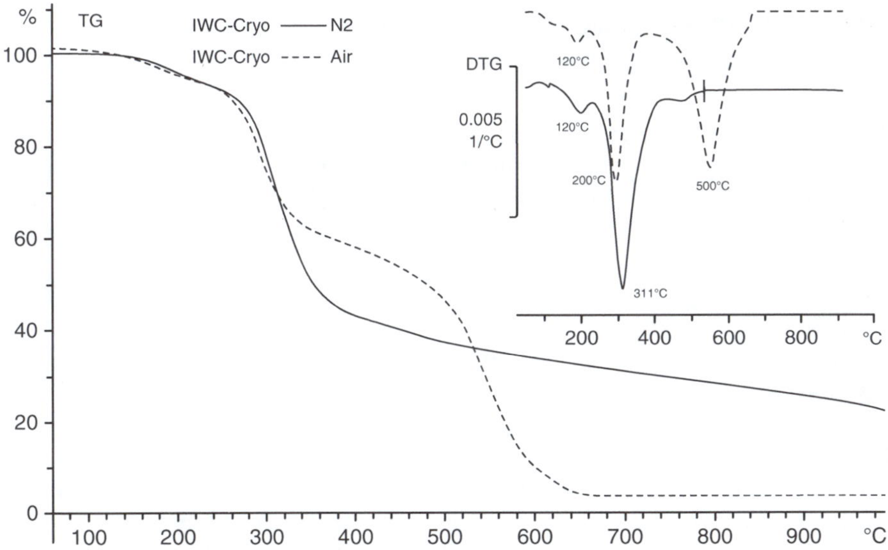

Fig. 10.20 TG tests of IWC-Cryo foams; *broken lines*—air atmosphere and *full lines*—N_2 atmosphere

Table 10.12 PUR foams IWC-Cryo thermal decomposition process

	Atmosphere	
Parameters	Air	N_2
Stage I		
Temperature range (°C)	147–226	149–225
Residue (%)	93.4	93.7
DTG_{max} (°C)	190	193
Stage II		
Temperature range (°C)	226–374	224–375
Residue (%)	60.0	52.8
DTG_{max} (°C)	298	311
Stage III		
Temperature range (°C)	374–1,000	375–1,000
Residue at 1,000 °C (%)	3.65	22
DTG_{max} (°C)	554	–
Mass residue at 450 °C (%)	54	50

At the temperatures 320–550 °C the mass losses are smaller in the case of the decomposition in the air atmosphere. The decomposition in the air media at the temperatures higher than 500 °C is due to the decomposition products oxidation.

10.8 Thermal Stresses and Strains in Foams' External Cryogenic Insulation

The adequacy of PUR and PIR foams to applications in cryogenic technique can be estimated several safety coefficients. According to the investigations [13] the ability of ET SOFI to withstand thermal strains can be characterised by safety coefficient K_{S1}:

$$K_{S1} = \varepsilon_{11}{}^{77} / \Delta l_{11}{}^{77-296}, \tag{10.4}$$

where $\varepsilon_{11}{}^{77}$—PUR foams' ultimate strain at 77 K (Temperature of liquid nitrogen LN_2) perpendicular to foams' rise direction (Rise direction is parallel to axis O3), %; $\Delta l_{11}{}^{77\text{-}296}$—the relative thermal contraction of PUR foams perpendicular to foams' rise direction, when cooling it from 296 to 77 K, %. The coefficient characterises how many times the measured ultimate tensile strain of foams at 77 K is bigger in comparison with the foams' relative thermal contraction, cooling foams from room temperature to 77 K. The PUR foams' contraction $\Delta l_{11}{}^{77\text{-}296}$ is measured in a fast cooling process in liquid nitrogen medium, similar to the conditions, when the tank is filled with LH_2 or LO_2. In all cases, the coefficients of thermal expansion are calculated using the contraction characteristics at the temperature difference $\Delta T = 219$ K.

In order to avoid appearance of defects in the cryogenic insulation, the complex of physical (Especially thermal) and mechanical properties should ensure that the tensile strength of PUR or PIR foams at 77 K would be at least 1.5 times higher than the thermal stresses that appear in the thermal insulation due to its coefficient of thermal expansion (CTE) perpendicular to foams' rise direction being considerably higher than that of aluminium alloys used to produce the ET [32, 45].

There are several methods for evaluation of the resistance of cryogenic insulation against thermal stresses that appear when filling LH_2 in the ET or discharging it from the ET, e.g. the safety coefficient K_{S2} determined by the (10.5), [32]:

$$K_{S2} = \frac{\sigma_T}{\Delta T E_T \alpha_F / 1 - \nu}, \qquad (10.5)$$

where σ_T—foams' tensile strength at 77 K; E_T—foams' tensile modulus of elasticity at 77 K; ν—foams' Poisson′s coefficient at 77 K; ΔT—temperature range from the initial state of the system and after filling in the cryogenic agent (LN_2); α_F—average CTE of foams' in the temperature range ΔT.

The coefficient K_{S2} characterises how many times the tensile strength σ_T of foams at 77 K is bigger than the foams' thermal stresses, when cooling foams from temperature of the initial state of the system and after to that filling in the cryogenic agent (LN_2). We propose a modification of the (10.5) for determination of the effective safety coefficient ${K_{S2}}^e$:

$$ {K_{S2}}^e = \frac{\sigma_T}{\Delta T E_T (\alpha_F - \alpha_M) / 1 - \nu}, \qquad (10.6)$$

where α_M—average CTE of metal in the temperature range "Room temperature … 77 K". The coefficient ${K_{S2}}^e$ characterises how many times the tensile strength σ_T of foams at 77 K is bigger than the foams' thermal stresses, when cooling foams from room temperature to 77 K, taking into account the thermal expansion of the metal substrate in the same temperature range.

In order to choose the most appropriate PUR or PIR foams by analysis of their safety coefficients, comprehensive experiments are carried out filling and discharging cryogenic agent into tanks of different size, volume and shape. Only after cyclic thermal loads appearing in these processes it is possible to estimate the influence of thermal stresses on foams as well as formation of the defects in insulation.

Comprehensive experiments on the effect of thermal stresses on the cryogenic resistance of different foams have been carried out [1]. Candidate insulation foams for LH_2 aircraft systems were selected on the basis of information available in publicly accessible information sources. The foams chosen for testing were selected on the basis of availability, properties and the aim to include candidates of different chemical types. The selected insulations included polyurethane, polyisocyanurate, polybenzimidazole, polymetacrylimide foams etc. with densities ranging 30–90 kg/m^3. All of the insulations tested were commonly available materials; none of them had been developed specifically for LH_2 service. The choice of test materials did not imply recommendations or endorsement of any material by NASA or *Bell Aerospace Textron*.

In Anthony et al. [1] an experimental evaluation of closed cell foams as cryogenic insulation for LH_2 tanks under thermal conditions representing airline type operations is provided. The original objective was to determine if any commercially available foams materials could survive more than a few hundred mission thermal cycles. In addition it was possible to assess the process of degradation and to identify failure modes. The PUR foams insulations exhibited the best cyclic life and an excellent thermal performance: two insulations of unreinforced PUR foams attained 4,400 thermal cycles (Equivalent to approx. 15 years of airline service) without serious thermal or structural degradation instead of the initial goal of 2,400 ones. The addition of fibre glass reinforcement or flame-retardant materials to an insulation degraded thermal performance and/or the life of the foams material. PIR foams exhibited worse performance under thermal cycles.

Thermal stresses in ET cryogenic insulation foams. Thermal stresses appearing in a cryogenic insulation material, when filling or discharging LH_2 from ET, are among the most important factors, influencing the system's safety and workability. These stresses should be assessed at the following levels: (a) in foams' insulation throughout its whole thickness; (b) in the contact surface "Foams—ET wall" and (c) between the spray-on foams layers (Interlayer adhesion).

Experiments were carried out with the aim of choosing potentially best cryogenic insulation materials for LH_2 tanks, intended for applications in aviation. One of the most important areas is the thermal protection for the tanks. Studies of subsonic and hypersonic vehicle applications indicate that low-density polymeric foams are attractive candidates for LH_2 tank insulation if their reliability and life are adequate. It indicated that (1) tanks should be kept at the liquid hydrogen temperatures (Except during overhaul periods) to reduce the possibility of fatigue failures from thermal stresses, (2) boil-off during ground-hold has a major influence in defining insulation thickness, and (3) system cost is a direct function of insulation thickness.

Analysing the available data on the thermal stresses in cryogenic insulation, the following can be concluded:

1. A PUR or PIR foams' layer, which adheres to the ET surface, about 5 mm in thickness, is subjected to thermal stresses of tensile nature.
2. The upper layer of the ET foams mainly performs the function of thermal insulation and thermal protection, and is subjected to the cryo-pumping effect as well as the action of heat flows. The ET upper layer foams in plane are is subjected to thermal stresses of compressive nature, although these stresses are about 2–3 times smaller than the thermal in-plane stresses of tensile nature for the foams layer, which adheres to the ET cold surface [1].

10.9 Cryo-Pumping Processes

Cryo-pumping occurs when a void in a material or structure is at a temperature low enough to densify a contained volume of gases. As the gases are condensed, a vacuum is created. If there is a pathway to the surface or other voids, additional gases are pumped into the void. When the material or structure is heated, the densified gases expand and are expelled from the void. During this expansion phase of cryo-pumping, the most damage in the material can be incurred. Cryo-pumping can be problematic for cryogenic materials and structures if the gases are densified in a void, then the void is rapidly heated, but the flow of gases from the void is restricted, causing a rapid pressure build-up.

Cryo-pumping and cryo-ingestion effects of cryogenic insulation. Cryo-pumping and cryo-ingestion processes are among the main reasons, which cause a whole range of hazardous phenomena in cryogenic insulation systems (A considerable decrease in thermal insulation capacity, ice formation, cracks formation etc.) [7–9, 12]. Cryo-pumping has long been theorised as one of the processes contributing to foams loss from larger areas of coverage. If there are cracks in the foams, and if these cracks lead through the foams to voids at or near the surface of the liquid oxygen and liquid hydrogen tanks, then air, chilled by the extremely low temperatures of the cryogenic tanks, can liquefy in the voids. After launch, as propellant levels fall and aerodynamic heating of the exterior increases, the temperature of the trapped air can increase, leading to boiling and evaporation of the liquid, with concurrent build-up of pressure within the foams.

Cryo-ingestion follows essentially the same scenario, except it involves gaseous nitrogen seeping out of the intertank and liquefying inside a foams void or collecting in the Super Lightweight Ablator. The intertank is filled with nitrogen during tanking operations to prevent condensation and also to prevent liquid hydrogen and liquid oxygen from combining [12].

Cryo-pumping index test. The test specimen consists of a 28.0 × 30.5 cm substrate with a 28.0 × 28.0 × 2.54 cm square block of cryogenic insulation bonded on its upper surface [46]. Thermocouples are located through-the-thickness of the cryogenic insulation, on the tank wall (inner) surface; in the bond line, and on the upper cryogenic insulation (external) surface. The through-the-thickness thermocouples are located at approximately 0.3175 cm (1/8 in.) intervals to measure temperature change inside the specimen during the test. The specimen is attached over a cryogenic chamber that can hold either LH_2 or LN_2. Cryogen flows into the chamber until steady-state temperatures are achieved (10–30 min). Once steady-state temperature has been achieved, the cryogen supply is turned off and the inner chamber is heated to room temperature. The cryo-pumping index CP_{index} is defined as:

$$CP_{index} = (T_{CP} - T_{SS})/T_{SS}, \qquad (10.7)$$

where T_{CP} is the lowest temperature after the cryogen supply has been turned off, and T_{SS} is the lowest steady-state temperature while the cryogen is flowing into the chamber.

The foams designed for cryogenic insulation should have a closed cell structure and no defects (Cracks, voids, delamination from the substrate etc.). There is no problem to obtain PUR or PIR foams with a high (>90 %) closed cells content, which ensures a low thermal conductivity coefficient at 20 K. The defects may appear during the foams' service time due to the technological inaccuracies and favour the cryo-pumping effect [7–9, 12].

If the foams' structure has defects, there is a possibility that air from the environment may be condensed in the thermal insulation, as the tank is filled with LH_2.The reason for the cryo-pumping effect can be the foams' low safety coefficient, too, as a result cracks and/or material's peel-off from the insulated surface may occur. The cryo-pumping effect is favoured also by the foams' hydrophility and the high relative moisture of the environment, as a result water vapours can condense in the foams' pores, and frost or ice may be formed on the foams' inner surface [12]. The diffusion characteristics of the PUR and PIR foams are dependent mainly on the production conditions of the thermal insulation, the type of the blowing agent and the structure of the foams (Porosity, cell size, content of closed cells, their orientation, cell wall thickness, etc.), but depend little on the chemical structure of the polymeric matrix.

10.10 Defects of External Tank Spray-On Insulation Foams

The operating environment of the ET is a complex combination of thermal conditions, aerodynamic heating, aerodynamic and acoustic loads and mechanical loads [8]. Internally, the ET shell structure contains various amounts of cryogenic liquid fuels, while externally the ET is exposed to ambient air temperature and pressure prior to launch and aerodynamic heating and pressure, vibro-acoustic loading and external pressure gradients during ascent. In Table 10.13 the major failures and defects are summarised.

Typical spray-on PUR foams' defect—delamination along the layers' knit line is presented in Fig. 10.21. The primary ET TPS failure modes appear to be a result of the consequences stemming from the SOFI application process and/or the overall ET system level design. An important point for the development of ETI is the possibility to check failures and defects during production, application and operation. Suitable nondestructive testing (NDT) methods have to be identified for that.

Table 10.13 The major defects and failure modes of spray-on foams ET insulation [7]

Delamination from the substrate	Separation between foams and substrate resulting in a layer of broken cells remaining on the primed aluminium surface Due to: thick spray passes; a substrate temperature that is too cold; surface contamination
Delamination along the knit line	Separation along the foams-to-foams interface Due to: inadequate overlap time, excessive thickness per foams pass, or contamination
Divoting	A large-scale, cohesive failure mode that has been observed on several flights Due to: the entrapped gas within the foams cellular structure, cryo-ingestion of condensed liquid nitrogen from the intertank, and cryo-pumping of energy sources into local voids in the SOFI or substrate debonds near the substrate surface
Rollover	During foams application, air pockets can be covered by foams that rebound off of adjacent surfaces Due to: gun over spray, regions with closely packed geometric features and 90° corners
Cracks	Break in foams exhibiting no material loss Due to: stress build-up/physical inducement. Large difference in coefficients of thermal contraction between foams and substrate—greater thermal contraction of foams causes it to crack/split
Porosity	Concentrations of elongated cells Due to: a high spray temperature/excessive heat of reaction
Voids	Absence of foams Due to: air pockets close to substrate when applying foams; conditions where air becomes trapped during foams application process; moisture present causing chemicals in foams to react forming carbon dioxide which is an excess volatile
Formation of elongated cells	Presence of cells with excessive expansion Due to: thick foams applications where blowing agent escapes during foams' rise

Fig. 10.21 Delamination along the knit line of PUR multilayered foams

10.11 Applications of Foams' Insulation in Space Technologies

10.11.1 Space Shuttle

Spray-on foams insulation has been developed for use on the cryogenic tanks of space launch vehicles beginning the 1960s with *Apollo* programme and further developed for the *Space Shuttle* programme. The *Space Shuttle* consists of three major components. The Orbiter carries the crew into space and returns to land on Earth. Two Solid Rocket Boosters provide the majority of the thrust needed to lift the Orbiter into space. Holding these major components together is the ET. The primary purpose of this tank is to carry the liquid propellants used by the Orbiter's three main rocket engines. The tank is the only part of the system that is not reused. After the main engines of the Orbiter have been shut down, the ET is released to reenter the atmosphere where it burns up. The *Space Shuttle* ET tank has three major sections as shown in Fig. 10.22: the liquid-oxygen (LO_2) tank, the intertank and the LH_2 tank [9]. The ET is coated with a spray-on cryogenic insulation that performs two major functions: (1) to insulate to maintain propellant quality during fill and ground-hold, and (2) to act as an ablative TPS during the ascent phase.

The *Space Shuttle* ET is covered with multiple urethane foams that basically belong to two categories of urethane foams: polyurethane and polyurethane isocyanurate (PIR) foams. The foams "systems" are comprised of parts A & B from raw materials originating from multiple sources. Part A is the isocyanate, and Part B is the polyol, catalysts, surfactants and the blowing agent. The Parts A & B are then proportionately spray-on or "poured" at a specified ratio onto the aluminium substrate of the ET and, in an exothermic reaction, form a protective closed cell foams insulation structure (In the USA polyisocyanate is denoted as component A and system of polyols as component B. In Europe the demotions are vice versa).

In ET insulation, the PUR or PIR foams spray-applying method has received wide acceptance [7–9]. PUR and PIR foams are applied using automated or manual spraying operations. In the case of automatic spraying of PUR or PIR, different technical solutions are possible. Thus, the insulation of the *Space Shuttle* ET with the PIR foams NCFI 24-124 is performed automatically, spraying several foams layers.

Technological processes of the Space Shuttle external tank insulation. To develop optimum technological conditions for new ET cryogenic insulation materials, it is necessary to analyze the experience reached when developing *Space Shuttle* ET insulation materials. The structural analysis technology has evolved significantly over the past 30 years. The study [9] had three primary objectives. The first objective is to evaluate the past and present analysis tools and modelling procedures used to analyze the ET TPS. The second objective is to identify the analysis tools and modelling procedures that may be applicable to ET TPS, including near-term capabilities to be pursued. The third objective is to assess the current understanding of failure modes for complex loading environments; to evaluate the capability of analytical tools, test databases and rationale; to

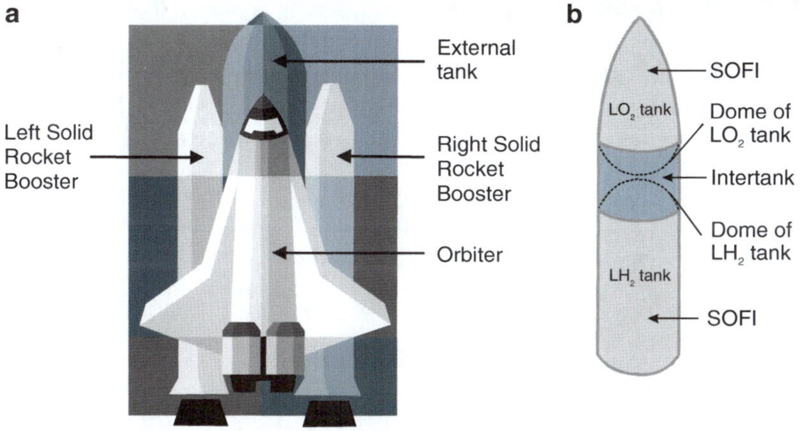

Fig. 10.22 a) "*Space Shuttle*" components and b) Components of the ET and locations of spray-on foams' thermal insulation (SOFI)

recommend analysis and testing needed; to identify near-term procedures and rationale to be implemented; and to identify long-term capabilities to be developed.

Automated PUR and PIR foams spray process and application. An automated spray process was used to cover the acreage surfaces with the two-part foams. Prior to the spraying process, all metallic surfaces were primed and cleaned. Nonmetallic surfaces were coated and dried, and wires were coated with an epoxy primer. Temperature and humidity of the work area and surface to be sprayed were measured and controlled. The parameters controlling the computer automated spray were the surface temperature of the part, ambient temperature, humidity and atmospheric pressure [9].

Tests used to qualify cryogenic insulation for a reusable launch vehicle. The tests include testing of thermal conductivity, glass transition temperature, radiant-heat panel test, LO_2 compatibility test, vertical burn test, cone calorimetry test, oxygen index test, open cell content testing and moisture absorption etc. [9].

The thermal properties of the cryogenic insulation of the *Space Shuttle* are evaluated in Semmes [47]. In the previous foams formulation efforts conducted by NASA, researchers were successful in identifying industry available foams formulations meeting *Shuttle* requirements with some physical attributes exceeding the existing foams performance.

Manual PUR and PIR foams spray process and application. A hand-spray process can be used to cover the closeout areas with the two-part foams. Prior to the spraying process, all metallic surfaces have to be primed and cleaned. Nonmetallic surfaces have to be coated and dried. Temperature and humidity of the work area and surfaces to be sprayed are measured and controlled. Other controlled parameters included the surface temperature of the part, ambient temperature, humidity and atmospheric pressure. The PIR cryogenic insulation NCFI 24-124 is sprayed on the acreage of the ET and is different from the polyurethane cryogenic insulation, BX-265 that is sprayed on the closeouts and areas not covered by the acreage cryogenic insulation. Other closeout foams, BX-265 were used on other super-light weight external tanks. There are significant areas of the ET left bare during fabrication and assembly at the NASA Michoud Assembly Facility (MAF). Each major tank component (LO_2 tank, LH_2 tank and intertank) is individually machine sprayed with NCFI 24-124, then assembled. A fair amount of the exposed areas, such as where the two propellant tanks are mated to the intertank, bipod, longerons, feed lines and instrumentation ramp, are secondarily or manually sprayed with BX-265 at MAF or Kennedy Space Center.

10.11.2 Launch System **Ariane**

The re-ignitable upper stage for the *Ariane 5ME* currently under development requires an innovative tank concept. To minimise the loss of cryogenic propellants like LO_2 and LH_2 during ground and during extended flight operation an effective insulation system is one of the major challenges. The both tanks have been arranged in such a way that the lower LO_2 tank is quasi-encapsulated by the upper LH_2 tank with a common bulkhead in between. The tank insulation has to fulfil the following main tasks: (1) to minimise the (aerothermal) heat fluxes to the upper LH_2 tank, (2) to minimise the heat input to LH_2 tank from the upper and lower inter stage cavities and equipment and (3) to provide an effective insulation at the common bulkhead to avoid LO_2 sub cooling and to prevent from heat fluxes to the LH_2 [13].

In order to meet the complex requirements various insulation solutions for the *Ariane* launchers including cryogenic closed cell foams compatible with aerothermal fluxes are developed. Beyond its primary function of heat leak reduction towards the propellants, the insulation of ET walls must also protect the aluminium structure from aerothermal fluxes during ascent. The closed cell PUR foams supplied by company "AIR LIQUIDE" for the *Ariane* launchers meets the following stringent requirements: low density, low thermal conductivity, suitability for bonding on cryogenic aluminium structure (no cracks), tightness (closed cell) to prevent cryo-pumping on launch pad, resistance to aerothermal fluxes during launcher ascent, easy shaping by curing and machining. "AIR LIQUIDE" offers closed cell foams for the *Ariane* launchers with the following properties: density 44 ± 5 kg/m^3 before curing, 50 ± 5 kg/m^3 after curing, thermal conductivity 33 mW/(m K) at room temperature, maximum temperature 90 °C in continuous use, 450 °C in 16 s transient [10].

10.11.3 Space Shuttle **Buran**

The surface (1,523 m^2) of the LH_2 external tank of the Soviet space shuttle's *Buran* rocket *Energy* was insulated with PUR foams *Ripor 2 N* and "PPU-17" [11, 48]. *Ripor 2 N* was spray-on applied on the cylindrical surfaces of the LH_2 tank, while the tank was situated in a horizontal position and rotated around its axis. The spray process was automated and it was characterised by the temperature range 20–25 °C both of the substrate and the environment. The cryogenic insulation was applied in one layer, thickness 50–60 mm, then mechanical processing of the external surface of the foams followed. Due to monolayer character there were no knit lines in the cryogenic insulation of the tank that could act as initiators of defects and faults. PUR foams "PPU-17" were applied for insulation of the intertank. The dome-shaped surfaces were insulated manually and the defects were detected by NDT methods. The component A of *Ripor 2 N* formulation comprised polyether and polyester polyols, a foaming agent Freon 113 ($CFCl_2$–CF_2Cl), flame-retardant TCEP and a catalyst package. The polymeric matrix had a small cross-link density ~600 [11] and the density of foams 40–45 kg/m^3. The ablative material PPU-306 was applied for thermal protection of LO_2 tank [48].

10.12 Conclusions

1. Wide applications of liquefied hydrogen not only in space technology, but in power industry and transport, could be forecasted for the twenty-first century. In this connection, efficient and safe cryogenic insulation materials, e.g. PUR and PIR foams will be necessary.
2. The main factors influencing the cryogenic resistance of external tank thermal insulation are analysed: (a) Thermal stresses and strains, dependent on the complex of physical and mechanical properties; (b) Thermal and diffusion (Cryo-pumping) properties dependent on cellular structure of the SOFI material, these factors being connected to the technological process of the SOFI production and (c) Defects in the SOFI.
3. Mechanical properties of rigid PUR foams in dependence of polymer matrix' molecular weight per branching unit M_c have been investigated. With M_c values increasing from 360 to 1,150, (a) the PUR foams' tensile strength perpendicular to rise direction and the corresponding ultimate strain increases, while elasticity modulus decreases both at temperatures 296 and 77 K and (b) the compressive strength parallel to rise direction decreases at 296 K, but grows at 77 K.
4. The capacity of PUR and PIR foams to resist thermal stresses depends greatly on the chemical structure and macromolecule's architecture of the polymeric matrix. The laboratory results revealed that PUR foams with high safety coefficient values can be obtained, if the polymer matrix' molecular weight per branching unit is in the range 600–900. Conventional PIR and PUR foams are not suitable for insulation of liquefied hydrogen ET.
5. The functions of the bottom, middle and upper layers of SOFI material are analysed. Influence of the thickness of mono- and two-layer foams as well as technological parameters on the SOFI material IWC-Cryo cellular structure and mechanical properties is characterised in the range 77–296 K.

6. Technological processes of spray-on PUR and PIR foams cryogenic insulation, cryo-pumping and the main defects of ETI are analysed as well as applications of foams' insulation in space technologies (*Space Shuttle*, *Ariane* and *Buran*).

References

1. Anthony FM, Colt JZ, Helenbrook RG (1981) Development and validation of cryogenic foam insulation for LH2 subsonic transports. NASA contractor report CR-3404, February, pp 1–69. http://ntrs.nasa.gov/archive/nasa/casi.ntrs.nasa.gov/19810011784_1981011784.pdf. Accessed 31 Aug 2012
2. Taylor AH, Jackson LR (1981) Structural concepts for a mach 5 cruise airplane LH fuselage tank. J Aircraft 18:655. 81.70.242.211/. . ./AIAA%20Journal%20of%20Aircraft. Accessed 31 Aug 2012
3. Banyay GA (2006) Examination of polymeric foams as an on-vehicular HPR hydrogen storage media. MS thesis, 109p. http://www.oucom.ohiou.edu/ou-microCT/Downloads/2006_Greg_Banyay_MS_thesis_foam.pdf. Accessed 31 Aug 2012
4. Abe A, Nakamura M, Sato I, Uetani H, Fujitani T (1998) Studies of the large scale sea transportation of the liquid hydrogen. Int J Hydrogen Energy 23:115–121
5. http://www.aerospaceweb.org/question/investigations/q0122.shtml. Accessed 31 Aug 2012
6. Matienzo LJ, Shah TK, Gibbs AJ, Stanley JR (1985) Thermal protection system for the spaca shuttle external tank: application of instrumental methods of analysis. J Therm Insul 9:30–45
7. Weiser ES, Nemeth MP, Clair TL (2004) Assessment of technologies for the space shuttle external tank thermal protection systems and recommendations for technology improvement. Part 1: materials characterization and analysis. NASA/TM-2004-213238, July, pp 1–19. http://ntrs.nasa.gov/archive/nasa/casi.ntrs.nasa.gov/20070013736_2007011218.pdf. Accessed 31 Aug 2012
8. Knigth NF, Nemeth MP, Hilburger MW (2004) Assessment of technologies for the space shuttle external tank thermal protection systems and recommendations for technology improvement, part 2, NASA TM-213256, August, pp 1–58. http://ntrs.nasa.gov/archive/nasa/casi.ntrs.nasa.gov/20040129726_2004135086.pdf. Accessed 31 Aug 2012
9. Gates TS, Johnson TJ, Whitley KS (2005) Assessment of technologies for the space shuttle external tank thermal protection systems and recommendations for technology improvement, part 3, NASA TM-2005-213778, July, pp 1–47. http://ntrs.nasa.gov/archive/nasa/casi.ntrs.nasa.gov. Accessed 31 Aug 2012
10. AIR LIQUIDE. Space cryogenic insulation. www.dta.airliquid.com; http://www.airliquideadvanced-technologies.com/file/otherelement/pj/insulations6933.pdf. Accessed 31 Aug 2012
11. Stirna U, Alksnis A, Tukums P, Yakushin V et al (1982) Certificate of USSR 1159309 (A patent). A method for obtaining polyurethane
12. (2003) Columbia Accident Investigation Board (CAIB) Mishap Report, Chapter 3. 1:49–84 http://klabs.org/richcontent/Reports/Failure_Reports/columbia/caib_report/caib_report.htm. Accessed 24 Jan 2013
13. Fischer WPP, Stirna U, Yakushin V, Cabulis U (2010) Cryogenic insulation for LOX and LH2-tank application. In: AIAA-2010-6295, 40th international conference on environmental system, Barcelona, Spain, 11–15 July 2010, CD-Rom:1–18
14. Ionescu M (2005) Chemistry and technology of polyols for polyurethanes, chap 21. Rapra Technology Limited, UK, 535p
15. Fedors RF (1974) A method for estimating both the solubility parameters and molar volumes of liquids. Polym Eng Sci 14:147–154
16. Соголова ТИ, Демина МИ (1977) Температурная зависимость механических свойств полимеров различного химического строения в интервале температур от 4,2 до 300 К

[The temperature effect on mechanical properties of different type polymers in temperature from 4.2 to 300 K]. Mech Compos Mater 13:387–391 (in Russian)
17. Соголова ТИ, Демина МИ (1975) Механические свойства полимеров при 4,2 К и их зависимость от химического строения и условий физического структурирования [The mechanical properties of polymers at 4,2 K and it dependance from chemical structure]. Mech Compos Mater 11:771–783 (in Russian)
18. Щербакова ТС, Макушкин АП, Чудина ЛИ et al (1987) Хрупкое разрушение полимеров при криогенных температурах [Friable destruction of polymers at cryogenic temperatures]. Plast Mass 1:11–12 (in Russian)
19. Stirna U, Beverte I, Yakushin V, Cabulis U (2011) Mechanical properties of rigid polyurethane 1094 foams at room and cryogenic temperatures. J Cell Plast 47:337–353
20. Becker R (1978) Beziehungen zwischen der Glastemperatur und der Chemischer Struktur von Polymeren. Faserforschung Textiltechnic 29:361–395
21. Becker R (1977) Quantitative Bezieungen zwischen der Glastemperatur von Polymeren und Energiegrössen. Z Phys Chem 258:953–966
22. Oertel G (1993) Polyurethane handbook. Hanser, Munich
23. Perepetchko II (1977) Свойства полимеров при низких температурах [Properties of polymers at low temperatures]. Khimija, Moscow, 271p (in Russian)
24. Welte RE (1984) Calculation and measurement of reaction temperature in rigid polyurethane and polyisocyanurate foams. J Cell Plast 20:331–356
25. Krüger J, Müller M, Yakushin V, Cabulis U, Stirna U (2010) Polyurethane foam for thermal insulation at extremely low temperatures. Patent DE102010007713A1
26. Берлин АА, Шутов ФА (1980) Химия и технология газонаполненных высокополимеров [Chemistry and technology of foamed polymers]. Наука, Moscow, 508p (in Russian)
27. ISO 8873–1 (2006) Rigid cellular plastics – spray-on polyurethane foams for thermal insulation-part 1 material specifications
28. ISO 8873–2 (2007) Rigid cellular plastics – spray-on polyurethane foams for thermal insulation-part 2 application
29. ISO 8873–3 (2007) Rigid cellular plastics – spray-on polyurethane foams for thermal insulation-part 3 tests methods
30. ASTM C591 - 12a standard specification for unfaced preformed rigid cellular polyisocyanurate thermal insulation
31. Fesmire JE, Coffman BE, Meneghelli BJ, Heckle KW (2012) Spray-on foam insulations for launch vehicle cryogenic tanks. Cryogenics 52:251–261. doi:10.1016/j.cryogenics.2012.01018
32. Demharter A (1988) Polyurethane rigid foams, a proven thermal insulating material for applications between +130°C and −196 °C. Cryogenics 38:113–117
33. Timmerhaus KD (2007) Insulation progress since the Mid-1950s. Cryogenic engineering, International cryogenic monograph series, part 3. Springer, New York, pp 120–133
34. Lerch BA, Sullivan RM (2006) Thermal expansion of polyurethane foams. In: 43rd Annual technical meeting of the Society of Engineering Science. The Pennsilvania State University, University Park, 14 Aug 2006
35. Tarecpir polyisocianurate foams, Kingspan. www.kingspantarec.com. Accessed 31 Aug 2012
36. Stokes E (2006) Thermal expansion of three closed cell polymeric foams at cryogenic temperatures. Source of application. NASA Marshall Space Flight Center 202.118.250.135/nasa/STAR/star0621.pdf. Accessed 31 Aug 2012
37. Reed RP, Arvidson JM, Durcholz RL (1973) Tensile properties of polyurethane and polystyrene foams from 76 to 300K. In: Advances in cryogenic engineering 18, Proceedings of cryogenic engineering conference. Colorado University, Boulder, 9–11 Aug 1972. Plenum Press, New York, pp 184–193
38. Sparks IL, Arvidson JM (1985) Thermal and mechanical properties of polyurethane foams at cryogenic temperatures. J Therm Insul 8:198–232
39. Nadeau H, Kolakowski G, Ulrich H (1981) Polyisocyanurate foams for cryogenic, petrochemical and solar energy applications. J Therm Insul 5:93–112

40. Yakushin VA, Zmudj NP, Stirna UK (2002) Physico-mechanical characteristics of spray-on rigid polyurethane foams at normal and low temperatures. Mech Compos Mater 38:273–280
41. Thimm T (1983) Derzeitige Erkentnisse über thermischen und thermooxidativen Beanspruchung von Polyurethan elastomeren. Kautschuk Gummi und Kunststoffe, 1983, Bd.36, N4, s.257–268
42. Yakushin VA, Stirna UK, Zhmud NP (1999) Effect of flame retardants on the properties monolithic and foamed polyurethanes at low temperatures. Mech Compos Mater 35:447–450
43. Stirna UK, Tukums PS, Goba DzN (1986) Пластификация сшитых полиэфируретанов трис(β-хлорэтил)фосфатом (Plastification of polyesterurethanes by trichlorethylphosphate). J Latv Acad Sci (Chem Ser) 1:445–449 (in Russian)
44. Perepechko II, Maksimov AV, Stirna UK (1990) Dynamic viscoelastic properties of plasticized polyesterurethanes. Vysokomol Soedin B 32:607–612 (in Russian)
45. Komada H, Imoto K (1981) ML system cryogenic insulation with rigid urethane s. Int Progr Urethanes 3:99–111
46. Johnson TF, Weiser ES, Grimsley BW, Jensen BJ (2003) Cryopumping in cryogenic insulations for a reusable launch vehicle. In: 48th international SAMPE symposium and exhibition, Long Beach, CA, USA, 11–15 May 2003
47. Semmes EB (2012) Comprehensive Shuttle foams debris reduction strategies. Source of acquisition NASA Marshall Space Flight Center 202.118.250.135/nasa/STAR/star0707.pdf. Accessed 31 Aug 2012
48. Hendrickx B, Vis B (2007) Energiya-Buran: the Soviet space shuttle. Springer-Praxis, Chichester, 522 p

Chapter 11
Cryogenic Treatment of Materials: Cutting Tools and Polymers

Simranpreet Singh Gill and Harpreet Singh

Contents

Abstract Cryogenic treatment (CT), a supplementary process to conventional heat treatment process, is the process of deep-freezing materials at cryogenic temperatures to enhance the mechanical and physical properties of materials being treated. CT is one of the field in which the materials to be treated play a very imperative role in its technological development. Some of its successful applications in nonferrous materials defy the conventional reasoning that the only affect it has is to convert retained austenite to martensite. For example tungsten carbide cutting tools, electronics materials, some plastics, composites, and

S.S. Gill (✉)
Department of Mechanical Engineering, Beant College of Engineering and Technology, Gurdaspur 143 521, India
e-mail: ritchie_223@yahoo.com

H. Singh
School of Mechanical, Materials and Energy Engineering, Indian Institute of Technology Ropar, Rupnagar 140001, India

S. Kalia and S.-Y. Fu (eds.), *Polymers at Cryogenic Temperatures*,
DOI 10.1007/978-3-642-35335-2_11,

polymers show significant improvements that cannot be supported by the conventional theories as to why the process works. CT technologies have applications in wide range of areas including cutting tools, polymers, plastics, power industry, medicine, rocket propulsion and space simulation, food processing, to name but a few of them. The execution of CT on cutting tool materials increases wear resistance, hardness, and dimensional stability and reduces tool consumption and down time for the machine tool set up, thus leading to cost reductions. Similarly, improvements in wear resistance, hardness, dimension stability, crystallinity, tensile strength and elongation have been reported for polymers subjected to CT. The effects of CT on tool materials (steels and carbides) and polymers along with their applications are reviewed for manufacturing industry in this chapter. Although it has been confirmed that CT can improve the service life of tools and polymers, the degree of improvement experienced and the underlying mechanism remains ambiguous. The steps involved in CT are critical enough to account for the significant incongruity in post-treated performance of treated materials. If we look toward the next century, the topic of the-state-of the art and future developments in CT areas has several aspects to it. Firstly it is essential to look at the history of the development of CT. It is equally important to evaluate the current status of science of CT and to identify research and development trends in this area which can act as starting point for future developments. In this chapter, an attempt has been made to present the past, present, and future of CT technology for tool materials and polymers.

Keywords Cryogenic treatment • Mechanical properties • Metallurgical properties • Tool steels • Tungsten carbide • Polymers

11.1 Introduction

11.1.1 Historical Background of Cryogenic Treatment

The science of subzero treatments known as cryogenic treatment (CT), as we recognize it today, started in the late 1800s when Sir James Dewar (1842–1923) perfected a technique for compressing and storage of gases from the atmosphere into liquids. However some credit a couple of Belgians as being first to separate and liquefy gases. These compressed gases were super cold and any metal that came in contact with the ultra-low temperatures showed some interesting changes in their characteristics. Practical applications of CT go as far back as the 1930s when the Junkers Company in Germany used it on components of their Jumbo aircraft engines. CT had its US origins in the 1940s, but that was a very primitive process compared to today's procedures. In the 1940s scientists discovered that by immersing some metals in liquid nitrogen they could increase the wear resistance of motor parts, particularly in aircraft engines, giving a longer service life. At the time this was little more than dipping a part into a flask of liquid nitrogen, leaving there for an hour or two and then letting it return to room temperature. They managed to get the hardness they wanted but parts became brittle. As some benefits could be found in this crude method, further research into the process was needed. The applications at this stage were mostly military. Around the mid-1950s California-based McCulloch Chain Saw Company was the first to use CT on chain saw blade links thus pioneering the use of CT in nonmilitary applications. They started CT of chainsaw blades but kept it a secret so other manufacturers could not make better blades. Research related to CT was reported in the form of magazine article as far back as the July, 1957 edition of Tooling and Production Magazine. Later on several people picked up the thread in the early 1970s. The most notable and persistent was Dr. Randall Barron of Louisiana Technical University. Dr. Barron has written multiple research papers about the subject, as have some of his students. These papers are widely cited in the cryogenics industry [7, 50, 89].

Till around the 1960s, the lack of any significant technological breakthrough in CT restricted its applications. The prominent reason for this was problem of cracking and chipping of components after direct immersion into liquid nitrogen. Thus, the need of some way of ramping temperature down on a component in a controlled manner was felt. This tended to be expensive and cumbersome before microprocessor controls became available. It was the invention of the microprocessor-controlled temperature control that really made it possible to get repeatable results in CT. Another item that needed to be added to the mix was the tempering phase of the CT process. Often those who experimented in CT ran into the brittle behavior of the processed materials, and decided that the cold had crystallized the metal, and that all was lost. Another thing that hindered CT of materials was that intuitively, one would not think of materials changing due to cold. Archeological evidence shows that humans have been using heat on metals for over 75 centuries. Extreme cold has only been available for roughly 100 years. So it is almost intuitive that you use heat to change metals not cold. Another thing had to happen in order to make people think in terms of cold changing things. The changes had to be detectable. Even into the 1940s measurements were not precise enough to indicate changes caused by CT. CT processes were not controlled well enough to be able to say with certainty that the change was a direct result of the low temperature. A cutting tool may last twice as long, but that was easily put off to the effects of how it was grounded or the variance in cutting geometry. Given that there could be large differences between one piece and another, the benefits given by the CT process could always be attributed to the production differences. It is important to realize that CT process has been largely empirically developed. The reason is that it has not been researched deeply by large corporations or by government bodies with deep pockets. Even Dr. Barron's published works go largely to the results of the process rather than how the process works. The process is in need of research to optimize its results.

NASA led the way and perfected a method to gain the best results, consistently, for a whole range of materials. The performance increase in parts was significant but so was the cost of performing the process. Work continued over the years to perfect the process, insulation materials improved, the method of moving the gas around the process developed, and most importantly the ability to tightly control the rate of temperature change. Technology enabled scientists to look deeper into the very structure of materials and better understand what was happening to the atoms and how they bond with other carbons. They also started to better understand the role that temperature plays in the treatment of materials to effect the final characteristics. The microprocessor enabled a steady but continual reduction in size of the control equipment required as well as increasing the accuracy of that part of the process. Subsequently, the research about CT has been validated during the 1980s by the first result in machine tools [4, 6]. ASM International, the metallurgist's and materials scientist's professional society started a committee on the process in the late 1990s. This committee was founded when members started to complain about the peculiar claims made by some of the early companies that were formed to promote the process but these claims ultimately ended up as harm to the CT process. The committee also strove to create a database of research and articles about the process, and promoted sessions at ASM conferences that featured CT. It is only since the mid-1990s that the process has started to become a commercially viable treatment. From the 1990s, CT has also been applied to many different components i.e., motor racing parts [84], in particular gears and bearings [53, 69, 70, 75, 94], oil drills, gun barrels, knives, surgical and dental instruments [59], and even brass musical instruments [52], piano and guitar strings, baseball bats, golf clubs, cutting tools [39, 42], and polymers [49, 81, 103] too.

11.1.2 Applications of Cryogenic Treatment

11.1.2.1 Tool Steels

These steels can be classified in three major categories as cold-work, hot-work and high-speed steels. Cold-work tool steels respond significantly well to CT. These steels are commonly used for applications such as cold heading, blanking, and trimming where tooling that operates below 200 °C. Collins and Dormer [21] investigated the W and D series of the cold-work steel grades and found that CT reduced the toughness of D2 by nearly 40 % primarily due to the transformation of retained austenite. However, CT below −100 °C gave a clear improvement in toughness, although the reason for this improvement was not clear. CT was successfully applied to two different cold-work tool steels, X155CrMoV121 and X110CrMoV82, to improve their wear resistance [76]. Yugandhar and Krishnan [98], after cryogenically treating AISI O1 and O2 press tools and powder compacting tools, found that the tools worked very well as the life of the tools increased. The performance of D2 and D3 grades cutting tools and metal forming tools were improved to three times that of hardened and tempered tools. Also, Das et al. [30] successfully enhanced the wear resistance of D2 tool steel by a few folds by applying CT. The hot-work tool steels are processed at temperatures above 200 °C. They typically use these steels for forging, die casting, and aluminum extrusion dies. H-13 showed a 50 % increase in die life in one of the study by Paulin [75]. Performance of AISI H11, H12, and H13 grades metal forming and pilgering tools were found to the tune of 200 % to that of hardened and tempered tools and chisels made out of O1 and S1 grade steel improved their performance to 150–200 %. High-speed steels are dominantly used for cutting applications. Cohen and Kamody [18] found 42 % increase in tool life of cryogenically treated TiCN-coated M4 tools when compared to untreated coated tools in a broaching operation. The performance of M2 and M6 grades cutting tools and metal forming tools were improved to three times that of hardened and tempered tools [98]. da Silva et al. [32] evaluated the performance of the cryogenically treated M2 HSS tools by Brandsma rapid facing test and achieved 44 % improvement in tool life. CT increased the performance of the M2 HSS twist drills as the gain observed during drilling tests adopting catastrophic failure as the end of tool life criterion varied from 65 to 343 % depending on the cutting conditions used [37]. Gill et al. [43] did detailed study on the effects of CT on M2 HSS in laboratory conditions and found the results consistent with findings of previous studies.

11.1.2.2 Tungsten Carbide

This material also responds significantly well to cryogenic treatment. Seah et al. [85] implemented the CT on cobalt-bonded tungsten carbide (Co-WC) and found that the treated inserts were superior to those of the untreated as-received inserts at high cutting speeds. Stewart [88] successfully applied cryogenic treatment on C2 tungsten carbide inserts and recorded lower tool force while turning MDF thus enhancing the life of inserts. Arner et al. [2] examined five different style cryogenically treated tungsten carbide inserts in full production operations and confirmed very consistent changes in overall tool life; however, the amount of improvement was dependent on the tool style. Yong et al. [96] showed that CT no doubt improves the resistance to chipping of tools and, to a less significant extent, improves flank wear resistance. Work by Quek [80] also agreed with the findings of these authors. In another study, Yong et al. [97] cryogenically treated tungsten carbide milling inserts and found 28.9–38.6 % increase in tool life. After experimental evaluation of the comparative performance of cryogenically proceeded TiCN-coated carbide inserts and Kennametal Grade KC 990 inserts by gas infusion process (dry process), Bonilla et al. [13] also succeeded in enhancing the tool life. Reddy et al. [82] observed the improvement in life of normal and deep cryogenically processed carbide inserts by an amount of 9.58 and 21.8 %, respectively. Gill et al. [44] machined C60 steel to evaluate the performance of cryogenically treated tungsten carbide inserts under dry and wet turning conditions. They corroborated that the performance of cryogenically treated carbide inserts depends a lot on cutting conditions. The maximum gain achieved in tool life was 55 %. In other independent study, Gill et al. [40] studied machining behavior of TiAlN-coated inserts and concluded that CT can have negative effect on the performance of coated inserts. In recent study, Gill et al. [41] also verified the positive effects of CT on tungsten carbide in laboratory conditions.

11.1.2.3 Polymers

There are many reports related to the approaches used to modify the polymer surface ([17, 22, 47, 49, 60, 61, 64, 68, 74, 93, 102, 105]). Among these approaches, CT seemed to be a simple and interesting method. Significant changes in dimension, crystallinity, tensile strength, and elongation in ultra-high molecular weight polyethylene (UHMWPE) samples after CT has been reported by Trieu et al. [90]. In their paper Indumathi et al. [49] investigated the effectiveness of CT on the wear performance of a series of thermoplastic engineering polymers such as polyimide (PI), polyetherimide (PEI), polytetrafluoroethylene (PTFE), polycarbonate (PC), and polyurethane (PU). A series of composites of PI, PEI, and its copolymer with increasing amount of short glass fiber and solid lubricants was also investigated. The authors identified CT as to be an effective technique for enhancing the abrasive wear performance. Increase in hardness due to CT was found to be the most important parameter responsible for performance enhancement in the case of the PTFE. With the help of scanning electron microscope images, a rougher topography for the surface of the cryogenically treated PEI has been reported. In another study short carbon fibers were treated by cryogenic treatment and the mechanical and tribological properties of carbon fiber reinforced epoxy were investigated [100, 101]. The results showed that the mechanical and tribological properties of carbon fiber reinforced epoxy were improved after CT due to the enhanced fiber–matrix interfacial bonding. In a recent study Hybrid PTFE/Kevlar fabric was treated by CT [103]. The authors found that the wear resistance of the fabric/phenolic composite was improved after CT. Enhanced fiber/resin adhesion was considered to contribute to the improved wear resistance of cryogenically treated fabric/phenolic composite.

11.2 Process of Cryogenic Treatment

CT is not, as it is often mistaken for, a substitute for good heat treating, rather it is an add-on or supplemental process to conventional heat treatment to be done before tempering. However, it has been reported that some improvement can be obtained by carrying out the treatment at the end of the usual heat treatment cycle, i.e., on the finished tools [78]. A typical CT consists of a slow cool-down rate (2.5 °C/min) from ambient temperature to the temperature of liquid nitrogen (Fig. 11.1). When the material reaches approximately at −193.15 °C (80 K), it is soaked for an appropriate time (generally 24 h). Then the part is removed from the liquid nitrogen and allowed to warm at room temperature in ambient air. By conducting the cool-down cycle in gaseous nitrogen, temperature can be controlled accurately and thermal shocks to the material are avoided [16, 79].

A fundamental distinction among different CTs is given by the parameters of the cooling–warming cycle. Depending on the minimum temperature reached during

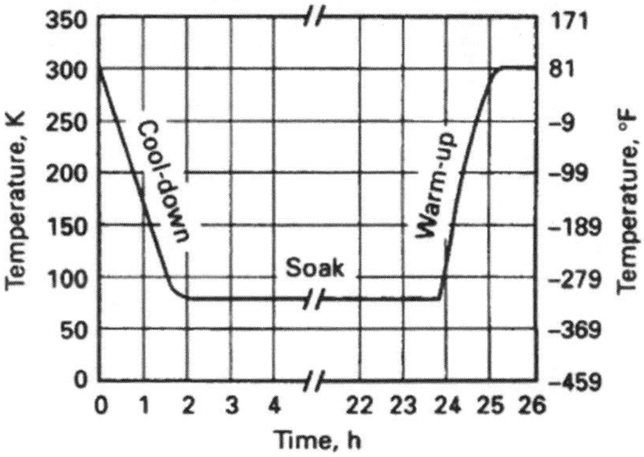

Fig. 11.1 Plot of temperature versus time for the cryogenic treatment process. Soaking temperature is −196 °C [16].

the cycle, Bensely et al. [9] proposed two categories: firstly shallow cryogenic treatment (SCT) in which the samples were placed in a freezer at −80 °C and then they are exposed to room temperature; secondly deep cryogenic treatment (DCT) in which the samples were slowly cooled to −196 °C, held-down for many hours and gradually warmed to room temperature. The typical process parameters of CT include rate of cooling, minimum soaking temperature (T_{soak}), soaking period, and rate of warming. It is not possible to recommend a single process for every tool material, nor even a single cycle for all components of the same material. Each component needs to be separately assessed and an individual process route devised for it that will depend on the combination of various mechanical properties required in service. The CT is just one step in that process and must be integrated into the processing route [87]. Literature reveals that to maximize the benefit obtained, CT should occur after quenching and before tempering [36, 55, 56, 70, 71, 73, 87, 92, 106]. Mohanlal et al. [71] and Zurecki [106] recommended the use of CT to be most effective if applied soon after quenching and before tempering. Kamody [55, 56] introduced a new Nu-Bit process, a rapid, deep cryogenic method, as an inline extended quench performed after conventional quenching and before a single, modified temper. Dymchenko and Safronova [36] advised to carry out low-temperature tempering after the CT of quenched steel in order to avoid austenite stabilization. Additionally, Vimal et al. [92] also confirmed that the cryogenic treatment should be done before tempering immediately after quenching to obtain maximum benefits. Stratton [87] proposed steps in the processing route for maximum wear resistance as heat to an austenitizing temperature, quenching, CT, reheat to room temperature, and finally tempering. Gill et al. [40, 41, 43, 44] subjected M2 HSS and carbide inserts to two tempering cycles to relieve the stresses induced by CT (Fig. 11.2). This was accomplished by increasing the temperature to +196 °C and then slowly reducing the temperature back to room temperature at the rate of 0.5 °C/min.

Ray [81] subjected the short beam shear (SBS) specimens of glass/epoxy laminates to two different treatment conditions. In first treatment, the samples were exposed to 40 °C for 5 min followed by immediate treatment at −40 °C. Whereas in second treatment, samples were first cryogenically treated at −40 °C for

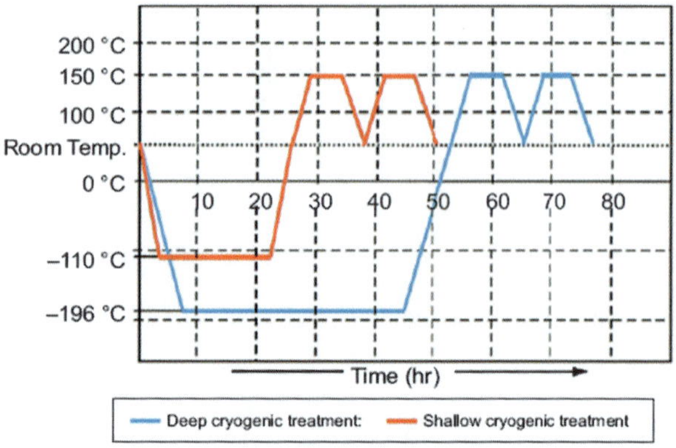

Fig. 11.2 Thermal treatment cycle used by Gill et al. [40, 41, 43, 44]

Table 11.1 Various kinds of heat treatment cycles studied by Alexandru et al. [1]

Route	Cycle
A	Quenching 1,230 °C
B	Quenching 1,230 °C + double tempering 560 °C
C	Quenching 1,230 °C + subzero (−70 °C)
G	Quenching 1,230 °C + subzero (−70 °C) + tempering 560 °C
H	Quenching 1,230 °C + tempering 560 °C + subzero (−70 °C)
M	Quenching 1,230 °C + tempering 560 °C + subzero (−70 °C) + tempering 560 °C
N	Quenching 1,230 °C + tempering 560 °C + subzero (−70 °C) + tempering 560 °C + subzero (−70 °C) + tempering 560 °C

5 min and then immediately exposed to 40 °C. They reported that CT does not have any positive effect if done after the tempering. Hence, it can be concluded from these studies that tempering may have stabilized the retained austenite in steels, making it more difficult to transform during CT if applied later. The effect of varying the sequence of tempering and CT for tungsten carbide has not been investigated till date and can be taken up for future studies.

In contrary to above results, some studies have proposed that CT can be performed after quenching and tempering without compromising the advantages of CT [1, 54, 72, 78, 99]. Alexandru et al. [1] established the influence of various CT cycles (Table 11.1) onto the material properties of steels by choosing different sequences of quenching, tempering and subzero treatments and concluded that CT should be applied after quenching and tempering. Like Alexandru et al. [1], Yun et al. [99] also investigated the microstructure of M2 high-speed steel after submitted this material to different CT cycles (Table 11.2) by varying the sequence of different cycle steps and proposed that the CT on M2 high-speed steel can be applied either after quenching and tempering or straight after quenching. Popandopulo and Zhukova [78] have also recommended tempering to precede CT. Another study also shows that, if CT is carried out after quenching and followed by the usual tempering cycle, its influence on the properties of steel is negligible [72]. Kalin et al. [54] also agreed that the subsequent tempering do not

Table 11.2 Different cycles applied to M2 high-speed steel [99]

Route	Cycle
A	Quenching from 1,250 °C + triple tempering at 560 °C
B	Quenching from 1,250 °C + 1 cycle 24 h subzero at −196 °C + triple tempering at 560 °C
C	Quenching from 1,250 °C + 1 cycle 48 h subzero at −196 °C + triple tempering at 560 °C
D	Quenching from 1,250 °C + 3 cycle totalling 48 h subzero at −196 °C + triple tempering at 560 °C
E	Quenching from 1,250 °C + triple tempering at 560 °C + 1 cycle 48 h subzero at −196 °C

have any significant impact on the fracture toughness of M2 high-speed steel when subjected to a range of tempering temperatures after CT.

It is evident from the review of the published literature that there is no consensus between researchers as far as process of CT is concerned. It may be concluded that as different materials react differently to the CT process, the need for optimization of the CT process was felt. Darwin et al. [23, 24] were the first to identify the main parameters involved in CT as cooling rate, soaking temperature, soaking time, heating rate, tempering temperature, and tempering time. The cooling rate is the rate at which the sample is cooled to the soaking temperature. Soaking temperature is the temperature at which the sample is held, while the soaking period is the time for which the sample is held at the soaking temperature. The heating rate is defined as the rate at which the sample is heated back to room temperature. Their results show that the significance of the parameters for improving mechanical properties of steels prevails in the following order of importance: (1) soaking temperature (72 %); (2) soaking period (24 %); (3) rate of cooling (10 %); (4) tempering temperature (2 %); and the parameter called tempering period is insignificant [23, 24]. Since the net contribution of the first three parameters is 98 %, these three parameters are currently being debated upon and hence considered for discussion in the following sections.

11.2.1 Minimum Soaking Temperature

The first users of CT applied soaking temperatures in the range of −80 to −100 °C for periods of about 30–60 min [45]. Later, Hallum [46] applied soaking temperatures in the range of −78 to −85 °C to the material and held at the temperature for several hours. In recent applications the materials are subjected to a controlled lowering of the temperature to −196 °C which is known as deep cryogenic processing [16]. The minimum soaking temperature is identified as the most significant factor [23, 24]. Bensely et al. [10] classified CT process in two types on the basis of minimum soaking temperature as SCT and DCT. They found that the amount of retained austenite in conventionally heat-treated (28 %), shallow cryogenically treated (22 %), and deep cryogenically treated (14 %) samples decreased with the lowering of soaking temperature. In another study, Bensely et al. [9] concluded that wear resistance of case carburized steel improves by 85 %

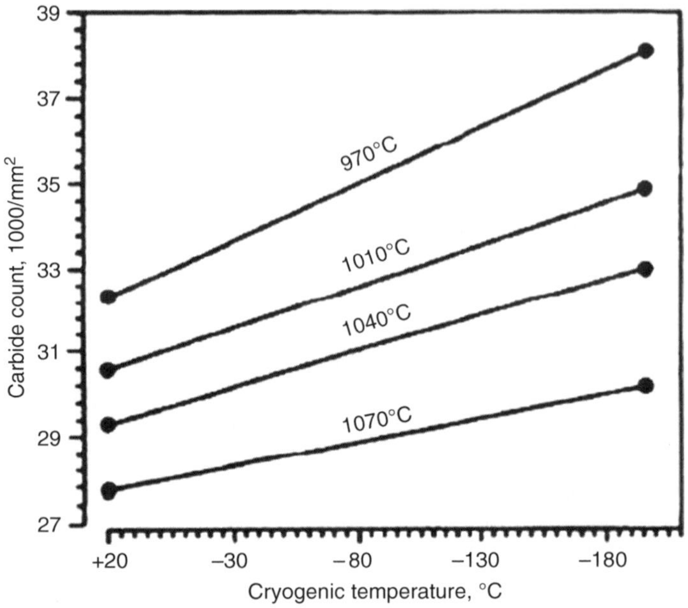

Fig. 11.3 Effect of cooling to subzero temperature on carbide number for D2 steel [21]

for shallow cryogenic processing (193 K) over conventional heat treatment and 372 % for deep cryogenic processing over conventional heat treatment. They also showed that wear resistance improvement of deep cryogenic process was 152 % higher than shallow cryogenic process. Barron [6], after cryogenically treating several materials including the M2 high-speed steel at −84 °C (maintaining it at this temperature for 24 h), observed a significant improvement on the wear resistance in sliding abrasion tests when compared to conventionally heat-treated steel (quenched and tempered). When the temperature of the CT was reduced further to −196 °C, the wear resistance was increased even more. The work of Mohanlal et al. [71] concluded that tempered samples when cryogenically treated at 133 K for 24 h yielded negative results but when treated at 93 K for 24 h the results were favorable. Hence, tempered samples if treated at still lower temperatures may yield still better results on par with untempered cryogenically treated samples. This also suggest to conclude that the stabilization of phases that would take place during tempering requires a sufficient degree of undercooling and time to get transformed to stable harder/tougher phase that offer better wear resistance. From experimental tests by Barron [5], it was found that the wear resistance of samples of 52,100, D2, A2, M2, and O1 tool steels soaked at −190 °C was approximately 2.6 times the wear resistance for tool steel samples soaked at −64.69 °C. Collins and Dormer [21] proved that the amount of carbides precipitated increases with decreased cryogenic temperature for D2 cold-work tool steel (Fig. 11.3).

It is interesting to note that using the lower austenitizing temperature does not offer the highest overall steel hardness; however, it does offer a significantly reduced wear rate compared to higher austenitizing temperatures. In contrary to tool steels, Seah et al. [85] observed no significant gain in wear resistance of tungsten carbide inserts by CT at −196 or −80 °C. However, all other studies on tungsten carbide performed under real machining conditions as well as laboratory conditions proved that tungsten carbide resist wearing more effectively if treated at −196 °C as compared to −80 to −110 °C [2, 13, 40, 41, 44, 80, 82, 88, 96, 97]. Also by immersing the hybrid PTFE/Kevlar fabric into liquid nitrogen, Zhang et al. [103] proposed that optimized CT process of hybrid PTFE/Kevlar fabric can improve the wear resistance of fabric/phenolic composite. They however warned that excessive treatment may negatively influence the tribological properties.

11.2.2 Soaking Period

The contribution of soaking period goes up to 30 % in improving the wear resistance of the steels [23, 24]. Das et al. [25–31] studied the effect of soaking period on the tribological behavior and carbide precipitation of D2 steel and suggest that the wear resistance increases by increasing the soaking period whereas hardness remains the same. In another study, Jinyong et al. [51] treated the Mo–Cr HSS by CT for 2 and 16 h as soaking period and showed that wear resistance is more in case of steel treated for 16 h. Mohanlal et al. [71] examined many treatment conditions (Table 11.3) for wear resistance study of M2, T1, and D3 tool steels and by comparing the results of soaking period of 6 and 24 h, they successfully proved that the mechanism causing improvement in wear resistance/tool life is essentially an isothermal process.

In two independent studies by Barron [8] and Dobbins [35], it has been shown that the soaking period is important to the final properties of the tool steels and soaking period of 20 h is enough as the atoms in the material require time to diffuse to new locations. Collins and Dormer [21] and Collins [19, 20] showed that soaking time affects the hardness, wear resistance, and carbide density of a variety of steels because diffusion occurs very slowly at cryogenic temperatures. Lomte and Chikalthankar [65] also confirmed that different soaking times during CT of AISI D2 steel considerably affect the extent of residual stresses which is a measure of durability of any tool steel. However, work by Moore and Collins [73] on the effect of soaking period on various steels shows that different materials react differently to variations in soaking period. In his patent, Kamody [55] asserts that the soaking period has no role in deciding the final condition of the material being processed and a soaking period of 10 min was recommended to allow the material to achieve thermal equilibrium before it is removed and reheated. By working on AISI M2 HSS and H13 hot-work tool steel, Molinari et al. [72] recommended that a more prolonged period does not have any significant effect and the maximum soaking period should not be more than 35 h. Das et al. [25–31] performed a series of wear tests on AISI D2 steel samples subjected to cryogenic treatment for different durations and established that critical time duration exists for achieving the best wear resistance properties. Hence, an optimum soaking time may, therefore, exist

Table 11.3 List of treatment conditions considered for wear resistance study [71]

Case	Purpose	Nomenclature followed
Hardened and tempered	Standard tool for comparing the tool life of cryotreated tools	Standard heat-treated
Cryotreated at 93 K for 24 h	Cryogenic treatment for tool steels	CT (93/24)
Cryotreated at 133 K for 24 h	To check the effect of lowering the temperature deep below cold treated levels	CT (133/24)
Cryotreated at 163 K for 24 h	To check the effect of lowering the temperature deep below cold treated levels	CT (163/24)
Cryotreated at 93 K for 6 h	To check the significance of isothermal holding time at the treatment temperature	CT (93/6)
Cryotreated at 93 K for 6 h and quenching in LN_2 for 2 h	To check the effect of reducing the temperature below 93 K	CT (93/6) + LN_2/2
Tempered and cryotreated at 133 K for 24 h	To confirm whether tempering is to be done before cryogenic treatment	TCT (133/24)
Tempered and cryotreated at 93 K for 24 h	To confirm whether tempering is to be done before cryogenic treatment	TCT (93/24)
Titanium nitride coated	To compare the benefits of cryogenic treatment with that of surface coatings	TiN
Cryotreated (CT(93/24))	To compare the benefits of cryogenic treatment with that of surface coatings	CT + TiN
Titanium nitride coated and then cryotreated (CT(93/24))	To compare the benefits of cryogenic treatment with that of surface coatings	TiN + CT

for each type of material, though the difference in properties due to different soaking periods may be quite small.

11.2.3 Rate of Cooling

With 10 % contribution, the rate of cooling is the third most significant parameters for improving mechanical properties [23, 24]. Literature review reveals the two major hypotheses proposed for rate of cooling are (1) high rate of cooling to achieve the CT temperature rapidly but without causing thermal shock to the material and (2) plodding rate of cooling so as to achieve the CT temperature in several hours [39, 42]. The literature contains many references into the need for low cooling rates and accurate cooling curve control down to liquid nitrogen temperatures [46]. Zhirafar et al. [104] cooled down and heated up the specimens slowly, to and from the cryogenic temperature (−196 °C), over a 2-h period with the temperature being monitored by a thermocouple attached to the specimen so as to achieve average heating/cooling rate of 1.8 °C/min in order to avoid thermal shocks from rapid cooling and heating. Bensely et al. (2007) processed the case carburized steel 815 M17 samples by slow cooling from room temperature to 77 K at 1.24 K/min and finally heating back to room temperature at 0.64 K/min. In another study, Preciado et al. [79] recommended slow cool down (~1 °C/min) from ambient temperature to liquid nitrogen temperature. Delaye and Limoge [33] proved that fast cooling rates allow the creation of moving defects because the defect equilibrium conditions change abruptly with temperature. As per Molinari et al. [72], one of the most critical parameters is the cooling rate which must not exceed 20–30 °C/h in order to prevent the rupture of the tool steels because of the cooling stress. Dobbins [35] also claims that the rate of cooling significantly affects the resulting material properties. He recommended 2 h time for cool down so that the full effect of the CT could be realized and slow warm-up period of 1 h to reach room temperature. In addition, many other studies also recognize the damage that can occur by rapid cooling in the form of thermal shock cracks [10, 16, 39, 42]. In contrary to the above-mentioned studies by different researchers, Kamody [55, 56] suggests that the rate of cooling has little, if any, effect on the final properties of the material being treated, and so should be as rapid as possible in order to minimize the treatment time, thus dipping the processing cost. He proposed that cooling should take place by submerging the material into liquid nitrogen, allowing the natural rate of cooling provided by the evaporation of the nitrogen to determine the cooling rate and the reheat rate to room temperature should also take as little time as possible. Kamody [55, 56] preferred a stream of warm air that can provide a warming rate of 1–6 °C/min as a medium to reheat the samples to room temperature. It is very much comprehensible that the rate of cooling still requires to be debated upon to achieve the desired results.

From the published literature, it can be found that a wide range of different values of CT process parameters has been used on different materials. Some general observations inferred are as follows:

1. In order to identify optimum conditions, every material is required to be treated at different soaking temperatures. In most cases soaking temperature range of 88–193 K is enough to select a specific soaking temperature.
2. Soaking period more than 36 h may not provide any significant improvements, thus usually soaking period of 24–36 h is enough to obtain results.
3. Cooling rate value is restricted in order to prevent microcracking in the material due to thermal shock and is usually varied from 0.3 to 3.0 K/min.
4. Tempering temperature, tempering period, and rate of warming are found out to be insignificant hence not closely controllable and little importance to this parameter is given in literature.

11.3 Effect of Cryogenic Treatment on Tool Materials and Polymers

11.3.1 Wear Resistant

It has been claimed by several researchers that CT enhances wear resistance of certain high alloy and tool steels ([5, 6, 9, 19, 21, 23, 24, 43, 62, 69–72, 75, 78,

Table 11.4 Different test parameters used to evaluate the wear resistance of cryogenically treated tool steels

			Parameter for wear test		
Tool Steel	Shape of counter body	Shape of sample	Load (N)	Sliding velocity (m/s)	Sliding distance (m)
M2, D3	Disc	Pin	20, 30, 50	0.18–0.60	324–1080
M2, H13	Disc	Disc	150	0.8	5000
M2	Disc	Pin	10	0.11	3.22
M2, M1, T2, T1, H13, D2, A10	Wheel	Pin	430	0.48	2160
M2	Ball	Disc	50	0.027	200
D2	Wheel	Block	21	0.50–3.62	200,400,600

Table 11.5 Increase in wear resistance by cryogenic treatment [75]

Type of steel (AISI)	Description	At −79 °C	At −190 °C
D2	Chromium steel	316	817
S7	Silicon tool steel	241	503
52100	Bearing steel	195	420
O1	Oil hardened cold-work die steel	221	418
A10	Graphite tool steel	230	264
M1	Molybdenum HSS	145	225
H13	Hot-work tool steel	164	209
M2	Tungsten/Molybdenum HSS	117	203
T1	Tungsten HSS	141	176
CPM 10 V	Alloy steel	94	131
P20	Mold steel	123	130
440	Martensitic stainless steel	128	121

79, 99]. Some typical results and wear test parameters reported by different investigators are given in Table 11.4.

Barron [6] and Pauline [75] reported improvement in wear resistance for cryogenically treated D2 steel over their conventionally treated counterparts by 8.2 times (Table 11.5).

Leskovsek et al. [62] reported that wear resistance of vacuum heat-treated high-speed steel, when subjected to cryogenic treatment, gets improved by an order of magnitude. Molinari et al. [72] examined the effect of deep cryogenic process on quenched and tempered AISI M2 and H13 steels and reported higher wear resistance. Bensely et al. [9] reported 372 % increment of wear resistance of deep cryogenically treated samples of case carburized En 353 steel over the conventionally heat-treated ones. The results obtained by them are represented as wear rate of pin as shown in Fig. 11.4 for various loadings of 60, 70, and 80 N.

Mohanlal et al. [71] studied the effect of cryogenic process on different steels and by machining tests and obtained nearly 110, 87, and 48 % increment in tool life for cryogenically treated T1, M2, and D3 tools, respectively, in comparison to the

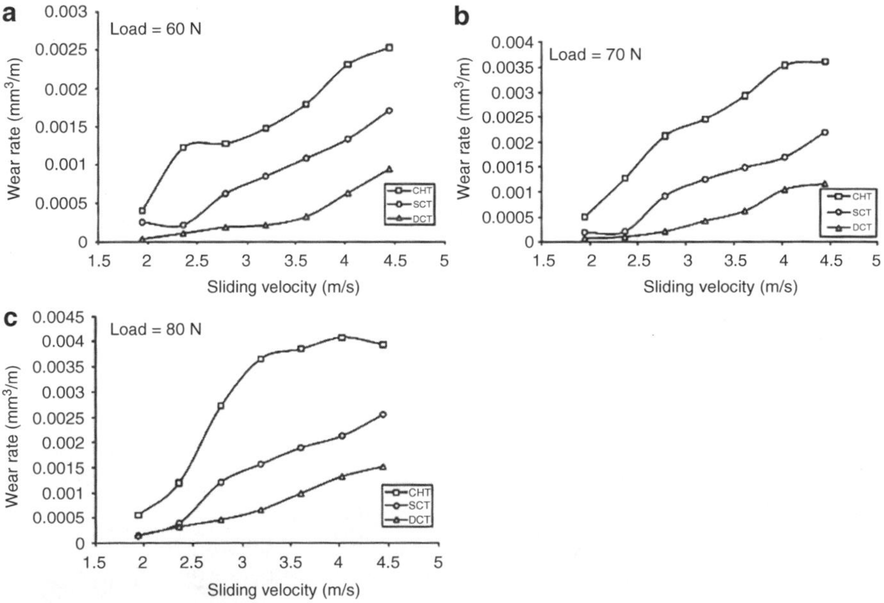

Fig. 11.4 Wear rate at (**a**) 60, (**b**) 70, and (**c**) 80 N load for three different treatments: conventional heat treatment (CHT), shallow cryogenic treatment (SCT), and deep cryogenic treatment (DCT) [9]

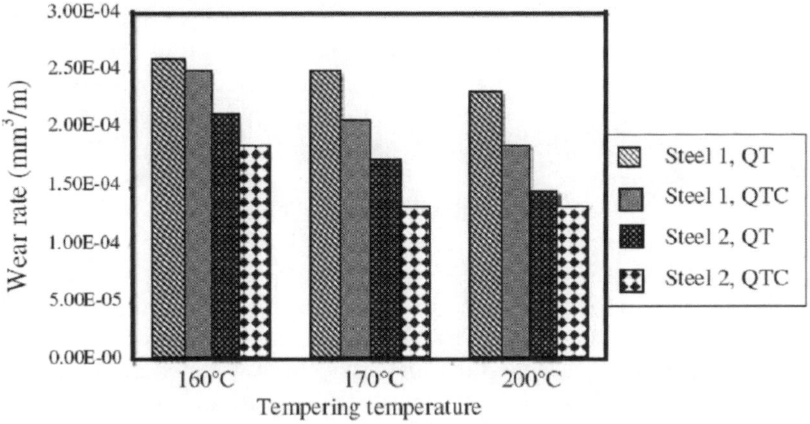

Fig. 11.5 Wear rate of heat-treated steels [79]

corresponding standard heat-treated tools. The effect of deep cryogenic process on the improvement of wear resistance of carburized steels was also reported by Preciado et al. [79] (Fig. 11.5).

Meng et al. [70] reported that cryogenic treatment improves the wear resistance of Fe–12Cr–Mo–V–1.4 °C tool steel by about 110–600 %, depending on the sliding speed of wear tests. Li et al. [66] showed that the hardness of the T12 steel increases slightly and the abrasive resistance increases by 50–60 % after cryogenic treatment. It was shown that cryogenic treatment can decrease harmful cracks and hence improve the wear resistance. Pellizzari et al. [77] successfully reduced the wear rate up to 42 % of AISI M2 HSS by coupling the DCT with conventional vacuum heat treatment. In a recent study, Firouzdor et al. [37] studied the influence of DCT on wear resistant and tool life of M2 HSS drills in high-speed dry drilling of carbon steel. Their experimental results indicated 126 % improvement in drill life which was attributed to enhanced wear resistance. Babu et al. [3] treated M1, EN19, and H13 tool steels at 0 °C, −20, −40, −80, and −190 °C and recorded from 315 to 382 % improvement in wear resistance depending upon the material. Gill et al. [43] also reported improvement in hardness as well as wear resistance of M2 HSS.

There are many studies which reported that CT also effect positively the another commonly used tool material tungsten carbides. Gill et al. [44] proved that cryogenically treated tungsten carbide inserts perform better in both interrupted and continuous machining. The extent of advantage gained was even better at higher cutting speeds. They also reported that cryogenically treated tungsten carbide inserts perform more consistently under interrupted machining mode compared with continuous machining mode. Seah et al. [85] also found that such treatment increases wear resistance of tungsten carbide inserts. Quek [80] concluded that cryogenically processed tungsten carbide tool inserts exhibit better wear characteristics than untreated ones at low turning speeds and feeds. Kao [57] also reported increase in abrasion wear resistance of sintered tungsten carbides after cryogenic treatment. Reddy et al. [82] tested carbide turning inserts subjected to normal and deep cryogenic processing and noted better tool life compared to untreated inserts in terms of flank wear. Yong et al. [96, 97], in two independent studies, noted that the cryogenic treatment of tungsten carbide inserts improves tool life (flank wear) performance and resistance to chipping in milling and turning to a certain extent (Fig. 11.6). They stated that tools under mild cutting conditions stand to gain from cryogenic treatment, but heavy duty cutting operations with long periods of heating of the cutting tool will not benefit from it.

Vadivel and Rudramoorthy [91] machined nodular cast iron to evaluate the performance of cryogenically treated coated carbide inserts and confirmed lower power consumption along with other benefits. However, Kim and Ramulu [58] performed drilling tests on thermoplastic PIXA-M composites by cryogenic-treated C2 grade carbide drills and found that the cryogenically treated drills wear 10 % faster than the conventional drills though the quality of the holes produced were better. In another recent study, Gill et al. [43] reported that a substantial decrease in tool life of deep cryogenically treated inserts as compared to untreated inserts indicating the destructive effect of deep cryogenic temperature on TiAlN-coated inserts which was further supported by VDI-3198 indentation tests by the authors.

There are some studies which claim improvement in wear resistance of polymers also by applying CT. Indumathi et al. [49] compared wear rates of treated and untreated polymers and composites samples under various loads and revealed that this technique has potential to increase the wear resistance of some polymers and

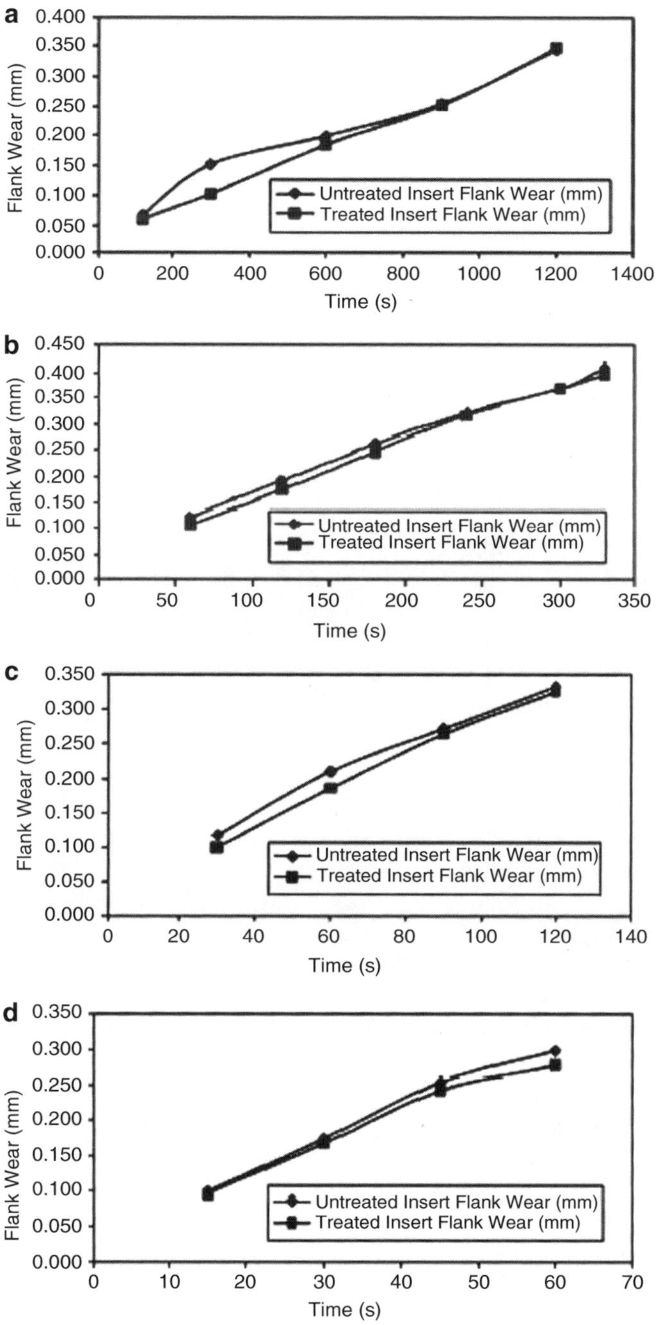

Fig. 11.6 Flank wear development for cutting speeds of (**a**) 150, (**b**) 200, (**c**) 250, and (**d**) 300 m/min, depth of cut of 1 mm, and feed rate of 0.1 mm/rev [96].

composites. The extent of influence of cryotreatment on wear resistance depended on the type of base matrix and the nature and amount of fiber or particulate filler. In two studies, the authors treated short carbon fibers by a cryogenic treatment and investigated the mechanical and tribological properties of carbon fiber reinforced epoxy [100, 101]. Their results showed that the mechanical and tribological properties of carbon fiber reinforced epoxy were improved after cryogenic treatment due to the enhanced fiber–matrix interfacial bonding. In another study, Zhang et al. [103] claimed that fabric/phenolic composite after cryotreatment exhibited a lower wear rate under the examined loads.

11.3.2 Metallurgical Properties

Several tool steels are vulnerable to the problem of retaining residual austenite after quenching to room temperature. This is the reason why some researchers believe that CT promotes transformation of retained austenite into martensite, and this can be attributed to the enhanced mechanical properties of the tool steels [3, 6, 12, 18, 32, 36, 45, 67, 73, 86, 95, 104]. Gulyaev [45] was the first to identify transformation of retained austenite into martensite with more stable structure as major outcome of CT on high-speed steels which was later on supported by many other researchers. Barron [6] did momentous research and proposed that the room temperature aging effect is reduced when the part is deep cryogenically treated. In nineteen's Dymchenko and Safronova [36], Moore and Collins [73], and Cohen and Kamody [18] examined specimens of various tool steels after treatment at below zero temperatures in their respective independent work and confirmed the transformation of retained austenite into martensite. Later on Blankinship [12] also accounted that CT improves the ability of tool steels to withstand abrasive wear by causing transition of relatively soft retained austenite to the harder and more durable martensite. Soundararajan et al. [86] proposed that the beneficial effects of CT are limited by the alloying elements as the alloying elements increase the content of retained austenite. Babu et al. [3] suggested that the retained austenite should be transformed to the maximum possible extent before any component or tool is put into service by CT. In more recent studies, Yang et al. [95], Da Silva et al. [32], and Zhirafar et al. [104] showed that during the CT, the secondary carbides precipitate in austenite, which promote the transformation of retained austenite to martensite. Li et al. [67] proposed an interesting fact that following the DCT process retained austenite transformed into martensite, however, not completely.

However, in many other studies, it has been reported that all the retained austenite content gets converted into martensite at temperature of cold treatment (−80 °C) whereas wear resistance of tool steels can be further enhanced by treating them cryogenic temperature of −196 °C [19, 21, 30, 69, 70, 78, 87]. To explain this phenomenon, another school of researchers attributed the improvement in mechanical properties of tool steels after CT to the inter martensite changes rather than to austenite–martensite transformations [11, 15, 34, 69, 70, 75, 83, 87, 98, 106]. Paulin

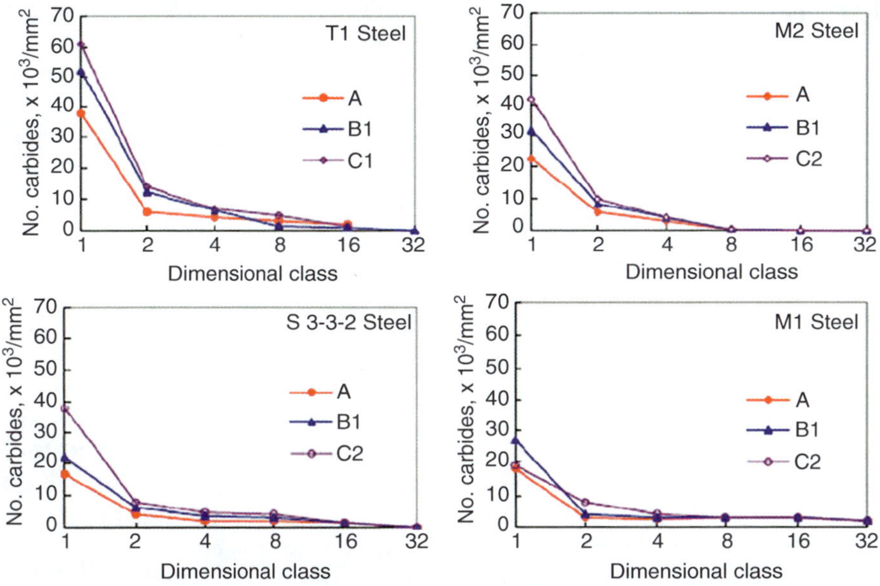

Fig. 11.7 Carbides density distribution on dimensional classes (Blancea et al., 2008)

[75] confirmed the precipitation of fine carbon carbides due to CT which results in tight lattice structures. Meng et al. [69, 70] have shown presence of fine η-carbides instead of the usual ε-carbides after CT. Biao and Jiyuan [11] specifically reported that transformation from residual austenite to martensite is not the cause of increase of wear resistance and toughness of tool steel. Satish Kumar et al. [83] endeavored to understand the basic mechanism of improvement in the wear resistance of cryogenically treated tool steels and proposed interesting concept of clustering of finely dispersed carbides in the supersaturated martensite matrix. Yugandhar and Krishnan [98] identified martensite decomposition and precipitation of fine η-carbides as the main mechanisms responsible for the beneficial effects of DCT. Zurecki [106], in an independent study, examined the effects of various heat treatment schedules including liquid nitrogen (−196 °C) and liquid helium (−269 °C) quenching on steel microstructure and confirmed the presence of large number of finely dispersed (Cr–V-depleted) dark carbides. In another latest study, Stratton [87] identified time-dependent decomposition of the primary martensite which occurs at cryogenic temperatures typically of −196 °C which causes some initial softening but nucleates numerous coherent nano-carbides. In another study, Bulancea et al. [15] carried out investigations on the qualitative parameters of the carbide phase and increase of tools durability was appreciated through the increased amount of small size carbide particles (<1 ηm) uniformly distributed in the martensite mass, increased at the expense of the residual austenite (Fig. 11.7).

In the most recent study, Dhokey and Nirbhavne [34] confirmed the segregation of heavy and relatively large size carbide particles in non-treated steel.

The improved mechanical properties of the tool steels can be attributed to carbide precipitation though the misconception is still prevailing. The most prominent findings that can be extracted from the above discussion include unique mechanism of martensite formations, time-dependent size of precipitated carbide particles, and importance of post-cryogenic treatment tempering. But many old and contemporary studies by different researchers documented that the improvement in mechanical properties of tool steels subjected to cryogenic treatment can be attributed to combined effect of transformation of retained austenite to martensite and precipitation of fine carbides rather than their respective individual effect [9, 10, 19, 21, 25–31, 37, 43, 48, 54, 62, 63, 71, 76, 79, 92, 99]. The results obtained by Collins [19] indicated two quite different phenomena, i.e., conversion of austenite into martensite and low-temperature conditioning of martensite which promoted the formation of large numbers of very fine carbide particles. Collins and Dormer [21] reported that all cryogenic treatments at temperatures below about $-80\ ^{\circ}\mathrm{C}$ resulted in negligible amounts of retained austenite in the microstructure, and the only significant microstructure variation that was observed by the authors was the size and distribution of carbides. A study by Yun et al. [99] showed that martensite decomposition and ultrafine carbide precipitation are important factors in the improvement of the mechanical properties. The findings of Mohanlal et al. [71] reinforced the idea that the reported improvements in wear resistance may be due to combined effect of martensitic transformation and precipitation of fine alloying carbides. In another study Leskovšek and Ule [63] noticed rod-like carbide precipitation in the investigated M2 steel. Pellizzari and Molinari [76] concluded that secondary hardening occurs through precipitation of carbides rather than through the decomposition of retained austenite. Huang et al. [48] studied the microstructure changes of M2 tool steel before and after cryogenic treatment and proposed that carbides with different sizes were homogeneously distributed in the cryogenically treated sample whereas the distribution of carbides in the non-cryogenically treated samples was inhomogeneous. Bensely et al. [9] identified the phenomenon of precipitation of fine carbides induced by DCT which causes the extra improvement in wear resistance. Preciado et al. [79] confirmed that the DCT of quenched and low-temperature tempered carburized steels improved the wear resistance probably due to the segregation of carbon atoms and alloying elements during the cryogenic cycle and the transformation of the retained austenite into martensite (Fig. 11.8). Kalin et al. [54] proved that DCT following vacuum hardening stimulated the transformation of austenite into martensite in addition to precipitation of rod-like carbide (Fig. 11.9).

Firouzdor et al. [37] reported volume fraction of carbides in the cryogenically treated (CT) and cryogenically treated and tempered (CTT) drills were approximately 10 % whereas for untreated drills were about 6 %. The effect of cryogenic treatment on the distribution of residual stress in the case carburized steel (En 353) was studied by Bensely et al. [10] and concluded that the large relaxation of residual stress in deep CTT specimens indirectly reflects about the amount of carbide precipitation. Vimal et al. [92] found that the austenite is completely transformed to martensite during cryogenic treatment. Authors found that cryogenic treatment

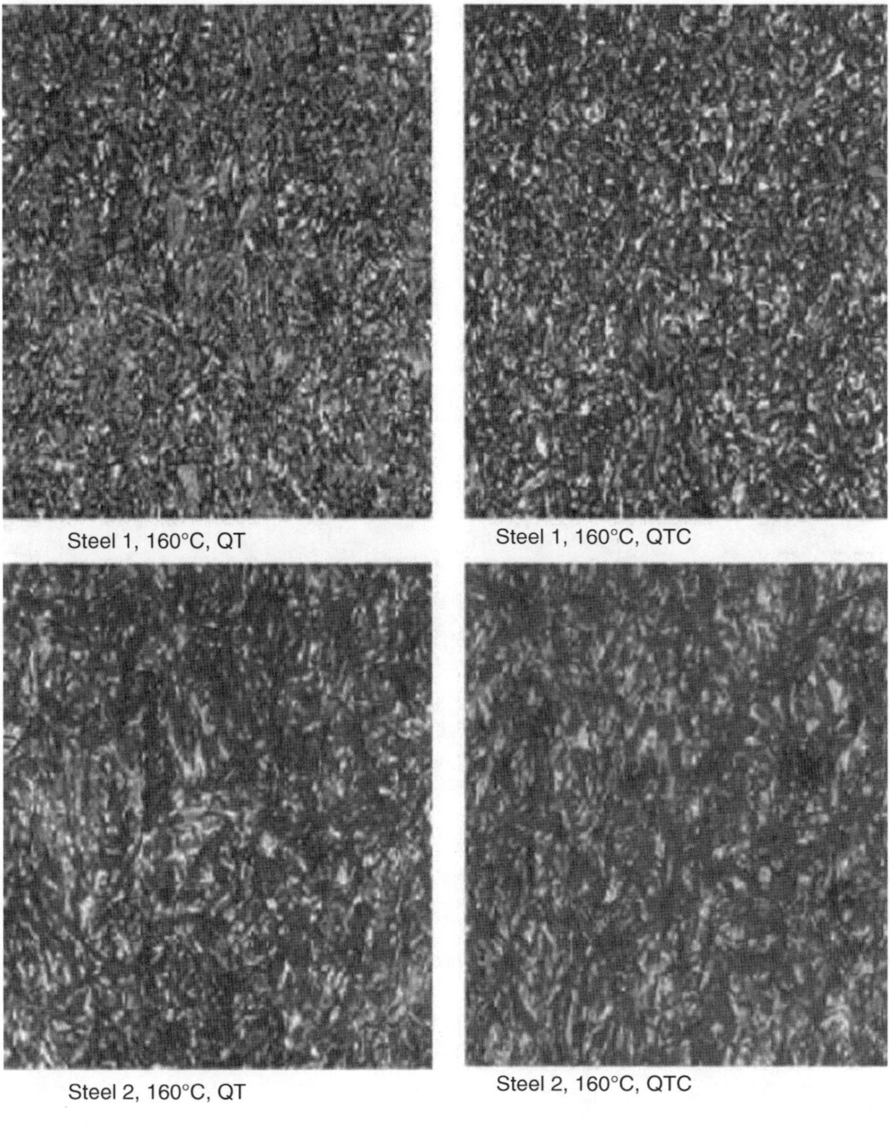

Fig. 11.8 Optical micrographs showing the 160 °C tempered steels with and without cryogenic treatment at 150 μm from the surface (1,000×) [79]

can overcome stabilization of retained austenite but carbide precipitation can be ensured only by post-cryogenic treatment tempering. Das et al. [25–31] have done series of investigations to assess the influence of cryogenic treatment on the mechanical performance and concluded that the improvement in mechanical properties of AISI D2 steel after cryogenic treatment can be attributed to the patently reduced amount of retained austenite and significant changes in the

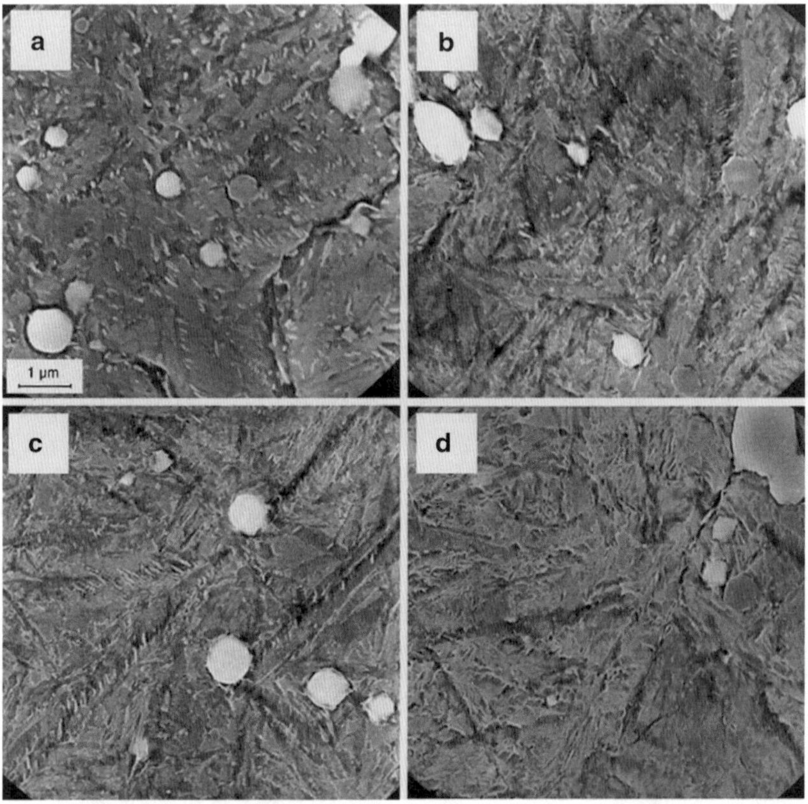

Fig. 11.9 Microstructure of AISI M2 high-speed steel specimens which were vacuum hardened, deep cryogenic treated at −196 °C, and finally single tempered at temperatures as follows: (**a**) 500 °C, (**b**) 540 °C, (**c**) 550 °C, and (**d**) 600 °C [54]

precipitation behavior of secondary carbides. In another recent study, Gill et al. [43] also concluded that any significant improvement after CT can be attributed to the combined effect of austenite transformations and precipitation of fine carbides.

Research efforts on the effect of cryogenic treatment on carbides and polymers remain relatively few and far between. It should be noted that there is no martensite phase in these materials; as such, any improvement in tool life or wear resistance would be due to other mechanisms [97]. Seah et al. [85] also support this hypothesis after conducting a study on cryogenically treated tungsten carbide inserts and confirmed an increase in the number of η phase particles after cryogenic treatment. Reddy et al. [82] observed a slight increase in grain size which increases the toughness. So far, few researchers have proposed other mechanisms that explain the effect of cryogenic treatment on tungsten carbide. Bryson [14] attributes the wear resistance, and hence the increase in tool life, of carbide tools to the improvement in the holding strength of the binder after cryogenic treatment. Stewart [88] and Gallagher et al. [38] concluded that cryogenic treatment might have an effect upon the cobalt binder by changing the phase or crystal structure so that more cobalt binder was retained during cutting. In relatively new studies, Gill et al. [40, 41] found that the hard phase particles of tungsten carbide are refined into their most stable form via the phenomenon of spheroidization after shallow and DCT. The authors established that the cryogenic treatment causes the crystal structure changes in both hard and soft binder phase of tungsten carbide material which may be responsible for the enhanced hardness and wear resistance properties along with precipitation of η-phase carbides. As far as effect of CT on microstructure of polymers is concerned, the research is at very initial stage and we can expect significant results in the near future.

11.4 Discussions and Recommendations

After a long struggle to seek legitimacy, it seems that CT will probably now become accepted and will be used extensively in industry. In recent times, we can find CT facilities in many countries all over the world. However, the growth in the use of these technologies is slower than expected. The science of CT is still not fully understood and this fact can explain the delay in the complete acceptance of CT technology. We are fortunate that this situation has already started to improve thanks to the dedicated efforts made by scientists and researchers. Nonetheless, there is an apparent need of improving the research activity in this field in order to achieve a more complete understanding of the mechanisms that are involved in the microstructural enhancement of cryogenically treated materials and also in search of new applications. From the literature, we can locate few research activities in this field; most of which has been done in USA and Europe. But during the last few years this situation has changed and, nowadays, one can find ever increasing number of studies concerning CT of materials from Asian countries also. China, India, and Singapore are leading the way from Asian countries. Hopefully, more new research activities will also be presented in the near future from other parts of the world also. Anyway, the empirical development of the technology will probably prevail in the development of new applications of the CT during the coming days. The future of CT of materials is expected to be very exciting and dynamic.

The topic of the-state-of the art and future developments in CT of materials areas looking toward the next century has several aspects to it. It is important to evaluate where we are today in this science and to look at research and development trends in this area which can give us some clues to future developments. Until now the use of CT in industry has been essentially focused on all kind of tools and consumables. This is perhaps the most palpable use for CT because the increase of cutting life of tools has a direct positive outcome in productivity and cost, something that is always interesting for the users. The increase in wear resistance is the prevailing CT effect when treating tools. But there are other treatment properties that will probably get more relevance in the coming years. The most prominent of them is increasing of fatigue life. Apart from the tools the process will be applied to the manufactured components in order to increase their service life and their reliability. The other most critical issue in CT of materials is going to be the treatment cost. Only the positive economics can pave the way for logical, creative, and innovative ideas in this field. To realize this forecast, we need to treat much more quantities of materials. CT is not going to be limited to a few types of materials, but a variety of metals, alloys, composites, and polymers are expected to find application in this technology. We are no longer limited by shape, density, size, composition rather we are only limited by our imagination, knowledge, and understanding of how to achieve the highest level of performance from CT of materials. We must not only continue to make incremental improvements in present applications of CT but identify whole new range of materials to achieve the highest performance in CT technology.

11.5 Conclusions

1. CT has significant favorable influence on the performance of cutting tool steels, carbides, and polymers. Hence, CT is a good alternative for having productivity enhancement.
2. Tempering should be followed by CT. Untempered tool materials have surface cracks, which is not beneficial to the life of cutting tools. The wear resistance ability of cutting tool materials can be maximized by incorporating CT in the heat-treating cycle in the following order: austenitizing, quenching, cryoprocessing, and tempering.
3. The three most significant parameters of CT identified as primarily affecting the wear resistance are soaking temperature, soaking period, and cooling rate. The CT cycle can be optimized by controlling these parameters for every specific material. Some useful indications for optimal CT parameters can be inferred from published works in case of steels. Determination of appropriate levels of the process parameters will result in maximum wear resistance as well as save time and energy involved in the process. However, detailed scientific studies still required for tungsten carbide and polymers.
4. Wear resistance and hardness improvement have been widely confirmed especially for tool steels and tungsten carbide. Beneficial effects of CT on toughness and fatigue behavior have also been reported in some studies.
5. The improvement in wear resistance and hardness by CT is attributed to the combined effect of conversion of retained austenite to martensite and precipitation of η-carbides in case of tool steels. The phenomenon responsible for improved wear resistance in carbide cutting tools is the combined effect of increased number of η phase particles and increase in bounding strength of binders used.
6. The phenomena of carbides precipitation in tool steels can be explained on the basis of martensite contraction which occurred due to thermal stresses during cooling, which leads carbon atoms to segregate near lattice defects.

7. The mechanisms like nano-martensite formation, dislocation pinning effect, contribution of retained austenite, and residual stresses are required to be investigated so as to justify the fatigue strengthening in some materials.
8. Considering the improvements in wear and hardness of PTFE, PI, and PEI, further investigations about the effects of CT on mechanical properties of polymers and composites could be a stimulating research topic.

References

1. Alexandru G, Ailincai G, Baciu C (1990) Influence of cryogenic treatments on life of alloyed high speed steels. Mém Étud Sci Rev Métall 4:283–388
2. Arner KA, Agosti CD, Roth JT (2004) Effectiveness of the cryogenic treatment of tungsten carbide inserts on tool wear when in full production operations. American Society of Mechanical Engineers, Mechanical Engineering Division, Anaheim, CA, Nov 13–19, pp. 31–40
3. Babu PS, Rajendran P, Rao KN (2005) Cryogenic treatment of M1, EN19, and H13 tool steels to improve wear resistance. J Inst Eng India MM 86:64–66
4. Baldissera P, Delprete C (2009) Effects of deep cryogenic treatment on static mechanical properties of 18NiCrMo5 carburized steel. Materials and Design 30:1435–1440
5. Barron RF (1974) Do treatment at temperature below −120 °F help increase the wear resistance of tool steels? Here are some research findings that indicate they do. Heat Treat 14–17
6. Barron RF (1982) Cryogenic treatment of metals to improve wear resistance. Cryogenics 22:409–413
7. Barron RF, Mulharn CR (1980) Cryogenic treatment of AISI T8 and C1045 steels. In: Cryogenic engineering conference. Madison, WI, Aug 21–24, 1979. Plenum, New York
8. Barron RF (1996) In: Proceedings of the conference of manufacturing strategies, vol. 6, Nashville, pp. 137
9. Bensely A, Prabhakaran A, Mohanlal D, Nagarajan G (2005) Enhancing the wear resistance of case carburized steel (En 353) by cryogenic treatment. Cryogenics 45:747–754
10. Bensely A, Venkatesh S, Mohanlal D, Nagarajan G, Rajadurai A, Junik K (2008) Effect of cryogenic treatment on distribution of residual stress in case carburized En 353 steel. Mater Sci Eng A 479(1):229–235
11. Biao G, Jiyuan C (1997) Property improvement and microstructure transformation of high speed steel due to deep cryogenic treatment. J Dalian Univ Technol 37(3):285–289
12. Blankinship S (2001) Ultra-cold could help freeze plant repairs. Power Eng 105:13–16
13. Bonilla C, O'Meara R, Perry L (2007) Evaluation of the comparative performance of cryogenically treated cutting inserts as a capstone design project. In: Proceedings of the ASEE Annual Conference and Exposition, pp. 9
14. Bryson WE (1999) Cryogenics. Hanser Gardner, Cincinnati, pp 81–107
15. Bulancea V, Condurache D, Gheorghiu DA (2008) Frozen high alloyed steels. Annals of the University of Craiova, Electrical Engineering Series, No. 32, 2008; ISSN 1842–4805, pp. 342–345
16. Carlson EA (1990) Cold treating and cryogenic treatment of steel. In: ASM Handbook, vol. 4 Heat Treating, ASM International, 10th Ed., Metals Park, OH, pp. 203–206
17. Cech V, Prikryl R, Balkova R, Grycova A, Vanek J (2002) Plasma surface treatment and modification of glass fibers. Compos A 33(10):1367–1372
18. Cohen P, Kamody D (1998) Cryogenics goes deeper. Cutt Tool Eng 150(7):46–50

19. Collins DN (1996) Deep cryogenic treatment of tool steels—a review. Heat Treat Met 23 (2):40–42
20. Collins DN (1998) Cryogenic treatment of tools steels. Adv Mater Process 154(6):H23
21. Collins DN, Dormer J (1997) Deep cryogenic treatment of a D2 cold-worked tool steel. Heat Treat Met 3:71–74
22. Combellas C, Richardson S, Shanahan MER, Thiebault A (2001) Modification of the adhesion properties of PTFE by a magnesium treatment. Int J Adhesion Adhesives 21 (1):59–64
23. Darwin JD, Mohan Lal D, Nagarajan G (2007) Optimization of cryogenic treatment to maximize the wear resistance of Chrome Silicon spring steel by Taguchi method. Int J Mat Sci 2(1):17–28
24. Darwin JD, Mohan Lal D, Nagarajan G (2008) Optimization of cryogenic treatment to maximize the wear resistance of 18 % Cr martensitic stainless steel by Taguchi method. J Mater Process Technol 195(1–3):241–247
25. Das D, Dutta AK, Ray KK (2008) On the enhancement of wear resistance of tool steels by cryogenic treatment. Phil Mag Lett 88:801–811
26. Das D, Dutta AK, Ray KK (2009) Correlation of microstructure with wear behaviour of deep cryogenically treated AISI D2 steel. Wear 267:1371–1380
27. Das D, Dutta AK, Ray KK (2009) Influence of varied cryotreatment on the wear behavior of AISI D2 steel. Wear 266(1–2):297–309
28. Das D, Dutta AK, Ray KK (2009) On the refinement of carbide precipitates by cryotreatment in AISI D2 steel. Philos Mag 89(1):55–76
29. Das D, Dutta AK, Ray KK (2009) Optimization of the duration of cryogenic processing to maximize wear resistance of AISI D2 steel. Cryogenics 49:176–184
30. Das D, Dutta AK, Toppo V, Ray KK (2007) Effect of deep cryogenic treatment on the carbide precipitation and tribological behavior of D2 steel. Mater Manuf Process 22(4):474–480
31. Das D, Ray KK, Dutta AK (2009) Influence of temperature of sub-zero treatments on the wear behaviour of die steel. Wear 267:1361–1370
32. da Silva FJ, Franco SD, Machado AR, Ezugwu EO, Souza AM Jr (2006) Performance of cryogenically treated HSS tools. Wear 261(5–6):674–685
33. Delaye JM, Limoge YJ (1993) Molecular dynamic study of vacancy like defects in a model glass: dynamical behaviour and diffusion. J Phys (Paris) 2:2079–2097
34. Dhokey NB, Nirbhavne S (2009) Dry sliding wear of cryotreated multiple tempered D-3 tool steel. J Mater Process Technol 209(3):1484–1490
35. Dobbins DB (1995) Cryogenic treatment can boost life. Metal forming, pp. 29 (May)
36. Dymchenko VV, Safronova VN (1993) Refrigeration treatment of quenched roll steel. Tyazheliy Mashinostroenie 9:29–32
37. Firouzdor V, Nejati E, Khomamizaden F (2008) Effect of deep cryogenic treatment on wear resistance and tool life of M2 HSS drill. J Mater Process Technol 206(1–3):467–472
38. Gallagher AH, Agosti CD, Roth JT (2005) Effect of cryogenic treatments on tungsten carbide tool life: microstructural analysis. Trans North Am Manuf Res Inst SME 33:153–160
39. Gill SS, Singh H, Singh R, Singh J (2010) Cryoprocessing of cutting tool materials – a review. Int J Adv Manuf Technol 48(1–4):175–192
40. Gill SS, Singh J, Singh H, Singh R (2011) Investigation on wear behaviour of cryogenically treated TiAlN coated tungsten carbide inserts in turning. Int J Mach Tools Manuf 51 (1):25–33
41. Gill SS, Singh J, Singh H, Singh R (2012) Metallurgical and mechanical characteristics of cryogenically treated tungsten carbide (WC–Co). Int J Mach Tools Manuf 58(1–4):119–131
42. Gill SS, Singh J, Singh R, Singh H (2011) Metallurgical principles of cryogenically treated tool steels – a review on the current state of science. Int J Adv Manuf Technol 54(1–4):59–82
43. Gill SS, Singh J, Singh R, Singh H (2011) Effect of cryogenic treatment on AISI M2 high speed steel: metallurgical and mechanical characterization. J Mat Eng Perf. doi:10.1007/s11665-011-0032-z

44. Gill SS, Singh R, Singh H, Singh J (2009) Wear behaviour of cryogenically treated tungsten carbide inserts under dry and wet turning conditions. Int J Mach Tools Manuf 49 (3–4):256–260
45. Gulyaev AP (1937) Improved methods of heat treating high speed steels to improve the cutting properties. Metallurgy (12): 65
46. Hallum DL (1996) Cryogenic tempering delivers better cutting tool durability. Am Mach, pp. 140–142
47. Huang HC, Ye DQ, Huang BC (2007) Nitrogen plasma modification of viscose-based activated carbon fibers. Surf CoatTechnol 201(24):9533–9540
48. Huang JY, Zhu YT, Liao XZ, Beyerlein IJ, Bourke MA, Mitchell TE (2003) Microstructure of cryogenic treated M2 tool steel. Mater Sci Eng A 339:241–244
49. Indumathi J, Bijwe J, Ghosh AK, Fahim M, Krishnaraj N (1999) Wear of cryo-treated engineering polymers and composites. Wear 225–229:343–353
50. Jha AR (2006) Cryogenic technology and applications. Butterworth- Heinemann, Oxford, UK
51. Jinyong X, Yanping L, Jianzhong W, Xiaoyun K, Yuan G, Zhong X (2007) A study on double glow plasma surface metallurgy Mo–Cr high speed steel of carbon steel. Surf CoatTechnol 201:5093–5096
52. Jones JN, Rogers CB (2003) Chilling trumpets: does it have an acoustic effect? In: 146th Acoustical Society of America Meeting [Online]. Available: http://www.acoustics.org/. Accessed Apr 14, 2012
53. Jordine A (1996) Increased life of carburised race car gears by cryogenic treatment. Int J Fatigue 18(6):418
54. Kalin M, Leskovsek V, Vizintin J (2006) Wear behavior of deep cryogenic treated high speed steel at different loads. Mater Manuf Process 21:741–746
55. Kamody DJ. US Patent 5,259,200, 9 November 1993
56. Kamody DJ (1999) Cryogenics process update. Adv Mater Process 155(6):H67–H69
57. Kao M (1984) The effect of cryogenic treatment of sintered tungsten carbide. Master's Thesis, Arizona State University, Arizona
58. Kim D, Ramulu M (2004) Cryogenically treated carbide tool performance in drilling thermoplastic composites. Trans North Am Manuf Res Inst SME 32:79–85
59. Kim JW, Griggs JA, Regan JD, Ellis RA, Cai Z (2005) Effect of cryogenic treatment on nickel-titanium endodontic instruments. Int Endod J 38(6):364–371
60. Klaas NV, Marcus K, Kellock C (2005) The tribological behaviour of glass filled polytetrafluoroethylene. Tribol Int 38(9):824–833
61. Krump H, Hudec I, Jasˇsˇo M, Dayss E, Luyt AS (2006) Physical–morphological and chemical changes leading to an increase in adhesion between plasma treated polyester fibres and a rubber matrix. Appl Surf Sci 252(12):4264–4278
62. Leskovsek V, Kalin M, Vizintin J (2006) Influence of deep cryogenic treatment on wear resistance of vacuum heat-treated HSS. Vacuum 80:507–518
63. Leskovšek V, Ule B (2002) Influence of deep cryogenic treatment on vacuum heat-treated high-speed steel. Heat Treat Met 29(3):72–76
64. Li F, Hu KA, Li JL, Zhao BY (2001) The friction and wear characteristics of nanometer ZnO filled polytetrafluoroethylene. Wear 249(10–11):877–882
65. Lomte SV, Chikalthankar SB (2007) Investigation of the influence of deep cryogenic treatment on residual stresses in HCHC D2 steels using X-ray diffraction. In: Proceedings of the AIST Steel Properties and Applications Conference, pp. 959–972
66. Li S, Chen C, Li X, Yuan Z, Liu X (2000) Effect of cryogenic treatment on the abrasive resistance of T12 steel. Mocaxue Xuebao 20(4):276–279
67. Li S, Xie Y, Wu X (2010) Hardness and toughness investigations of deep cryogenic treated cold work die steel. Cryogenics 50(2):89–92
68. Lu XC, Wen SZ, Tong J, Chen YT, Ren LQ (1996) Wettability, soil adhesion, abrasion and friction wear of PTFE (+PPS) + Al_2O_3 composites. Wear 193(1):48–55

69. Meng F, Tagashira K, Azuma R, Sohma H (1994) Role of eta-carbide precipitation in the wear resistance improvements of Fe-12Cr-Mo-V-1.4C tool steel by cryogenic treatment. ISIJ Int 34:205–210
70. Meng F, Tagashira K, Sohma H (1994) Wear resistance and microstructure of cryogenic treated Fe-1.4Cr-1C bearing steel. Scripta Metall Mater 31:865–868
71. Mohanlal D, Renganarayanan S, Kalanidhi A (2001) Cryogenic treatment to augment wear resistance of tool and die steels. Cryogenics 41(3):149–155
72. Molinari A, Pellizzari M, Gialanella S, Straffelini G, Stiasny KH (2001) Effect of deep cryogenic treatment on the mechanical properties of tool steels. J Mater Process Technol 118 (1–3):350–355
73. Moore K, Collins DN (1993) Cryogenic treatment of three heat treated tool steels. Key Eng Mater 86–87:47–54
74. Pappas D, Bujanda A, Demaree JD, Hirvonen JK, Kosik W, Jensen R, McKnight S (2006) Surface modification of polyamide fibers and films using atmospheric plasmas. Surf CoatTechnol 201(7):4384–4388
75. Paulin P (1993) Frozen gears. Gear Technol 10:26–29
76. Pellizzari M, Molinari A (2002) Deep cryogenic treatment of cold work tool steel. In: Proceedings of the 6th International Tooling Conference. Karlstad University, September, pp. 657– 669
77. Pellizzari M, Molinari A, Girardini L, Maldarelli L (2008) Deep cryogenic treatment of AISI M2 high-speed steel. Int J Microstruct Mater Prop 3(2–3):383–390
78. Popandopulo N, Zhukova LT (1980) Transformation in high speed steels during cold treatment. Met Sci Heat Treat 22(10):708–710
79. Preciado M, Bravo PM, Alegre JM (2006) Effect of low temperature tempering prior cryogenic treatment on carburized steels. J Mater Process Technol 176(1–3):41–44
80. Quek TW (2007) Machining of steel using cryogenically treated cutting tool inserts. Ph.D. Thesis, National University of Singapore, Singapore
81. Ray BC (2005) Effects of thermal and cryogenic conditionings on mechanical behavior of thermally shocked glass fiber/epoxy composites. J Reinf Plastics Comp 24(7):713–717
82. Reddy TVS, Sornakumar T, Reddy MV, Venkatram R (2009) Machinability of C45 steel with deep cryogenic treated tungsten carbide cutting tool inserts. Int J Refract Met Hard Mater 27 (1):181–185
83. Satish Kumar M, Mohan Lal D, Renganarayanan S, Kalanidhi A (2001) An experimental investigation on the mechanism of wear resistance improvement in cryotreated tool steels. Indian J Eng Mater Sci 8(4):198–204
84. Schiradelly R, Diekman FJ (2001) Cryogenics—the racer's edge. Heat Treating Progress 6:43–50
85. Seah KHW, Rahman M, Yong KH (2003) Performance evaluation of cryogenically treated tungsten carbide cutting tool inserts. Proc IME B J Eng Manufact 217:29–43
86. Soundararajan V, Alagurmurthi N, Palaniradja K (2004) On the enhancement of wear resistance of hardened carbon tool steel (AISI 1095) with cryogenic quenching. Trans Mater Heat Treat 25(5):531–535
87. Stratton PF (2007) Optimising nano-carbide precipitation in tool steels. Mater Sci Eng A 449–451:809–812
88. Stewart HA (2004) Cryogenic treatment of tungsten carbide reduces tool wear when machining medium density fiberboard. For Prod J 54(2):53–56
89. Timmerhaus KD, Reed RP (2007) Cryogenic engineering. Springer, New York
90. Trieu HH, Morris LH, Kaufman ME, Hood R, Jenkins LS (1997) Investigation of cryogenic treatment of UHMWPE. In: Proceedings of the Sixteenth Southern Biomedical Engineering Conference, Biloxi, MS, pp. 90–91
91. Vadivel K, Rudramoorthy R (2008) Performance analysis of cryogenically treated coated carbide inserts. Int J Adv Manuf Technol 42(3–4):222–232

92. Vimal AJ, Bensley A, Lal DM, Srinivasan K (2008) Deep cryogenic treatment improves wear resistance of En 31 steel. Mater Manuf Process 23(4):369–376
93. Wu GM, Chang CH (2007) Modifications of rigid rod poly (1, 4-phenylene-*cis*-benzobisoxazole) fibers by gas plasma treatments. Vacuum 81(10):1159–1163
94. Wurzbach RN, DeFelice W (2004) Improving component wear performance through cryogenic treatment. In: Lubrication excellence proceedings [Online]. Available: www.cryoplus.com. Accessed Apr 14, 2012
95. Yang HS, Wang J, Shen BL, Liu HH, Gao SJ, Huang SJ (2006) Effect of cryogenic treatment on the matrix structure and abrasion resistance of white cast iron subjected to destabilization treatment. Wear 26:1150–1154
96. Yong AYL, Seah WKH, Rahman M (2006) Performance evaluation of cryogenically treated tungsten carbide tools in turning. Int J Mach Tools Manuf 46:2051–2056
97. Yong AYL, Seah KHW, Rahman M (2007) Performance of cryogenically treated tungsten carbide tools in milling operations. Int J Adv Manuf Technol 32:638–643
98. Yugandhar T, Krishnan PK (2002) Cryogenic treatment and its effects on tool steel. In: Proceedings of 6th international tooling conference, Karlstad University, Karlstad, Sept, pp. 671–684
99. Yun D, Xiaoping L, Hongshen X (1998) Deep cryogenic treatment of high-speed steel and its mechanism. Heat Treat Met 3:55–59
100. Zhang H, Zhang Z (2004) Comparison of short carbon fibre surface treatments on epoxy composites: II. Enhancement of the wear resistance. Compos Sci Technol 64 (13–14):2031–2038
101. Zhang H, Zhang Z, Breidt C (2004) Comparison of short carbon fibre surface treatments on epoxy composites: I. Enhancement of the mechanical properties. Compos Sci Technol 64 (13–14):2021–2029
102. Zhang ZZ, Xue QJ, Liu WM, Shen WC (1997) Friction and wear properties of metal powder filled PTFE composites under oil lubricated conditions. Wear 210(1–2):151–156
103. Zhang ZZ, Zhang HJ, Guo F, Wang K, Jiang W (2009) Enhanced wear resistance of hybrid PTFE/Kevlar fabric/phenolic composite by cryogenic treatment. J Mater Sci 44:6199–6205
104. Zhirafar S, Rezaeian A, Pugh M (2007) Effect of cryogenic treatment on the mechanical properties of 4340 steel. J Mater Process Technol 186(1–3):298–303
105. Zou XP, Kang ET, Neoh KG, Cui CQ, Lim TB (2001) Surface modification of poly (tetrafluoroethylene) films by plasma polymerization of glycidyl methacrylate for adhesion enhancement with evaporated copper. Polymer 42(15):6409–6418
106. Zurecki Z (2006) Cryogenic quenching of steel revisited. In: ASM Proceedings: Heat Treating, pp. 106–113

Chapter 12
Current and Potential Applications of Cryogenic Treated Polymers

Paolo Baldissera and Cristiana Delprete

Contents

Abstract As the use of polymers in structural components and power transmission systems has grown in the last decades, the demand for improved mechanical performances of these materials is increasing. Besides fibre reinforcement, the research about thermal treatments and surface coatings of these materials is focused on obtaining more strength and resistance to fatigue and wear. In this direction, a contribution can be given by unconventional treatments such as cryogenics. The use of deep cryogenic treatment (DCT) on polymers is already claimed as miraculous by some companies. The analysis of peer-reviewed literature shows that it is not all marketing: many pure and reinforced polymers have shown DCT improvements in hardness and in wear resistance under controlled experimental conditions. Starting from these results, a general overview of potential DCT applications on polymeric products is given in order to suggest further developments and research areas.

Keywords Deep cryogenic treatment • Wear • Polymers • Mechanical behaviour

P. Baldissera (✉) • C. Delprete
Politecnico di Torino, Corso Duca degli Abruzzi, 24, Torino 10129, Italy
e-mail: paolo.baldissera@polito.it

S. Kalia and S.-Y. Fu (eds.), *Polymers at Cryogenic Temperatures*,
DOI 10.1007/978-3-642-35335-2_12,

12.1 Introduction

Polymeric materials have been rapidly increasing their market during the last decades as a consequence of the growing demand of lightweight components for a wide range of industries and applications. In addition, while polymers were excluded from structural applications in the past, they are now being appreciated also in this field thanks to the introduction of new classes of polymers reinforced by high-strength fibres (typically carbon, but also polymeric fibres such as Kevlar®, Nylon® and Vectran®).

The research on mechanical behaviour enhancement of polymers is focused on fibre reinforcement, thermal treatments and surface coating. Among these approaches, the use of low temperature such as those involved in deep cryogenic treatment (DCT) has shown some interesting results.

The DCT process has been widely studied and applied to metals and alloys during the last decades, with interesting results on hardness, wear resistance and fatigue properties of tool steels, carburized steels for structural applications, stainless steels and also cast iron. However, while the mechanisms besides DCT effects on metallic materials have been partly identified (retained austenite elimination, fine carbide precipitation, residual stresses reduction), the road of knowledge is still open in the case of polymeric materials. First of all, a large range of polymers need to be tested and considering the whole range of properties (mechanical, optical, electrical, etc.), then, the classes of polymers showing good results could be used for an in-depth analysis of the microstructural causes of the change. In order to focus the research on a more defined area, it can be useful to spot some specific polymers and products where the potential improvement given by DCT could produce a significant technological progress. In the following paragraphs, a general review of potential use of DCT on polymers is proposed starting from the current claims by DCT companies, passing through the peer-reviewed literature analysis and then trying to identify the possible outcome and perspective from the point of view of design and production of mechanical components.

12.2 Current Applications of DCT on Polymers

Referring to some DCT companies [1–4], polymers such as polyamide (Nylon®) and polytetrafluoroethylene (PTFE, Teflon®) can be positively influenced by cryotreatment.

In particular, ethylene methacrylic acid (E/MAA) copolymers (Surlyn®) is widely used for golf balls and many DCT companies claim improved durability to resist denting, a more pleasing impact sound, a better "feel" and accuracy. Also 2–4 % increase in hitting distance is claimed by many companies, with 5–10 yards farther landing measured with a mechanical swing machine by comparing 100 untreated vs. 100 cryotreated balls [4]. However, no hypotheses or explanations are given about the mechanism responsible for the improvement.

UHMW Polyethylene for implants is cited among DCT successful applications in [2], but without any detail or reference (which implant, which properties, which improvement). Looking at the scientific literature (see the following paragraph), the improvement is probably referred to hardness and wear resistance and thus involve the durability of implants where parts are subjected to reciprocal motion (hip and knees).

Besides the above mentioned no other applications of DCT on polymeric components were found by the authors either in literature or on the web. However, the promising results reported by few recent works in the scientific literature (see the following paragraph) should bring the DCT opportunity to the attention of many companies involved in market production with polymers and the range of practical applications is expected to grow in the next years.

12.3 DCT Effects on Polymers in Peer-Reviewed Literature

More than looking at the DCT business claims, a serious evaluation of the potential advantage from cryotreating of polymers must be based on published scientific results from experimental campaign. Unfortunately, only few papers are available in this research field [5–9]. However, the results they found give a valuable indication about further research direction and possible implications on the applicative side.

In [5], the effects of DCT on a series of thermoplastic engineering polymers and their glass fibres (GF) composites were investigated in terms of hardness and wear resistance. Polyetherimide (PEI) and PTFE have shown significant enhancement of the investigated properties: +30 % and +60 % respectively in abrasive wear performance, +22 % and +12 % respectively in Shore-D hardness. However, the positive effect of DCT on the respective GF-reinforced composites was confirmed only for some percent of fibre reinforcement while some others shown a degradation of their properties and without a clear relationship between the parameters. An increase in crystallinity of amorphous and semi-crystalline polymers was found through X-ray diffraction analysis of cryotreated materials, but the authors addressed the measured improvements in wear resistance to the hardness increase.

The increase in wear resistance of PTFE through DCT has been recently confirmed in [6] and the optimal treatment parameters were found through quantum chemical approach (−185 °C with 12 h of soaking time). The increment in crystallinity was identified as the main responsible for the wear performance enhancement.

In [7], DCT was found to induce significant improvements in wear resistance of carbon fibre reinforced (25 and 30 %) composites of high-temperature polymers. In particular, PEI and PTFE confirmed their good results with this technique, but also CF-reinforced polyamide (PA 6,6), polyetheretherketone (PEEK) and polyethersulphone (PES) showed significant wear performance enhancement. Also in this case, the authors confirmed the hardness increase as one of the major wear controlling factors.

Any significant or detectable changes in dimension, crystallinity, tensile strength and elongation of Ultra-High Molecular Weight Polyethylene (UHMWPE) samples was found by [8] after DCT. However, a more in-depth investigation involving hardness and wear testing would be worthwhile in consideration of the large use of such polymer in artificial joints prostheses, as it will be discussed later.

In [9], no significant improvements in adhesion quality of glass fibre to epoxy resin were measured. In addition, cryotreated specimens showed a greater water absorption rate during hydrothermal ageing, suggesting the formation of interfacial debonding or matrix cracking due to the different expansion/contraction coefficients of epoxy and glass fibre.

Another related field of investigation concerns the effect of DCT directly applied on fibres and fabrics before including them into polymeric matrix in order to create composite materials. In this case the modifications induced on the fibre seem to lead to a stronger bonding with the matrix, resulting in enhanced wear performance. In [10], hybrid PTFE/Kevlar® fabric was subjected to cryogenic treatment before incorporation into phenolic matrix. Although the friction coefficient in cryotreated specimens was found to be higher than in untreated ones, the increased roughness of the fabric fibres after 10 min exposure to DCT resulted in an enhanced fibre/resin adhesion by mechanical interlocking. As a consequence, the wear rate at high temperatures was strongly reduced (by 75 % at 150 °C and by 86 % at 180 °C).

According to [11], cryotreatment applied to carbon fibres lead to the removal of amorphous pyrolyzate deposits from the fibre surface, which lead to an increase of the fibre strength, but can induce a reduction of the fibre-matrix adhesion as the above mentioned deposit plays a positive role in such direction. However, in [12, 13] the cryotreatment of short carbon fibres resulted in an increase of the fibre surface roughness with the consequence of an improved adhesion strength of fibres ad epoxy bonding due to mechanical interlocking. Enhancements in both flexural strength and wear resistance at high sliding pressure were measure on the final composite for the cryotreated-fibre specimens. Similar results were also confirmed in [14] on a CF-reinforced polyimide (PI) composite.

In order to give an overall picture, the most significant ($\geq$15 %) effects of DCT on wear and hardness of some polymers are summarized in Table 12.1.

At last, for chemical and microstructure specialized scientist, some useful information could certainly arise from the literature analysis of an adjacent field of research concerning the behaviour and the microstructure of polymers at low temperature for cryogenic applications. Some extensive studies in this area are reported in [15–18], but an in-depth exploration of such topics is beyond the purposes of this review and is left to the most specialized readers.

12.4 Potential Applications of DCT on Polymers

Analysing the potential benefits from the application of DCT on polymeric materials, three main areas should be considered:

Table 12.1 Summarized effects of DCT on polymers from the available literature

Base material	Reinforcement	Reference	Wear change	Hardness change
PEI	–	[5]	+30 %	+12 %
	10 % GF		+15 %	+7 %
	30 % GF		−25 %	−13 %
	40 % GF		−35 %	−10 %
	30 % CF	[7]	+37 %	+32 %
PI	15 % graphite	[5]	+15 %	+15 %
	15 % graphite and 10 % PTFE		+58 %	+9 %
PTFE	–	[5]	+60 %	+22 %
	–	[6]	+55 %	Not reported
	25 % short CF	[5, 7]	+40 %	+15 %
PES	30 % CF	[7]	+28 %	+30 %
PA6,6	30 % CF		+35 %	+7 %
PEEK	30 % CF		+40 %	+9 %
Phenolic resin	PEEK/Kevlar® fabric	[10]	+75 % at 150 °C +86 % at 180 °C	Not reported

GF glass fibre; *CF* carbon fibre

1. Current uses of polymers where DCT can add some improvement such as extended life or better performance (applicative field consolidation).
2. New applications where polymers could be used by reaching a certain target in mechanical properties (applicative field extension).
3. New approaches to be explored by using the DCT during the manufacturing process (i.e. extreme rapid quench, cryogenic moulds, etc.) instead of considering it as a post-processing phase.

Both directions are worthy of investigation and can lead to significant advancements in design and production of parts for a wide range of applications.

In the following paragraphs, these two areas will be explored with the aim of giving specific suggestions for further exploration.

12.4.1 Applicative Field Consolidation Through DCT

One of the most interesting area where the above mentioned effects of DCT on polymers could produce effective benefit is the biomedical one. In particular, considering the wear resistance enhancement and the reduction of the friction coefficient, articular joint endoprosthetic implants should be the first candidate for DCT. The use of polymeric parts in the joint interface of these components has been introduced in order to lower the friction, resulting in a more comfortable articular motion with respect to metallic or ceramic contact. Typical uses of polymers for artificial joints include:

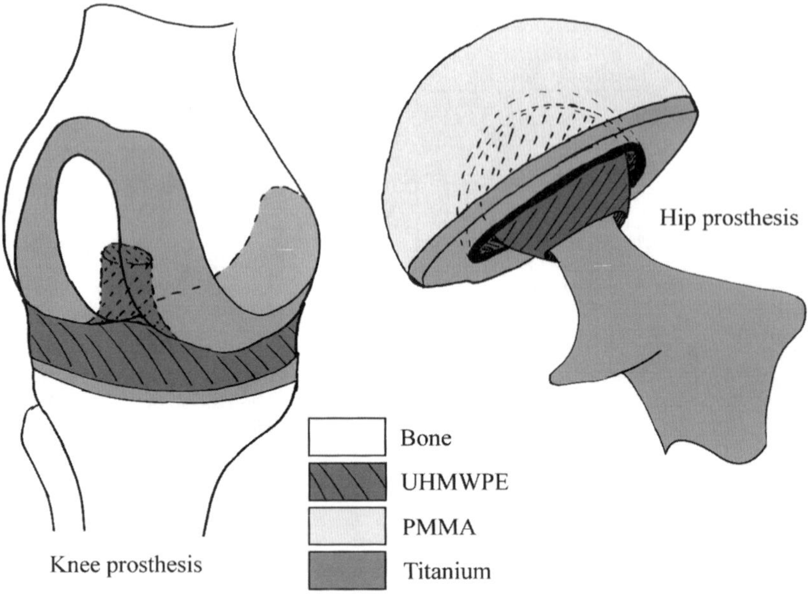

Fig. 12.1 Example geometries and materials for knee and hip joints prostheses

- UHMWPE is the most used material for the acetabulum in total hip prostheses and for spacers and plateau in knee prostheses [19].
- Also PA12 has been proposed for the acetabulum replacement [20].
- Polypropylene (PP) is often used finger joints prostheses [19].
- PTFE is suitable for both shoulder and knee joints prostheses [19].
- A special polyimide named PM-1® has been recently introduced on the market claiming extreme resistance to wear, fatigue and corrosion [19].
- PEEK has been recently introduced as substitute for titanium in prosthetic stems due to its elastic modulus much closer to that of the bone, which helps in avoiding the so called stress shielding phenomenon [21].

As an example, Fig. 12.1 shows the typical geometry and materials used for the construction of hip and knee joint artificial replacements.

One of the main limitations of prostheses is the durability under loading conditions and under the aggression of natural fluids and wear resistance is one of the weakest points of polymers. As a consequence, such kind of surgery is often delayed as much as possible in young aged patients, in particular for the highly stressed limb articulations (hip, knee and ankle). However, in front of a painful and incapacitating disease, delaying is not a nice option and a significant reduction in polymeric interfaces wear could relax the approach to the surgery of young aged patients. While PTFE cryotreating benefits on wear resistance were investigated [5–7], the other cited materials still need for an extensive DCT investigation.

Cryotreating of fluorocarbons (PTFE or TFE) and polyamides (Nylon®) is worthy of investigation for its potential drawback on wear resistance of gears, bearings, cams and bushings. In fact, as shown before, peer-reviewed literature has already pointed out the advantages of DCT in terms of hardness increase, friction reduction and wear life extension of such classes of polymers and it is expected that plastic bearings and gearings manufacturer will consider further investigation directly on the components.

Also thermosetting plastics such as epoxies have shown positive results from DCT in terms of wear resistance. In this case, the range of application of this class of polymers as matrix for composite materials is so wide that probably any mechanical, chemical or aesthetic properties which could be potentially affected by DCT is worthy of in-depth analysis.

DCT application on lenses and transparent aircraft enclosures made by acrylics (poly-methyl-methacrylate) should be investigated in order to check if the induced microstructural change could lead to some enhancement in the optical properties or at least in resistance to abrasion and scratch.

12.4.2 Applicative Field Extension Through DCT

Considering the already discussed case of fluorocarbons (PTFE or TFE) and polyamides (Nylon®), if the positive effects will be confirmed on the component itself, not only the current uses of plastic gears and bearings could be consolidated and improved, but also an extension of their potential application could be achieved. In fact, the enhancement of the wear performance can be used to give more reliability at the same load level or to apply higher loading condition at the current reliability level. The perspective in this case is an expansion of the market with the introduction of plastic bearings and gears also on products where they cannot be introduced at the present state of the art. In the last decade, polymers have started being largely applied in the manufacturing of gears for a wide range of products. In particular, Delrin® acetal resin, Zytel® nylon resin and Vespel® polyimide (all by DuPont) are used in a variety of small and medium size gearings in standard or epicyclic configuration (Fig. 12.2), mono and multi-stage systems. Some examples can be found in various tools such as electric screwdrivers and drills, car rear-view mirror positioners (Fig. 12.2), office automation equipments, household appliances, doors and barriers automatic movers, windshields, speedometers, rotary pumps and clocks. Recently, also PEEK® (by Victrex) has been successfully employed in the production of balance shaft gears for automotive engines [22], thanks to its good properties at relatively high temperature (155 °C). There are many advantages from the replacement of metal gears with polymeric materials: weight and inertia reduction, superior damping in case of moderate shock or impact loads, noise reduction. In addition, many of the cited polymers show self-lubricating properties that are essential in applications that do not permit external lubrication.

The international suppliers of polymeric materials usually distinguish three classes of applications for gears:

Fig. 12.2 Polymer planetary gears used in rear-view mirror positioners [24]

1. Power transmission, where high-strength materials are required.
2. Motion transmission, where high-durability materials are required.
3. Power and motion transmission, where an excellent balance between strength and durability is the key for a successful application.

In this case, the potential increase in mechanical properties of high-strength polymers through DCT could shift many of them from the first class to the third, with the result of consolidating and expanding the applicative range of polymeric gears.

Another industry that could be attracted by DCT-enhanced polymers is represented by the producers of high-end cycling components and frames. Such kind of business has been leading the composite market development in the last decades thank to the big production volumes that permitted a strong reduction of manufacturing prices. Until today, the typical composite material employed for such kind of production has been CF-reinforced epoxy resin, but the constant extreme research for lighter components without decreases in reliability level may lead to investigate special treatments and/or new materials. With the addition of some good marketing work, in which cryogenic treatment can play a role for its "exotic appeal", a new branch of applied research and development could easily find its start-up funds in cycling industries.

Also in the case of bearings, both rolling and plain (bushings or similar), the use of polymeric material is increasing. Frelon®, Nylon®, PEEK®, PTFE, Rulon®, UHMWPE and Vespel® are the most common polymers used by the bearing industries in pure or reinforced status, for the production of the whole component or as coating on a metallic base. Also in this case, the main advantages come from the lower friction coefficient, the self-lubricating behaviour and the resistance to corrosion. As the operating temperature can grow easily in plain bearing, any eventual increase in wear resistance given by DCT should be verified in the whole range of operating temperature nominally supported by each material (i.e. up to 249 °C for PEEK®, 260 °C for PTFE and Rulon®, 288 °C for Vespel®). In addition, as DCT is responsible for changes also in thermal expansion coefficient

Fig. 12.3 Bi-material bushings made by metal and inner plastic coating [25]

for metallic alloys [23], it should be interesting to analyze the DCT effects on this property also for polymers. Indeed, due to the extreme accuracy requirement in design and manufacturing of bearings, the thermal expansion coefficient plays a crucial role in the operating rage of such products (Fig. 12.3).

12.4.3 DCT Application in Manufacturing

All the cases proposed in the scientific literature concern the use of DCT as a post-processing on polymeric samples or components. However, some new approaches could be investigated by considering the introduction of DCT directly into the manufacturing processes. In general, quite all the production methods include a warming phase (near to melting temperature or at least over glass transition) in order to allow for the shaping and a cooling phase. The last operation is often obtained by air or water, but simple systems for cryo-cooling can be specifically designed for the most common processing methods. Some schematic proposal are reported in Fig. 12.4 (blow moulding) and Fig. 12.5 (extrusion), while similar systems could be designed for vacuum thermoforming, injection moulding and rotational moulding processes.

The introduction of a controlled cryogenic cooling in the manufacturing process could be interesting in order to achieve different mechanical characteristics of the final product. As it happens with metallic alloys, rapid cooling of molten polymers can freeze some microstructural configuration that are unstable at room temperature, without allowing any path for a transformation towards a more stable conditions unless the product is heated again. In addition, the glass transition temperature (T_g) of amorphous polymers increases with higher cooling rate and, through cryo-cooling, different T_g levels could be obtained for the same material, with the aim of having the desired mechanical behaviour at the operational temperature of the final component.

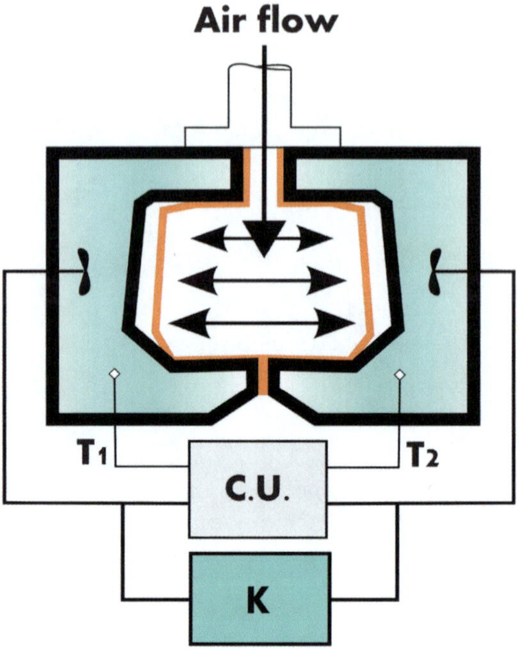

Fig. 12.4 Example of integrated DCT on blow moulding (*C.U.* control unit; *K* liquid nitrogen tank, T_1 and T_2 temperature feedback sensoring)

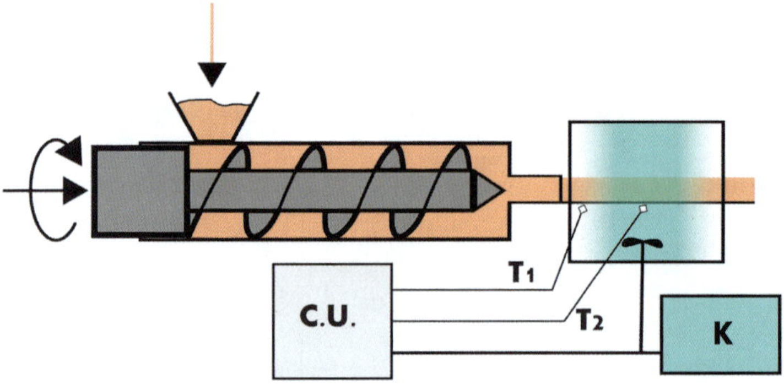

Fig. 12.5 Example of integrated DCT on extrusion processing (*C.U.* control unit; *K* liquid nitrogen tank; T_1 and T_2 temperature feedback sensoring)

At last, even if not directly applied on the polymer, some other advantages can arise from DCT on metallic moulds in terms of durability.

12.5 Conclusion

From the analysis of the few available peer-reviewed papers it emerges that cryogenic treatments can play a significant role for the enhancement of tribological characteristics of polymers. While DCT companies are mainly focused on metal treating, some of them claim positive effects also on polymers, but often focusing on specific products such as sporting goods. Given the increasing market for polymers in structural and power transmission applications, these first results are encouraging and should lead to a wide in-depth research during the next years. In particular, some field of application can have significant benefits:

- Biomedical prostheses for artificial joints could extend their operating life and reduce the need for surgical replacement due to wear.
- The growing production of plastic gearings could achieve significant improvements in durability of high-strength polymers.
- A wider range of bearings could replace their rolling parts with high-durability cryotreated polymers.

In addition, the potential enhancement in mechanical properties could attract other industries such as high-end cycling component makers, which are always looking for innovative processes to propose on the market.

Beyond the outlined potential applications, an extensive and rigorous research work has to be made in the next years in order to grow and consolidate the knowledge about DCT on polymers. In particular:

- Effects on mechanical properties other than hardness and wear resistance should be investigated, i.e. fatigue and toughness.
- The mechanical behaviour of cryotreated polymers should be investigated also at relatively high temperature for some specific materials such as PEEK®.
- Effects on optical properties need to be investigated for acrylics used in transparent aircraft enclosures.
- Effects of DCT as part of the manufacturing process (instead of the usual post-processing approach) could be worthy of investigation in order to verify benefits and drawbacks in terms of residual stresses, glass transition temperature, crystalline structure and mechanical properties of the final product.

References

1. http://www.300below.com/. Accessed 28 June 2012
2. http://www.cryogenicsinternational.com/. Accessed 28 June 2012
3. http://www.nitrofreeze.com/. Accessed 28 June 2012
4. http://www.cryotron.com/. Accessed 28 June 2012
5. Indumathi J, Bijwe J, Ghosh AK, Fahim M, Krishnaraj N (1999) Wear of cryo-treated engineering polymers and composites. Wear 225–229:343–353

6. Pande KN, Peshwe DR, Kumar A (2012) Optimization of cryo-treatment parameters for PTFE by quantum-chemical approach and its evaluation through mechanical, thermal and structural characterization. Trans Ind Inst Metals. doi:10.1007/s12666-012-0139-4
7. Indumathi J, Bijwe J (2006) On the feasibility of using cryo-treatment as a tool to enhance the abrasive wear behaviour of solid-lubricated polymeric composites. J Synth Lubr 17–2:123–134
8. Ray BC (2006) Adhesion of glass/epoxy composites influenced by thermal and cryogenic environments. J Appl Polymer Sci 102:1943–1949. doi:10.1002/app. 24488
9. Trieu HH (1997) Investigation of cryogenic treatment of UHMWPE. Proceedings of the 1997 Sixteenth Southern Biomedical Engineering Conference, Biloxi, MS, USA, pp. 90–91
10. Zhang Z, Zhang H, Guo F, Wang K, Jiang W (2009) Enhanced wear resistance of hybrid PTFE/Kevlar fabric/phenolic composite by cryogenic treatment. J Mater Sci 44:6199–6205. doi:10.1007/s10853-009-3862-4
11. Rashkovan IA, Korabel'nikov YG (1997) The effect of fiber surface treatment on its strength and adhesion to the matrix. Compos Sci Technol 57:1017–1022
12. Zhang H, Zhang Z, Breidt C (2004) Comparison of short carbon fibre surface treatments on epoxy composites I. Enhancement of the mechanical properties. Compos Sci Technol 64:2021–2029. doi:10.1016/j.compscitech.2004.02.009
13. Zhang H, Zhang Z (2004) Comparison of short carbon fibre surface treatments on epoxy composites II. Enhancement of wear resistance. Compos Sci Technol 64:2031–2038. doi:10.1016/j.compscitech.2004.02.010
14. Zhang X, Pei X, Mu B, Wang Q (2008) Effect of carbon fiber surface treatments on the flexural strength and tribological properties of short carbon fiber/polyimide composites. Surf Interface Anal 40:961–965. doi:10.1002/sia.2846
15. Glaeser WA, Kissel JW, Snediker DK (1974) Wear mechanisms of polymers at cryogenic temperatures. Polym Sci Technol 5B:651–662
16. Yano O, Yamaoka H (1995) Cryogenic properties of polymers. Prog Polym Sci 20–4:585–613
17. Hübner W, Gradt T, Schneider T, Börner H (1998) Tribological behaviour of materials at cryogenic temperatures. Wear 216–2:150–159
18. Theiler G, Hübner W, Gradt T, Klein P, Friedrich K (2002) Friction and wear of PTFE composites at cryogenic temperatures. Tribol Int 35:449–458
19. Shirbhate DV (2008) Polymers used in artificial joints. http://www.scribd.com/doc/7365609/Polymers-used-in-Artificial-joints. Accessed 28 June 2012
20. Bougherara H, Bureau M, Campbell M, Vadean A, Yahia L'H (2007) Design of a biomimetic polymer-composite hip prosthesis. J Biomed Mater Res A 82A-1:27–40
21. Smit L (2009) http://medicaldesign.com/materials/designing-polymers-prosthetics-20091201/. Accessed 28 June 2012
22. http://www.victrex.com/en/victrex-library/case-studies/detail/detail.php?id=95. Accessed 28 June 2012
23. Bensely A, Paulin P, Nagarajan G, Mohan Lal D (2009) Influence of cryogenic treatment on the dimensional stability of gear steel-EN 353. http://www.300below.com/book/36-rs-dimensionalstability-en353/4-research.html. Accessed 28 June 2012
24. Gideonro (2009) http://en.wikipedia.org. Accessed 28 June 2012
25. David Benbennick (2005) http://en.wikipedia.org. Accessed 28 June 2012

About the Editors

Dr. Susheel Kalia is Assistant Professor in the Department of Chemistry, Bahra University, Solan (H.P.), India. After completing his Ph.D. at Punjab Technical University, Jalandhar, India, he began working as Assistant Professor in Singhania University (Rajasthan), Shoolini University (Solan) and Bahra University (Shimla Hills) where he has been for the past 5 years. Kalia has around 40 research papers in international journals along with 80 publications in national and international conferences and seven book chapters. Kalia's other editorial activities include work as a reviewer and member of editorial board for various International Journals. Additionally, he has edited a number of books such as *Handbook of Biopolymers and Their Application*—Wiley-Scrivener Publishing and *Cellulose Fibers: Bio- and Nano-Polymer Composites*—Springer, Heidelberg. He is also a member of a number of professional organizations, including the Asian Polymer Association, Indian Cryogenics Council, the Society for Polymer Science, Indian Society of Analytical Scientists, and the International Association of Advanced Materials. Presently, Kalia's research is in the field of biocomposites, nanocomposites, conducting polymers, cellulose nanofibers, inorganic nanoparticles, hybrid materials, hydrogels, and cryogenics.

S. Kalia and S.-Y. Fu (eds.), *Polymers at Cryogenic Temperatures*,
DOI 10.1007/978-3-642-35335-2, © Springer-Verlag Berlin Heidelberg 2013

Dr. Shao-Yun Fu joined his current position as a full professor at the Technical Institute of Physics and Chemistry, Chinese Academy of Sciences (CAS) in 2002. He was awarded his Ph.D. by CAS in 1993 and since then has held positions as a Humboldt Fellow at Leibnitz Institute of Polymer Research, Dresden and visiting scholar at the Center for Advanced Materials, Nanyang Technical University (Singapore), University of Sydney (Australia), and University of California at Riverside (USA). He has also worked as Senior Research Fellow at City University of Hong Kong. Since June 2012, Fu is a visiting professor at Tohoku University, Shinshu University, Chiba University of Japan, and the Technical University of Kaiserslautern, Germany. Fu has published over 170 articles in journals, one monograph entitled *Science and Engineering of Short Fibre Reinforced Polymer Composites* together with Prof. Yiu-Wing Mai (University of Sydney) and Dr. Bernd Lauke (Dresden, Germany) published by Woodhead, Cambridge, UK. He has also contributed three chapters in English books and three chapters in Chinese books.

Index

DOI 10.1007/978-3-642-35335-2, © Springer-Verlag Berlin Heidelberg 2013

Printed by Printforce, the Netherlands